ACCEPTANCE TESTING OF RADIOLOGICAL IMAGING EQUIPMENT

Co-sponsored by:

American Association of Physicists in Medicine
American College of Radiology
and
Society for Radiological Engineering

Edited by:

Pei-Jan Paul Lin, Editor-in-chief
Robert J. Kriz
Keith J. Strauss
Phillip L. Rauch
Raymond P. Rossi

Library of Congress Catalog Card Number: 82-72451
International Standard Book Number: 0-88318-400-1

Published by the American Institute of Physics, Inc.,
335 East 45 Street, New York, New York 10017

Printed in the United States of America

ACCEPTANCE TESTING OF RADIOLOGICAL IMAGING EQUIPMENT

The symposium entitled "Acceptance Testing of Radiological Imaging Equipment" was held on October 1 and 2, 1981. The co-sponsoring organizations were:

American Association of Physicists in Medicine

American College of Radiology

Society for Radiological Engineering

The symposium coordinators listed below served as the first round reviewers and editors for the papers submitted to this symposium so that all articles were assured of their quality and integrity for publication in the proceedings:

Pei-Jan Paul Lin | Robert J. Kriz
Keith J. Strauss | Phillip L. Rauch
Raymond P. Rossi

Publication of the proceedings at the time of the symposium was well received by all attendees, and served as a reference point for the Diagnostic Radiology Committee and the Publications Committee of AAPM in starting the review process for consideration of publication by AAPM. The symposium coordinators would like to express their sincere appreciation and thanks to the Diagnostic Radiology Committee and the Publications Committee of AAPM in streamlining the review process so that the publication was made available in the shortest possible time. The following people were involved in the AAPM review process; the symposium coordinators are greatful for their assistance:

From the Diagnostic Radiology Committee

Gary T. Barnes | Maynard High
Pei-Jan Paul Lin | Mary Ellen Masterson
William D. McDavid | Gerald Shapiro

From the Publications Committee

John R. Cameron | Arthur G. Haus

Upon completion of the review process the proceedings were reorganized to include pertinent comments and suggestions. Thus, the proceedings became available through AAPM.

FOREWORD

Introduction

A series of three symposiums have been organized, including this one, within the past four years. The first symposium entitled "Workshop in Purchase Specifications and Performance Evaluation of Radiologic Imaging Equipment" was held in May 1978. It was sponsored jointly by the Midwest Chapter of the American Association of Physicists in Medicine (AAPM), Northwestern University Medical School, and Northwestern Memorial Hospital. The second symposium entitled "Establishment and Administration of a Maintenance and Service Program for Radiologic Imaging Equipment" was held in October, 1979. The sponsoring organizations were the Midwest Chapter of AAPM, the American Hospital Radiology Administrators Midwest Region (AHRA), and the Society for Radiological Engineering (SRE). The third symposium, "Acceptance Testing of Radiological Imaging Equipment" is the most comprehensive and technical of these three symposiums, and will be the conclusion of this series designed for diagnostic radiological imaging. The sponsoring organizations are AAPM, SRE, and the American College of Radiology (ACR).

As one of the designers in the organizing and planning of these symposiums, I have been very fortunate working with many of my colleagues and friends in the radiology community. It has been extremely satisfying to me as I was able to enjoy the successful outcomes of these symposiums with everyone who has attended these meetings. I shall remember the support and cooperation given to me by those not only in the academic but also in the industry during the entire course of organizing these symposium meetings.

One of the reasons, I believe, that these symposiums have been well accepted by the attendees is the publication and distribution of the proceedings at the time of registration. All authors spent a great deal of time and effor in preparing their articles as the articles will be printed and made available to the attendees when the paper is being presented; which no doubt has contributed in maintaining the high level of quality. The proceedings for the first symposium are available from the National Technical Information Services (NTIS Publication Number 287-985). The proceedings for the second symposium, unfortunately are out of print and cannot be obtained from regular publication sources at this time.

The proceedings to the third symposium (this proceedings actually) is scheduled to be reviewed by AAPM for consideration as one of the AAPM Publications. For this reason no extra copies of the proceedings have been printed. If everything goes well the Acceptance Testing of Radiological Imaging Equipment Monograph will be available for those who could not attend this symposium. This proceeding should be available in approximately one year.

Editorial Comments

The papers in this proceeding have been reviewed by the symposium coordinators a few times; many of them were returned to the authors for rewriting. About two thirds of the papers have been retyped to conform with the standard format and/or corrections of typographical errors. Although every precaution has been made to ensure the accuracy of the papers, the symposium coordinators cannot be held responsible for the correctness or accuracy of the contents. The information supplied herein and/or opinions expressed are soley the responsibility of the author or authors. Additional changes and corrections may, in turn, be required upon the review process by AAPM.

Acknowledgement

The high level of integrity and quality of these papers can be attributed to the effort and time spent by the authors in preparing their articles. I would like to express my sincere thanks to the authors for their contributions to this symposium.

I would like to express my sincere thanks to the coordinators who assisted me in organizing this symposium. I could not have accomplished the many tasks involved in the various aspects of arranging this symposium without their assistance and encouragements.

The technical support and financial assistance provided by the following manufacturers has not only ensured the quality of this symposium, but has also made it possible to have this symposium in the first place. On behalf of the symposium coordinators, I would like to express my sincere thanks to the following manufacturing companies for their generous contributions:

E.I. duPont de Nemours & Company, Inc.

Eastman Kodak Company

General Electric Company

The Machlett Laboratories, Inc.

N.V. Optische Industrie de Oude Delft

Oldelft Corporation of America

N.V. Philips Gloeilampenfabrienken

Philips Medical Systems, Inc.

Picker-Dunlee Corporation

Picker International

Siemens Corporation

Finally, I would like to thank the symposium attendees for their continued support and interest in this highly technical and specialized field of diagnostic radiology.

Pei-Jan Paul Lin, Director
Acceptance Testing Symposium

CONTENTS

QUALITY CONTROL OF IMAGE INTENSIFIERS IN THE MANUFACTURING PROCESS

Aart A.G. Beekmans, Jacques A. den Boer, Johan W. Haarman, and Bart van der Eijk
N.V. Philips' Gloeilampenfabrieken, Eindhoven, The Netherlands

Abstract

A new understanding of image quality of X-ray image intensifiers is explained taking into account the total X-ray system in which the image intensifier performs its function.
It turns out that the low frequency drop is more important than the contrast ratio, the modulation at 1 lp/mm is more important than resolution and the MTF - behaviour and DQE of the image intensifier should be optimized taking into account the other parameters in a complete X-ray system which influence image quality.
Based on these ideas a new system of image quality control of image intensifiers in the manufacturing process is introduced. This paper discusses both the basis and the organisation of this quality control.

Introduction

In the context of this paper quality control is interpreted as image quality control of X-ray image intensifiers. It is clear that we cannot deal with all the different systems for quality control of image intensifiers which exist on the world. In this paper we confine our selves to the system that is in use in the manufacturing process of X-ray image intensifiers of Philips Medical Systems. This system is of course based on our understanding of the phenomena that control image quality.

Since we feel that our understanding is an extension of what many people may be accustomed to we have to start with an explanation of what our understanding is. Then the image quality control system in our understanding process is dealt with, which ensures a guaranteed level of image quality of our X-ray image intensifiers.

Image quality

Until now many people believed that image quality of X-ray systems is determined primarily by resolution, contrast ratio and noise behaviour (expressed in exposure per frame) of the imaging subsystem. A subsystem in this context is an X-ray image intensifier connected to a detector such as a film camera or a video camera.

Resolution

Let us first examine the meaning of resolution. Resolution in its usual sense is the resolution obtained by reading from film with the eye images of funkphantoms. In an observer test we asked 17 observers to read 100 funkpha ntom images in 5 sessions.

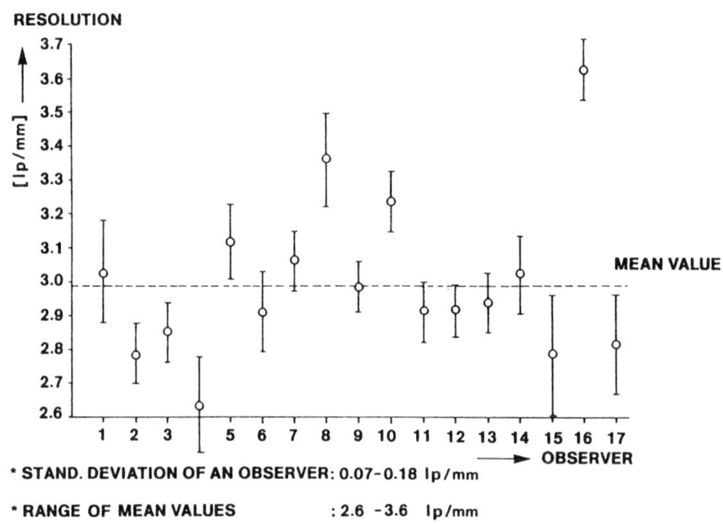

Fig. 1. Observer test.

Fig.1 shows the resolution observed by reading a single 100 mm funkphantom image. The readings are indicated on the vertical axis and the different observers are indicated on the horizontal axis. This experiment shows that, although each observer is rather consistent in his readings, the discrepancy between different observers is very large. From this experiment it is clear that resolution measurements with a funkphantom is not a good absolute measuring method. However it may be useful as a relative method.

It is quite understandable that resolution measurements yield non reproducing results.

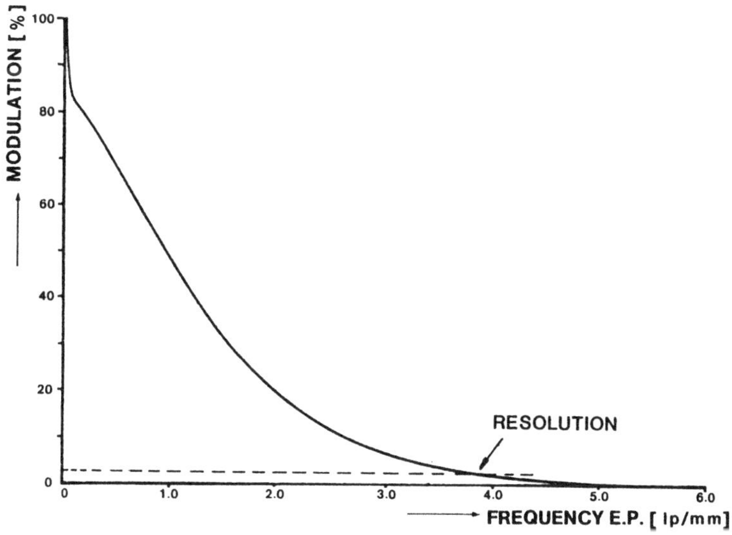

Fig.2. Determination of the resolution of an X-ray Image Intensifier.

Fig.2 shows a MTF-curve of a modern X-ray image intensifier. The intersection of the horizontal hatched line with the tail of the MTF-curve indicate the resolution. It is obvious that even very small changes in vertical position of this line, due to personal differences or due to somewhat different experimental conditions,will yield very different results.

Even if resolution could be measured in a reliable and easy way, one wonders what its significance is.
It indicates more or less one point of the MTF-curve in a region of spatial frequencies in which we know that there is no medical information. The reason for this is that smaller details of medical relevance never modulate the X-radiation by more than a few percent. The only use that can be made of resolution measurements is, in our opinion, to compare MTF-curves that are of the same shape. In that case we know that the MTF-curve with the highest resolution will have higher modulation transfer for every spatial frequency if other parameters (low frequency drop) are the same. In that case one must be sure that the same observer makes the comparison and that all exposure parameters as well as the film development parameters are kept constant. It is clear that resolution measurements can have only a remote relation with image quality.

Low frequency drop and contrast ratio

Let us now examine the meaning of the value of contrast-ratio. According to a measuring method standardised by the NEMA the contrast ratio is measured in the following way.
A lead disk is positioned in front of the image intensifier, covering 10% of the input area of the intensifier. The light intensity is measured in the center of the image of the disc, applying an appropriate exposure rate. Then the lead disc is removed and the light intensity is measured again at the same position with the same exposure rate. The contrast ratio is the ratio of those two light output measurements; it can be expressed in a percentage.

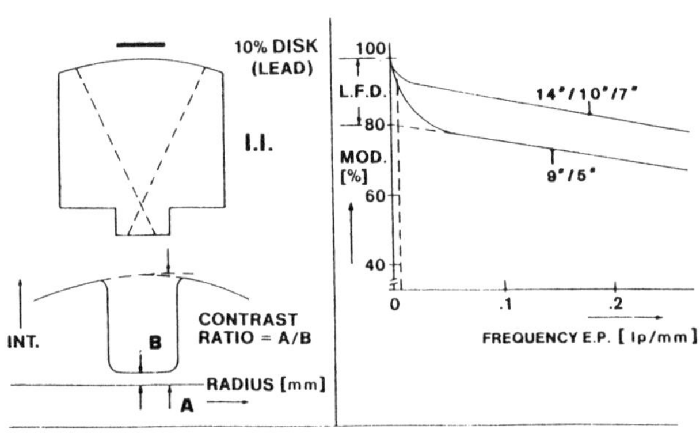

Fig.3. Contrast ratio and low frequency drop.

In fig.3 (left part) this measuring setup is explained graphically.

We prefer the concept of the so called low frequency drop. We define the low frequency drop as the intercept of the vertical axis by the extrapolated straight part of the MTF-curve directly following the fast decrease at low frequencies. The low frequency drop is expressed in a percentage being the difference between 100% on the vertical axis and the modulation value of the intersection point (see fig.3).

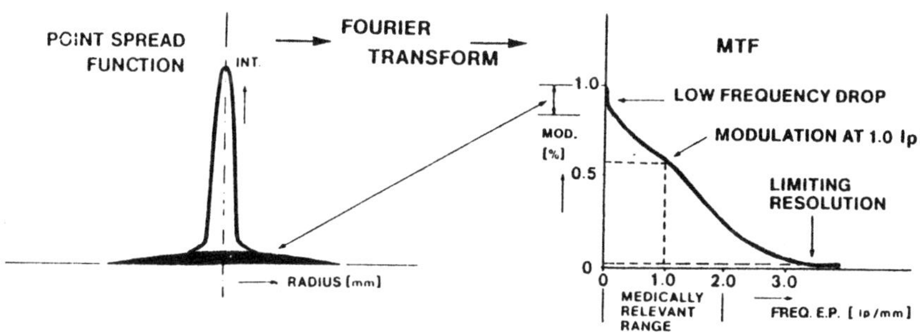

Fig.4 Point spread function and MTF. Relation between "tails" of the point spread function and the low frequency drop.

As indicated in fig.4 the low frequency drop is related directly to the tails of the point spread function.
These tails are the result of scattering of X-rays and light in the image intensifier, which causes X-ray quanta and light quanta to be transformed into phenomena that do not carry any information. This effect causes the modulation transfer to be decreased for all spatial frequencies. (In addition it adds to the noise in those parts of the image where the quantum flux is low, decreasing the contrast-to-noise ratio of details).

The low frequency drop is a measure of the amount of energy that is being scattered and in the same time indicates the amount by which the modulation is decreased at higher spatial frequencies.

I.I. Tubes	Mode	Contrast Ratio Measured	LFD
9"/5" (1973)	9"	14% ± 0,5	30% ± 0.5
9"/5" (1975)	9"	7%	20%
9"/5" (1979)	9"	6%	18%
9"/5" (1981)	9"	6%	12%
14"/10"/7"(1979)	10"	4%	6%

Table I. Large area contrast loss versus low frequency drop.

Table I summarises the situation for a family of Philips II's. If we look at this table we see that the values of the low frequency drop do not correlate very well with the contrast ratio values. Especially if we compare the contrastratios of the 9"/5" image intensifiers (9" mode) and the contrast ratio of the 10" mode of the 14" image intensifier we do not see very much difference. Our experience has shown that the subjective appreciation that the users of the image intensifiers have for the contrast behaviour of different tubes, is related much more to differences in low frequency drop than to differences in contrastratio.

It is interesting to inspect more closely what relation exists between contrast ratio and low frequency drop. Mathematical analysis reveales that in principle all points of the MTF-curve contribute to the intensity in the middle of the image of the lead disc that is used in the measurement of the contrast ratio. This means that the contrast ratio is a kind of average of all points of the MTF-curve that are weighted differently. The relative influence of each point of the MTF-curve on the contrastratio is dependent on the shape of the MTF-curve, especially at low spatial frequencies. This means that, if the MTF-curve of all image intensifiers would have the same shape, the contrast ratio would have a fixed relation with the low frequency drop.

If we look at the two MTF curves in the right hand part of fig.3 we see two different shapes. The upper curve is the MTF of the 10" mode of our 14" image intensifier. The lower curve is the MTF of the 9" mode of our 9"/5" image intensifier manufactured in 1975. The figures in table I also indicate that there is no fixed relation between the contrast ratio and the low frequency drop.

If the contrast ratio is a kind of weighted average of all the points of the MTF-curve, we should be able to find one point on the MTF-curve that correlates with the contrast ratio for a particular tube. Calculations and measurements have shown that for all MTF-curves, we have seen so far, the contrast ratio is within measuring accuracy equal to the difference between 100% and the modulation at a spatial frequency of 1.5 lp over the diameter of the input of the image intensifier. This is true also for other brands of image intensifiers than the tubes made by Philips. For the 9" mode of a 9"/5" image intensifier this spatial frequency is 0.006 lp/mm. In fig.3, this spatial frequency is indicated by the vertical hatched line. Graphically we can find the contrast ratio by subtracting from 100% the modulation that belongs to the intercept of this vertical line and the MTF-curve.
In the same way the contrast ratio of the 10" mode of our 14"/10"/7" image intensifier is found. We see that the contrast ratio is apparently very unsensitive to the drop in the MTF-curve beyond the spatial frequency that belongs to 1.5 line pairs per input diameter.

For these reasons we have the opinion that the low frequency drop is a much better concept for indicating low frequency behaviour in a single parameter of image intensifiers than the concept of the contrast ratio as measured with the measuring method standardised by NEMA.

DQE and noise behaviour

Now we will deal with noise. Instead of the undefined qualification of required exposure per frame, which is dependent on personal preference, we prefer the concept of the detective quantum efficiency. It is defined as the ratio of the squares of the signal-to-noise ratios at the output and the entrance, at a given radiation quality :

$$DQE = \frac{(S/N)^2 \text{ output}}{(S/N)^2 \text{ input}} \qquad (1)$$

Since the $(S/N)^2$ of the input is proportional to the input exposure, we can write :

$$DQE = \frac{\text{Constant}}{\text{input exposure}} \times (S/N)^2 \text{ output} \qquad (2)$$

This DQE definition includes all conversion processes in the image intensifier, not only those occurring at the input side of the tube.
The higher the DQE the better the signal-to-noise ratio for the same input exposure or the higher the DQE the lower the exposure for the same signal-to-noise ratio. (It is important that the measuring method is not influenced by temporal effects due to a slow timeresponse of the phosphors in the tube. Neglection of a correct correction for time lag in the phosphors will lead to erroueously high values of DQE. Temporal lag in phosphors can in some cases reduce the noise in fast sequence imaging such as cine and video, but then it will also cause movement unsharpness. In static imaging laggy phosphors will not reduce noise).

In order to minimise the noise one would be inclined to maximise the DQE. But in normal cases an increase in DQE causes a decrease in MTF-performance. So there is a trade-off between DQE and MTF which is dependent on many parameters that control the performance of a complete X-ray system as we will see. This means that we have to optimize a complete X-ray system in order to find the optimum trade-off between DQE and MTF.

Fig.5 shows the relation between kV and patient thickness in a cine-system with an Optimus M200 as generator (isowatt kV-mA control). The two curves referring to two tubes with different DQE's are obtained by setting the input exposures so that the same noise impression in the output images is obtained. The lower input dose for the high DQE tube means a lower kV for the same X-ray tube. A lower kV yields a better X-ray contrast for medical details. If the noise remains the same then the higher input modulation of details yields a better detail contrast-to-noise ratio for these details. This is of course only true if all other parameters are kept constant. However, we know already that the MTF-behaviour is negatively affected if the DQE is increased.

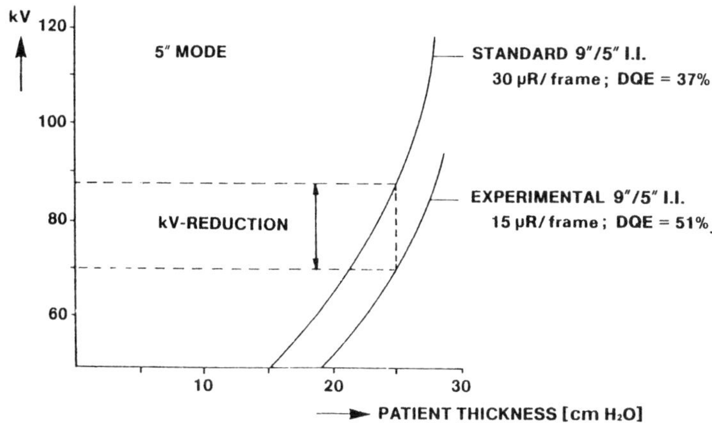

Fig.5. Relation between kV and patient thickness for a cine system with isowatt kV-mA control (Optimus M200) for two X-ray image intensifiers having different DQE.

A decrease in the MTF-behaviour decreases the contrast obtained from details. The smaller the detail the larger the decrease in contrast if the MTF deteriorates. On the other hand this means that also noise is imaged not so well; fluctuations caused by noise being the smallest details to be imaged by the image intensifier. This can be deducted from the exposures given in fig.5. The high DQE is less than 1.5 times the normal DQE, whereas the dose rates applied to the two tubes differ by a factor 2. We can calculate the ratio between the total average intensity in an aperture (S, signal) and the standard deviation of the statistical fluctuations in intensity in the same aperture (N, noise). (We reserve the word contrast (C) for the difference in intensity behind a detail with the surroundings of that detail).
Without going into mathematical details we state that the signal-to-noise ratio can be calculated (all other parameters being constant) as

$$(S/N)^2_{out} = \frac{n\tau\pi\rho^2 \, \mathrm{DQE}}{4\int_0^\infty \frac{1}{\nu} MTF^2(\nu) J_1^2(2\pi\rho\nu)\, d\nu} \qquad (3)$$

In this formula $n\tau$ represents the incident photon fluence and ρ is the radius of a circular aperture with which the noise is observed. $J_1(\nu)$ is the Bessel function of the first order, ν being the spatial frequency. (see also the appendix)

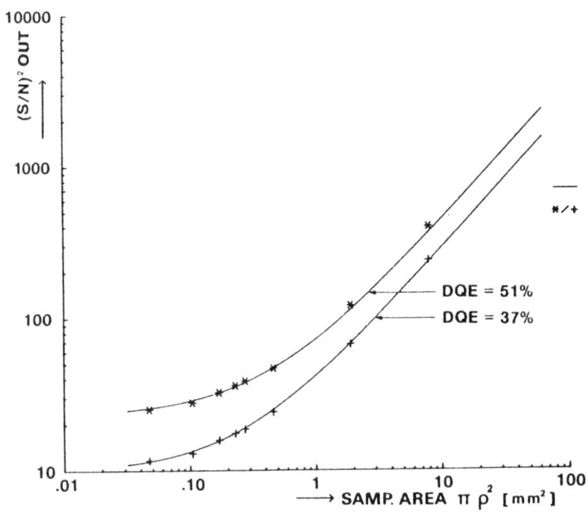

Fig.6. Signal-to-noise ratio as a function of sampling area for two X-ray image intensifiers having different DQE.

Fig.6 shows experimental values and calculated curves for the signal-to-noise ratio for two tubes with the same input exposures having different DQE's and MTF-curves.
The measuring points confirm the calculated curves. For a sampling area that corresponds to the area of the image of a single X-ray quantum (about 0,3 mm^2) the difference in the squares of the signal-to-noise ratio's is about 2. In other words: if the input exposure of the high DQE tube is one half of the exposure on the tube with the normal DQE the signal-to-noise ratio in the output will be the same, for an aperture which has the same area as an image of a single X-ray quantum. Apparently this is the size of the aperture with which the eye of a radiologist scans a medical image in order to judge the acceptability of the noise.

Systems response in relation to tube parameters.

What happens in a cine-system if the DQE of the image intensifier is increased from 37% to 51% and the exposure per frame is reduced by a factor 2? We already mentioned the decrease of X-ray tube voltage associated with an increase in radiation contrast of details and the counter acting decrease in detail contrast because of the decrease in MTF behaviour.
Fig.7 shows these effects in more detail.

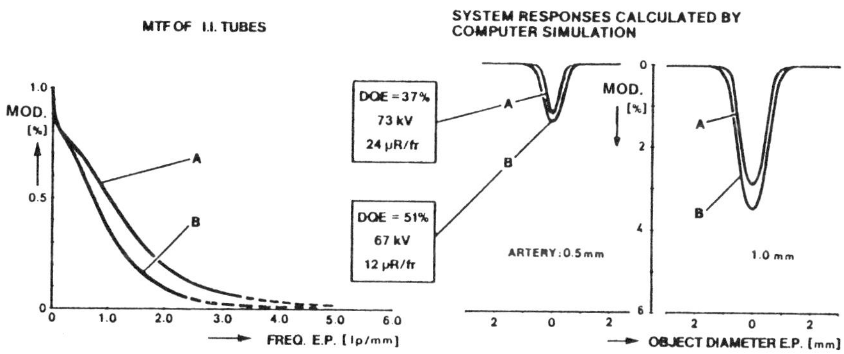

Fig.7. Contrast of two arteries filled with 50% angiografin in the image produced in a cine system with two X-ray image intensifiers with different DQE and MTF behaviour.

In the left hand part we see two different MTF-curves belonging to two X-ray image intensifiers A and B. In the right hand part we see the response of a cine system for two sizes of arteries both imaged with the two image intensifiers. The size of the arteries is 1.0 and 0.5 mm respectively; both are filled with 50% angiografin. The responses are calculated by computer simulation taking into account kV-mA control, focal spot size, filtration, focus-detector distance, patient thickness, scattered radiation, grid, absorption of iodine and all parameters involving the imaging subsystem except the properties (especially the gamma) of the film. The noise in the final image is in all cases the same. The cine system we are dealing with, is a completely normal off the shelf cine system.

Contrast visibility threshold for noiseless images

If we look at the responses we see that in the case of the 1 mm artery the response obtained by image intensifier B (high DQE) is the largest. In the case of an artery of 0.5 mm the response of the system with tube B is still larger than the response of the system with tube A, but the difference is much smaller. The absolute value of the response is somewhat larger than 1%. Even with a gamma of the film of 1.5, the chance that this artery will be seen is very small. The reason for this is the fact that the human eye has a threshold for the detection of contrast differences of about 2%. For an artery of say 0.25 mm the calculated response of tube B (high DQE) will be smaller than the response of tube A. This response will certainly not be seen by the human eye.
If the assumption that the small arteries will not be filled with more the 50% contrast medium is true, then the combination of MTF and DQE of tube B will be better than the choice that has been made for tube A. This is only true if there are no other limits which influence the detectability of details.

Contrast-to-noise ratio; detectability threshold

A phenomenon that effects the detectability of details is the contrast-to-noise ratio. This is an other fundamental limit, which becomes more important with the advent of new imaging techniques such as digital fluoroscopy and digital radiography.

In order to explain the influence of the contrast-to-noise ratio we have to look at fig.8

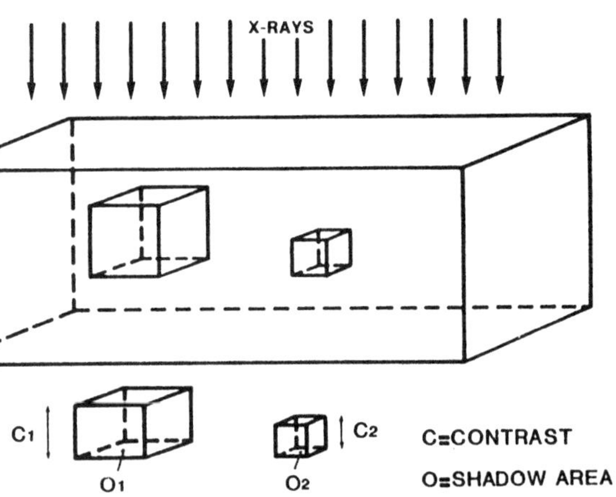

Fig.8. Contrast and shadow area of two cubical details in a rectangular water phantom.

In a rectangular "patient" two cubical details are shown. The two details are composed of the same material that absorps the X-ray quanta at a rate that differs from that of the surrounding material. If the details are relatively small the contrast indicated by the length of the arrow C increases linearly with the characteristic dimension.
When we define the noise of the shadow image of a detail as the relative standard deviation of the statistical uncertainty of the signal level integrated over the area of the image of the detail, then noise depends on the amount of quanta that are contributing to the shadow image :

$$N(oise) = \frac{1}{\sqrt{Q}} \tag{4}$$

where Q is the number of X-ray quanta that contribute to the image of the detail. Q is proportional to the exposure on the detector and to the area of the shadow image. This means that the noise in the shadow image of a cubical detail is inversely proportional to its characteristic dimension (see also Motz and Danos [1]).
Both effects are shown graphically in fig.9. The hatched lines give the noise in the shadow images of details as a function of the characteristic dimension for different dose levels. The solid lines present the contrast as a function of the characteristic dimension of details for different materials.

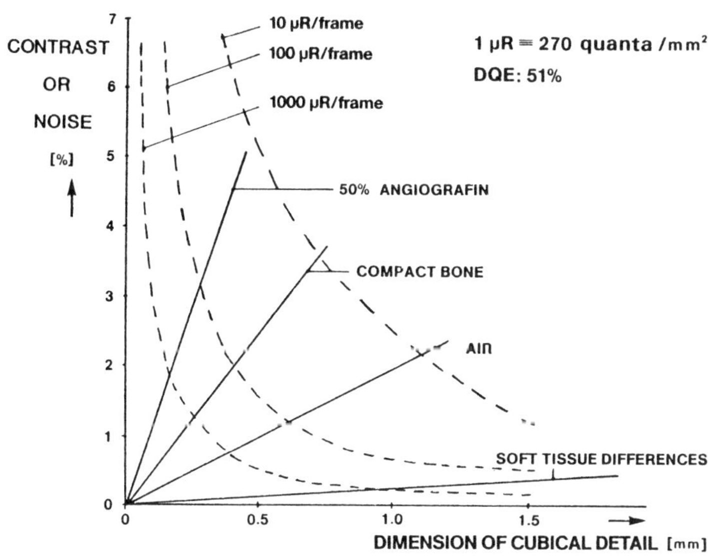

Fig.9. Contrast detectability for cubical details.

When will these details be detectable? These details will be detected if the observed contrast is interpreted as being an object and not just a statistical variation of the signal level due to noise. The noise as defined in formula (4) is the standard deviation of the statistical contrast variations. So whether an object will be detected is a statistical question. If the eye is used as detector and if the contrast threshold is no problem Rose[2] has found that the contrast-to-noise ratio should be 5 for the detection to be possible. Others have suggested values as low as 2 (e.g. Schnitzler[3]).

(One may ask why the eye is again assumed to be the detector. If other detectors are used it is possible to detect contrast differences which are below the visibility threshold of the eye. The contrast-to-noise ratio is not improved by such detectors. The final medical image is evaluated by some one who looks to it with his eyes and then the contrast-to-noise ratio is roughly the same as in the output of the detector).

Detection of details in an X-ray system

Accepting the lower value, the cubical details in fig.9 can only be detected if the contrast exceeds the noise by a factor 2.

So for example the smallest cube filled with 50% angiografin surrounded with water, that can be seen on 100 mm exposures using 100 μR per exposure is only about 0.4 mm. If other parameters like focal spot and movement unsharpness, scattered radiation and MTF of the detector are taken into account a more realistic number of 0.5 mm should be reckoned with.

If we go back to the cine system where exposures per frame of about 10 μR are used we have a different situation. Then cubical details filled with 50% angiografin can only be detected if they have dimensions of at least 0.7 mm.

However if we look at angio images we are looking for blood vessels which are not cubical but long objects. For those objects it is reasonable to assume for the eye that detection can be based on an integration area that has a length 3-10 times its width.

If we assume that the factor 10 is true then we can use the noise curve for 100 μR/frame if we read the numbers at the horizontal axis as width of the blood vessel. In that case we can conclude that blood vessels filled with 50% angiografin can only be detected if their width is minimal 0.4 mm. In reality effects like focal spot, MTF of the detector and the like play a role, and the detectability limit will be at least 0.5 mm.

Optimization criteria

What criteria have to be used in order to optimise an X-ray system and especially the image intensifier in it? Not knowing what the size and the shape of the relevant medical details will be and not knowing exactly how perception processes influence diagnosis we have to choose a "safe" criterion. From the previous discussion we know that if the DQE is increased and the MTF-behaviour accordingly decreased as compared to "normal" tubes the contrast is more improved for larger details than for small details. If we go to ever smaller details there will be a dimension below which the contrast of the detail will be lower for the high DQE tube than for the "normal" tube. If these details are not detectable for other reasons than the low contrast alone, then the high DQE tube will certainly give better results in practice than the "normal" tube. Consequently the "safe" criterion means that we have to find the highest possible DQE (and the MTF that belongs to it) which ensures that the smallest detectable medical details are visible in the final image. A detectable detail means in this context a detail that has in the final image a contrast-to-noise ratio of at least a factor of 2. A visible detail is a detectable detail of which the contrast in the final image exceeds the contrast detectability threshold of the eye.

It will be clear that the optimization is very difficult because all the parameters that play a role in an X-ray system are interrelated. Moreover the optimization is dependent on several boundary conditions imposed upon the optimization process by applicational circumstances. Therefore we make an estimation of the optimal image intensifier parameters in the following way.

First we have to choose for which application we want to optimize. We choose to optimize the image intensifier for cineradiography. Second we have to choose the exposure per frame. The exposure per frame is dependent on the acceptance by the doctor of the noise impression. Although this can vary according to personal preference we know that these variations are not very large. The reason is that doctors set the exposure per frame so that most of the noise falls just below the contrast detectability threshold of the eye, depending on viewing conditions. With this condition the exposure per frame is of course also dependent on the DQE and the MTF of the image intensifier as we have seen (fig.6). In order to be able to proceed with our optimization process we have to assume an exposure per frame, not exactly knowing what the DQE and the MTF will be at the end of the optimization process. However, we know that the outcome of the optimization process will not depend strongly on the choice of the exposure per frame.

Consequently we assume an exposure of about 10 μR per frame. Next we have to choose medical details on which the optimization process will focus. We choose bloodvessels filled with 50% angiografin, being the details with the largest contrast per characteristic dimension (fig.9) and consequently being the smallest details that will have to be visible in the final image. Further we can estimate the smallest bloodvessels that will be detectable.
If we take into account the influence of the size of the focal spot, movement unsharpness and stray radiation the smallest diameter will as discussed before be between 0.4 and 0.5 mm, say 0.45 mm. Now we can optimize the DQE and the MTF of the image intensifier assuming a reasonable thickness of the patient of for instance between 25 and 30 cm. We optimize by choosing the highest DQE and resulting MTF of the tube that yields a contrast that is just visible in the final image. The result of such an optimization is shown in fig.7.
We found that a DQE of about 51% and a MTF curve as shown in fig.7 (tube B) fulfills the optimization criteria. With this combination of DQE and MTF the detectability limit of the smallest detectable details coincides with the visibility limit (contrast detectability threshold of the eye). So in the example of fig.7, tube B is of optimal design. If we compare the performance of tube B (high DQE) and tube A ("old" tube) we see that the performance in an X-ray system is the same for the smallest detectable details and that for larger details tube B gives better results.

Applicational results

In the previous section we have shown the results of a purely theoretical exercise. Many boundary conditions have been assumed which depend on practical circumstances. Therefore only an extensive program of application tests can give the right answer. This is even more necessary if we take into account that one image intensifier can be used for more than one application. In a three channel system one tube can be used for 105 mm photography, for cineradiography and for fluoroscopy.

We have found that tubes like tube B in fig.7 are giving good results in applicational situations. A large increase in image quality for cine and fluoroscopy is experienced as compared with the results obtained with tube A (fig.7). The results for 105 mm photography are of equal quality or better than the results obtained with tube A. This is true even if no use has been made of the possibilities of kV reduction (if one wants to be able to see through the barium contrast medium) and if even then the exposure per frame has been reduced by approximately a factor 2.

Apparently we have optimized the tube staying on the safe side for cine application to such an extend that the results for 105 mm photography are not less than for the "old" tube (tube A fig.7). How far we have stayed on the safe side is still subject of study.
Maybe that in some future the optimization point will shift into the direction of even higher DQE and lower MTF behaviour. However we do not expect such a shift, if any, to be very large.

How to characterise image quality of X-ray image intensifiers

The chosen optimum implies that the smallest details that can be detected in an X-ray system have a characteristic dimension of roughly 0.5 mm even with an optimized image intensifier. Therefore we have the opinion that the modulation for a spatial frequency of 1.0 lp/mm (the modulation factor) is much more important than the resolution as measured with a funkphantom. In the same time the DQE is also important for the ability of the image intensifier to detect small medical details. Consequently we have to know both the modulation factor and the DQE in order to be able to judge an image intensifier on its ability to detect small medical details (in an X-ray system). This clearly illustrates that not a high DQE or high characteristic modulation in itself is a guarantee for a good image quality.

It would be attractive to have a kind of figure of merit with which the performance on image quality of different tubes can be compared. In this figure of merit the influence of DQE and modulation factor must be incorporated.
Such a figure of merit depends on the applicational circumstances, on the X-ray system in which the image intensifier is used and on personal preference. Therefore it will be very difficult to define a figure of merit of an image intensifier that has universal significance. Maybe it is more appropriate to try to characterise the performance of a complete X-ray system not bothering with the image intensifier itself.
However, in order to be able to judge the performance of a complete X-ray system it is necessary to know at least the following parameters of the image intensifier in that system: the low frequency drop, the DQE and the modulation factor.
We have the feeling that all manufactures should provide guaranteed numbers for these parameters to the customers.

Conclusions on image quality.

An X-ray image intensifier should be optimized for its use in an X-ray system. In order to be able to judge an image intensifier on its performance in an X-ray system other parameters have to be known than the manufacturers provided until now.

The DQE and the modulation factor in combination determine the ability to detect small medical details in an X-ray system. The resolution as measured with a funkphantom is not at all indicative for this aspect.
Image intensifiers with maximised resolution will not be able to detect small details as well as an optimized tube if used in the real situation, that is in a complete X-ray system.

An optimized (high) DQE will allow for lower doses delivered to the patient, the doctors and radiographers during the examination.

The low frequency drop is much more indicative for the low frequency contrast behaviour of image intensifiers than the old fashioned contrast ratio as measured with the measuring method standardised by the NEMA.

In order to give the users of image intensifiers the opportunity to judge the image performance of the tubes as used in complete X-ray systems the manufacturers should provide accurate numbers for the DQE, modulation factor and low frequency drop.

Image quality control in the manufacturing process of image intensifiers.

General quality control

From the previous sections it is clear that DQE, modulation factor and low frequency drop are the most important image quality parameters. In this part of the paper we will describe our industrial control system for these parameters. This does not mean that other quality aspects are not important. For a good understanding it should be mentioned that in the cause of our manufacturing process of X-ray image intensifiers many parameters are checked or adjusted such as sensitivity, conversion factor, focus voltage, image diameter, image displacement, high tension stability, pump action, ion spot, optical prefocussing and mechanical dimensions.

Apart from these measurements and adjustments we conduct design tests confirming to IEC 68: chock, bump and vibration tests, cold, warm and cyclic damp heat tests and tests on low air pressure and temperature change.

Life and shelflife tests are conducted on a sampling basis. During life tests the image intensifier receives 300 R (sometimes 1200 R) in 1000 hours simulating at least 5 years of intensive use.

All these tests and also practice of every day show that all image quality parameters of our image intensifiers are constant during many years of use. Only the conversion factor, which is not an image quality parameter as long as it is exceeds a certain limit, shows a change between 0 and 20% during lifetests and/or for 5 years of shelflife.

Control of DQE

In order to control the DQE one could think of measuring it in all X-ray image intensifiers.
However, the equipment to measure the DQE is sophisticated and the measurements are very complex and time consuming. Therefore the direct measurement is only done as a check on a sampling basis.
For the control of the DQE in the manufacturing process indirect measurements are used. We know the sources that contribute to the DQE and we know that nearly all contributions are independent on the manufacturing process.
An exception of course is the physical thickness of the CsI as deposited on its carrier. The thickness of this layer together with the attenuation of the carrier of the CsI layer is measured of every input screen before it is mounted in the tube. This is done by measuring the attenuatio n for X-rays with a well defined beam quality. The measuring setup is shown in fig.10.

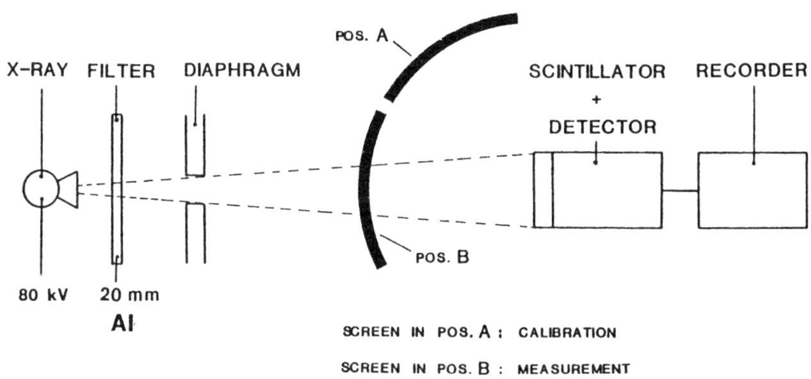

Fig.10. Measuring of the attenuation of X-ray in an input screen.

The quantum flux at the input of the detector is measured with the input screen in the beam and subsequently with the input screen removed from the beam. The ratio of the two values is a measure for the DQE of the image intensifier. The accuracy of the method results in a value of the DQE that is guaranteed to be within 1% equal to the specified value.

Control of the modulation factor

It is very expensive to measure the MTF of each image intensifier in order to measure and control the modulation factor (see fig.11) in the manufacturing process. Therefore a specific measuring set up has been developed for the measuring of the modulation factor [4] (fig.12).

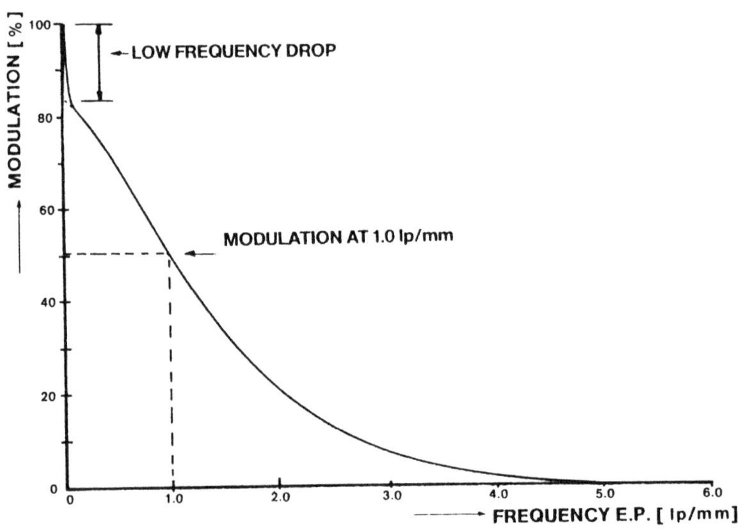

Fig.11. The modulation at 1.0 lp/mm (the modulation factor)

The measuring method is rather simple. A slit in a lead diaphragm in the entrance plane is irradiated by X-rays. The image intensifier forms an image of the slit on its output screen. This image is optically projected on a pattern that in its general form consists of transparant and opaque diamond shaped elements (see fig.12).

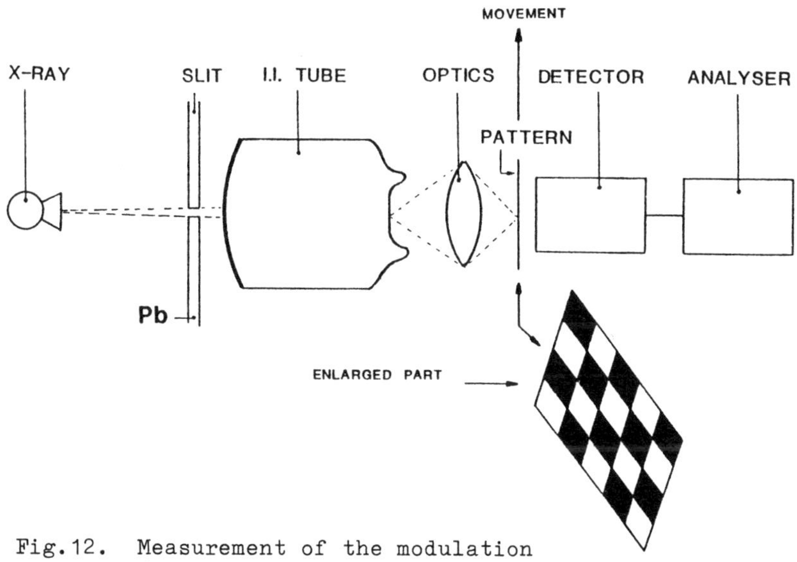

Fig.12. Measurement of the modulation factor at a single spatial frequency.

The amount of light from the slit image passing through the pattern is modulated by moving the pattern in a direction perpendicular to both the slit and the axis of the tube.
In this way the slit image is spatially filtered by a triangular wave filter. The output signal of the detector is modulated in time according to the movement of the filter. This signal is filtered with a sinusoidal spatial filter with the same spatial ground frequency as the triangular filter. After filtering the signal is proportional to the modulation transfer of the tube at a spatial frequency that corresponds with the spatial frequency of the triangular filter. By multiplying the derived signal with a calibration factor we find the modulation factor.
The absolute measuring accuracy is about 1%.

The low frequency drop is taken into account with this measurement by analysing the entire output screen in the optical projection.

So also those parts of the output screen that lie outside the image of the slit contribute to the eventual signal via scattering of X-rays and light in the image intensifier.

Each image intensifier is measured in this way and the measured number is indicated on the measuring protocol that is provided to the customer in most countries. A tube that does not fulfill a preset criterion in this respect is destroyed.

A complete MTF measurement of the manufactured image intensifiers is done on a sampling basis. An important spin off is that it provides a crosscheck on the measuring accuracy of the MTF and the modulation factor.

<u>Control and measurement of uniformity of modulation factor over the area of the tube</u>.

Due to manufacturing tolerances it is possible that the MTF is not the same everywhere on the active area of the image intensifier. A way to check this is to measure the modulation factor on several locations on the area of the image intensifier. This, however, is not economical since the modulation factor measurement, although less complicated than a MTF measurement, is still time consuming and expensive. Therefore an other method is used.

A special phantom has been designed which is used in a measurement of the uniformity of MTF performance of the tube. The phantom contains a number of funkphantoms distributed over the entire area of the tube placed in radial as well as in tangential direction. The most peripheral position is located at 90% of the radius of the image.
An X-ray image is made of this phantom on a 100 mm photo. This photograph is made after the optimal electrical and optical focussing adjustment is obtained with the aid of an X-ray photometer[4)]. The X-ray photometer is an instrument that optimizes in an objective way the image sharpness for 5 equidistant points on the radius of the image simultaneously. (The Philips service organisation in the field uses the same instrument for installation and maintenance. This ensures that the measurement in the factory is done with the same focussing adjustment as in the field).

The limiting resolution is read from the photograph. The resolutions measured in the image are not allowed to differ more than 0.3 lp/mm as compared with the mean value of all readings.

Previously in this paper we have explained why a resolution measurement is not a correct absolute measurement. We have also indicated that it may be used as a relative measurement. This is exactly the way in which the resolution measurement is used.

For every tube two of these photographs of the special phantom are made. One is kept in the file of the factory and the other is inserted in the measuring protocol that goes with the tube. This photograph can in principle be used to compare the result of a 100 mm photograph made after installation of the X-ray system with the result that has been achieved in the factory. Again here the resolution measurement is only used as a relative method. (We have to keep in mind that a resolution check done in this way only checks the correct adjustment of the imaging subsystem containing the tube, shield, high tension generator and the optics.
It does not say anything about the other (system) parameters that control the image quality performance of an X-ray system).

Control of luminance distribution

The luminance distribution in the output screen is controlled for each tube produced.
The entrance window of the image intensifier is irradiated with filtered X-radiation.
After calibration of the intensity distribution of the X-rays in the entrance plane the luminance distribution of the output window is measured. This is done by scanning with a pinhole over the diameter of the image of the output window produced by an optical system (see fig.13).

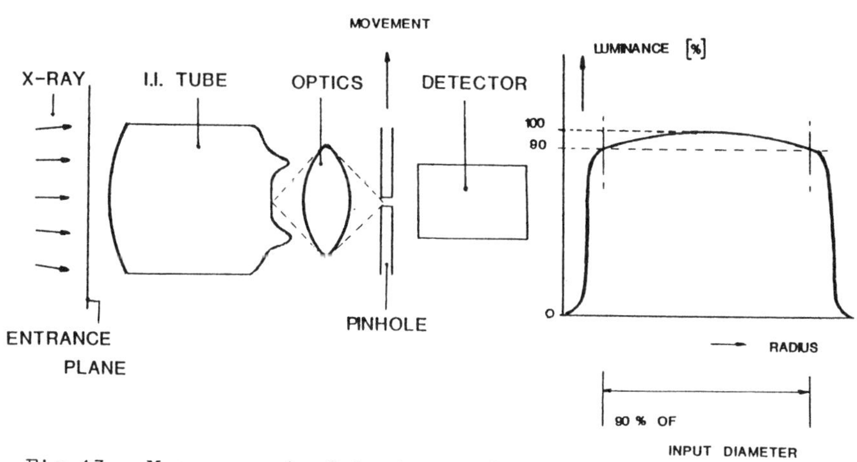

Fig.13. Measurement of luminance distribution.

On a sample basis also isoluminance curves are plotted over the entire output screen in order to have a check on the accuracy of the measuring system.

The previous measurement takes care of effects that vary relative slowly over the area. There are other nonuniformities in the image that have smaller dimensions. Very small nonuniformities in the image with high contrast are called spots. They are in general caused by local faults in input or output screen or by loose particles in the tube. Because these faults cannot be avoided completely there must be a method to decide whether these defects are acceptable. This is done by a system with which each spot scores penalty points, depending on its size and its location. The total summation of penalty points has to be below a certain number which is dependent on the type of image intensifier. Spots that are larger than a specified size are not allowed. If the score is higher than the predetermined number the tube is destroyed.

Apart from spots sometimes large sized nonuniformities of low contrast may show up in the image. Also disturbances like stains, stripes or rings with low contrast may be sometimes visible. These defects are generally caused by sensitivity spread or process faults. In the production control, these defects are compared with a series of photographs that give examples of such defects that are just acceptable. If the defects are not worse than the defects on the standard photographs these nonuniformities will not be visible in a normal medical image. Tubes that do not comply to this test are destroyed.

<u>Control of the low frequency drop</u>

Extensive experience has shown that the low frequency drop is a design parameter and does not vary because of variations in the manufacturing process.
As, already indicated the complete MTF, including the low frequency drop, is measured on a sample basis. In this way we continue to keep an eye on the low frequency drop. We have never found an image intensifier that had a low frequency drop differing more than 1% from the specified values. It is very likely that the small differences we do find are resulting from spread in the measuring procedure rather than from real differences in low frequency drop from tube to tube.

Conclusions

The most important parameters of an X-ray image intensifier that control the image quality performance of an X-ray system to a large extend are the DQE, the modulation factor (the modulation transfer at 1.0 lp/mm) and the low frequency drop. These parameters are effectively under control in the manufacturing process of the image intensifiers of Philips Medical Systems. The DQE is controlled with an accuracy of 1%. The low frequency drop is a design parameter that does not vary due to variations in the manufacturing process. It is constant with an accuracy of 1%. The modulation factor can vary due to differences in the processing during manufacturing. Therefore this parameter is measured for every tube. The test method has an absolute accuracy of 1%. Every tube that does not meet its specification in one or more aspects is rejected.

All important image quality parameters are guaranteed. In addition individual values are indicated on a measuring protocol that goes with the image intensifier. Measures are taken to check in the field if the imaging subsystem is optimally adjusted. This is important for the reproduction in the field of the potential imaging performance of the subsystem.

References

1. Motz, J.W. and Danos, M, Image information content and patient exposure, Medical Physics 5, 8(1978).

2. Rose, A, The sensitivity performance of the human eye on an absolute scale, Journal of the optical society of America, 38, 196 (1947).

3. Schnitzler, A.D., Image-detector model and parameters of the human visual system, Journal of the optical society of America, 63, 1357 (1973).

4. Leunen, H.A.J.van, An instrument for the measurement of the modulation transfer factor of image transfer devices, Sira conference april 1981.

5. Proper, J.; Schuur, J.; Monté, G.L.A., Method of and apparatus for adjusting an image intensifier chain, US patent no. 4272782.

6. Albrecht, C., Noise Sources in Image Intensifying Devices, Proceedings Diagnostic Radiologic Instrumentation, C.C.Thomas, Springfield, Illinois, U.S.A. (1965)

Appendix

The signal-to-noise ratio on the output of an image intensifier depends on :

a. the signal-to-noise ratio of the input signal
b. the detective quantum efficiency including its correction for the long-decay behaviour of the phosphors applied.
c. the blurring of the image intensifier (because it smeares out" the peaks of the random signal).

Albrecht[6)] has described the effect of blurring on the signal-to-noise ratio of an image intensifier.
This appendix will give a brief explanation of equation(3) in which the relation between (S/N), MTF and DQE is given.

Albrecht has shown that the statistic of a blurred output signal S_{out}, generated from domain A of the image plane of an image intensifier, can be described by

$$N^2_{out} = \sigma^2(S_{out}) = C n \tau \iint_{-\infty}^{\infty} MTF^2(\nu_x, \nu_y)\, \mathcal{F}^2[f_A(x,y)]\, d\nu_x\, d\nu_y \qquad (I)$$

where C indicates the efficiency of the conversion processes of the image intensifier

n is the number of X-ray quanta incident per mm^2 and second

τ is a time interval

$F[f(x,y)]$ is the two-dimensional Fourier transform of a two-dimensional function f (x.y).

MTF (ν_x, ν_y) is the two-dimensional MTF with the coordinates ν_x, ν_y in the Fourier domain.

$f_A(x,y)$ is the aperture function with

$$f_A(x,y) = 1 \quad (x,y) \in A$$
$$= 0 \quad (x,y) \notin A$$

The mean value of this output signal S_{out} can be derived in a similar way and is given by

$$\overline{S}_{out} = C.\, n \tau A \qquad (II)$$

Note that the mean value $\overline{S}_{out}$ is not affected by the blurring of the system.

The square of the signal-to-noise ratio of S_{out} follows from (I) and (II).

$$\left(\bar{S}/N\right)^2_{out} = \frac{C n \tau A}{\iint_{-\infty}^{\infty} MTF^2(\nu_x,\nu_y)\, \mathcal{F}^2\left[f_A(x,y)\right] d\nu_x\, d\nu_y} \qquad \text{(III)}$$

For a circular domain with radius ρ equation III can be rewritten as

$$\left(\bar{S}/N\right)^2_{out} = \frac{C n \tau A}{2\int_0^{\infty} \frac{1}{\nu}\, MTF^2(\nu)\, J_1^2(2\pi\rho\nu)\, d\nu} \qquad \text{(IV)}$$

where J_1 is the Bessel-function of the first kind and the first order, ν is the spatial frequency and under the assumption of :

a. the conditions of isoplanatism are met
b. the system has rotational symmetry
c. $\rho \ll R_{IP}$ (R_{IP} is the radius of the image plane)

The constant C in the formula is determined by the measuring conditions of the DQE-measurement.
In this measurement a small beam of X-ray quanta incidents on the centre of the input of the image intensifier and the output signal of the entire image field is analysed.
The measured values are measurement time interval dependent because of the temporal characteristics of the phosphors. After correction for the long-decay behaviour of these phosphors (the square of) the signal-to-noise ratio can be approximated by

$$(S/N)^2_{out} = 2C.n\tau B \qquad \text{(V)}$$

where B is the cross section of the incident beam.
Due to the Poisson distribution of the incident quanta equation V can be rewritten as

$$(S/N)^2_{out} = 2\ C\ (S/N)^2_{in}$$

or

$$C = \tfrac{1}{2}\ \frac{(S/N)^2_{out}}{(S/N)^2_{in}} \qquad \text{(VI)}$$

According to (1) and (VI)

$$C = \frac{1}{2}\,\mathrm{DQE} \qquad \text{(VII)}$$

Combination of (VII) and (IV) gives

$$(S/N)^2_{out} = \frac{\eta\tau . \pi\rho^2 . \mathrm{DQE}}{4\int_0^\infty \frac{1}{\nu}\,\mathrm{MTF}^2(\nu)\,J_1^2(2\pi\rho\nu)\,d\nu} \qquad \text{(VIII)}$$

DESIGN AND FUNCTIONAL DESCRIPTION OF OPTICAL IMAGE DISTRIBUTORS FOR IMAGE INTENSIFIER SYSTEMS

Arend J.W.A. Vermeulen, Aart A.G. Beekmans, and Johan W. Haarman

N.V. Philips' Gloeilampenfabrieken, Eindhoven, The Netherlands

Abstract

Design of optical image distributors for image intensifier systems implies cost effective guarding of tolerances as well as trade offs between conflicting design targets. The performance is often limited by the state of the art in technology.

Introduction

Optical image distributors are needed to relay the opto electronically demagnified X-ray shadow image at the output screen of an image intensifier to film - or TV cameras. Although the concept of use of optical image distributors for this purpose exists for more than 20 years the increasing quality of X-ray image intensifiers and the growing need for cost effective production make challenging and continuous demand for improvement of these systems.

In the next sections we will summarize the design targets and explain the design philosophy which often leads to trade offs between conflicting design targets.

Design targets

Requirements set to an image distributing system are:

1. Minimum loss of diagnostic information
2. Acceptable loss of luminous power
3. Maintenance of illuminance distribution
4. Minimum deterioration of object shape
5. Magnification to required size
6. TV-monitoring during operation of film cameras
7. Illuminance control and phototiming
8. Fast switching between channels
9. Suited for different image intensifier types
10. Suited for different cameras
11. Compact design
12. Insensitive to temperature variations
13. Cost effective manufacturing.

Structure of optical distributing systems.

The relay of the visual information is performed by conventional refractive optics. The required magnification varies from near unity up to five. The rules of optical design necessitate a nearly symmetrical optical design for these imaging ratios.
This in combination with the requirement of multi channel operation makes it obvious to split up the optics into a tandem combination of two lenses, applied at infinite conjugates, see fig.1.

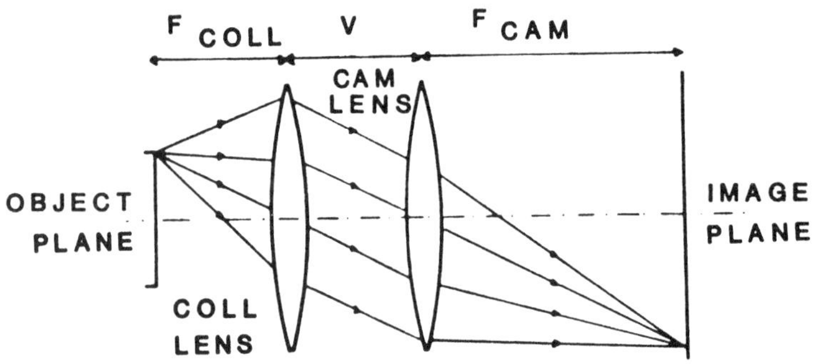

Fig.1 Schematic view of lens tandem combination

The image intensifier output screen acts as the object plane for the optics. Optically it coincides with the focal plane of the first tandem component, denoted as collimating lens. Each point in the object plane gives rise to a set of parallel rays with an inclination to the optical symmetry axis determined by the distance between the center of the image and the object.
The second tandem component is denoted as camera lens. It converts these sets of parallel rays back into a real image located in the focal plane.

The demand for simultaneous TV-monitoring during camera operation can be met by inserting a partially reflecting, partially transmitting mirror between the tandem components, see fig.2. By this mirror the sets of parallel rays can be relayed simultaneously to two channels.

By changing the position of the mirror one can switch over to different channels for other film cameras, if needed.

Via a prism a small fraction of the parallel rays is guided to a photodiode in order to supply a driving signal for the illuminance control and phototiming in the X-ray system. Pre adjusted diaphragms or, in recent designs, motor driven iris diaphragms control the illuminance in each camera channel.

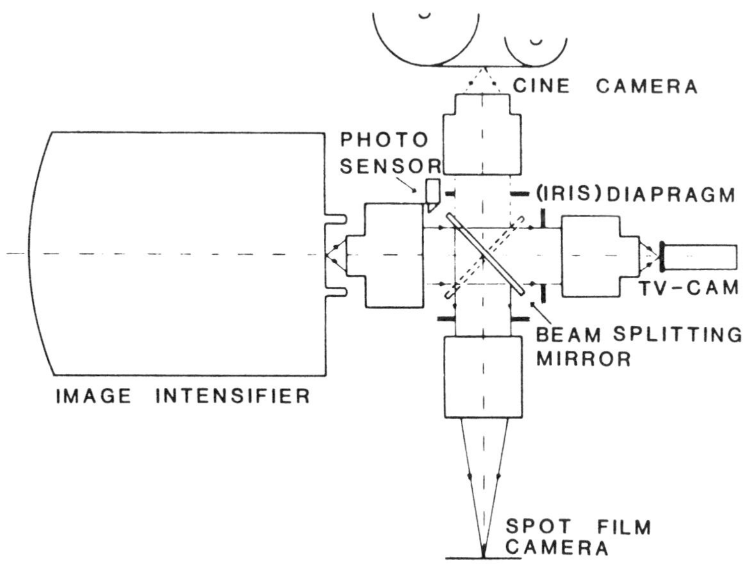

Fig.2 Structure of an optical image distributor system.

Design considerations

Apart from the housing,the image distributor consists of three functional units: the optics, the beam splitting mirror and the sensor for phototiming. In this section it will be explained in which way several design targets are conflicting and why accurate guarding of tolerances is necessary.

1. Optics

A. Focal length of the collimating lens:

Compact design requires a short focal length. By applying a short focal length the object plane is located very close to the entrance pupil of the optics. Rays from the edge of the object field enter the optics at a large field angle φ . This field angle is defined by the arc tangent of the ratio between object height, r, and focal length of collimating objective, f_{coll}. If φ assumes large values the design conflicts with the demand for maintenance of illuminance distribution. This can be seen in the following way :

Towards the edge of the object plane the distance between object plane and optics increases. This causes a decrease of the effective solid angle subtended by an object point and the entrance pupil of the optics. On top of the illumination decrease that results from a smaller effective solid angle, ray obliqueness also effects illuminance by decrease of the effective areas of both a light emitting and a light receiving surface element, see fig.3.

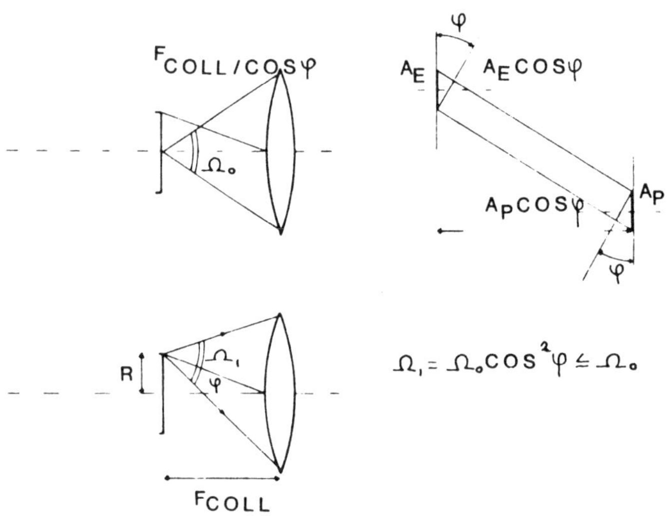

Fig.3 Illustration of photometric vignetting note that $\Omega_1 < \Omega_0$ by the increasing distance at the edge of the field.

As shown in this figure the projections of the surface elements in the object plane and in the plane of the lens entrance pupil on the oblique beam direction are effective for the illuminance transfer.
The over all effect results in an angular illuminance dependence given by $\cos^4\varphi$.
It is denoted as photometric vignetting.
For the practical design a maximum field angle ranging from 7° to 9° is found as the optimum compromise between compact design and maintenance of illuminance distribution.

Different image intensifier types involve output screen diameters ranging from 15 to 35 mm. The design target which requires that an image distributor system has to be suited for different image intensifier types can be met if an unique maximum field angle, φ_{max}, is specified. This specification requires for each output screen diameter the design of a collimating lens with such a focal length that the field angle at the edge of the object field equals φ_{max}. This implies for a specific output screen diameter, d, a focal length $f_{coll} = \frac{\frac{1}{2}d}{tg\varphi_{max}}$.

In this way the subsequent components of the image distributor system become format independent, see fig.4.

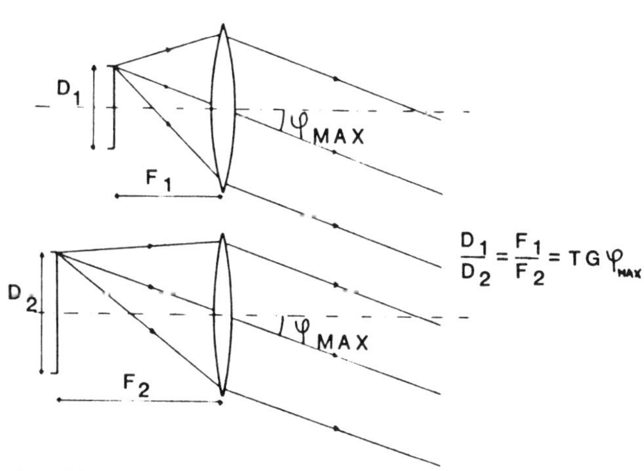

Fig.4 Illustration of format independence.

B. Focal length of camera lens :

The focal length of the camera lens, f_{cam}, is determined by the design target which demands for mangnification to a required size. For a required image diameter, d', the required focal length is given by : $f_{cam} = \frac{\frac{1}{2}d'}{tg\,\varphi_{max}}$

C. Choice of lens diameters :

Acceptable loss of luminous power necessitates a lens with a large diameter. Increase of the lens diameter causes increase of cost and lens aberrations. Lens aberrations conflict with the demand for minimum loss of diagnostic information.
Therefore the lens diamaters have to be carefully adapted to allow the required receptor response to be reached at the lowest operational light level.
If an illuminance, E_{req}, is required at a specific minimum brightness, B_{min}, the aperture angle, u, follows from

$$\sin^2 u = \frac{1}{\pi\tau\eta} \cdot \frac{f^2_{cam}}{f^2_{coll}} \cdot \frac{E_{req}}{B_{min}}$$

Here τ and η stand for the transmission of the optics and the fraction of light relayed by the beam splitting mirror respectively.

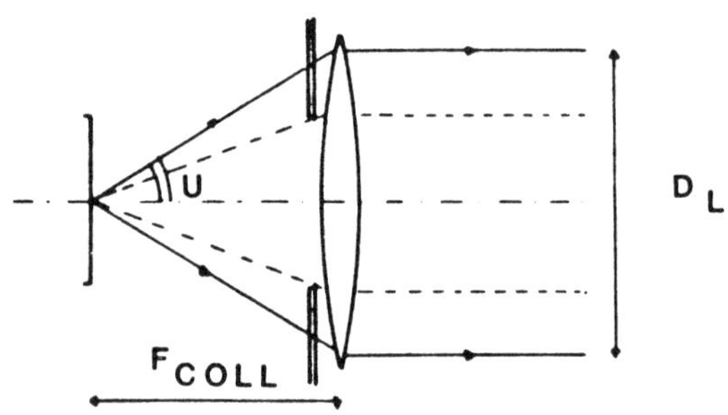

Fig.5 Defenition of aperture angle u. Note that the light flux is limited when stopping down a lens.

The aperture angle, see fig.5, for given lens diameter, d_l, is defined by

$$u = \text{arc tg}\left(\frac{\frac{1}{2}\,d_l}{f_{coll}}\right)$$

From this relation it is evident that a minimum value allowed from sensitivity considerations for the aperture angle also implies a minimum value, d_{min}, for the lens diameter.

The camera lens need not have a diameter larger than this minimum diameter. However, designing the collimating lens at this minimum diameter conflicts with the demand for maintenance of the illuminance distribution. This is caused by the separation of the tandem components at a vertex distance, v, that leaves room for the beam splitting mirror, see fig.6.

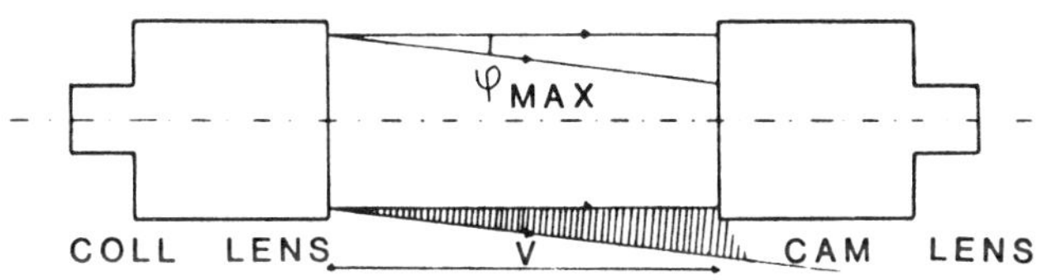

Fig.6. Geometric vignetting by separation of tandem components.

Rays from a central point leave the collimating lens parallel to the optical axis, so they are all able to enter the camera lens. Rays from an off axis point however are inclined to the optical axis. By this inclination they move aside, which makes it impossible for a fraction of the rays to enter the camera lens. By the design of a collimating lens with a diameter that exceeds the camera lens diameter by more than $2\ v\ tg\ \varphi_{max}$ this so called geometric vignetting can be completely avoided, see fig.7. In practice some geometric vignetting is allowed.

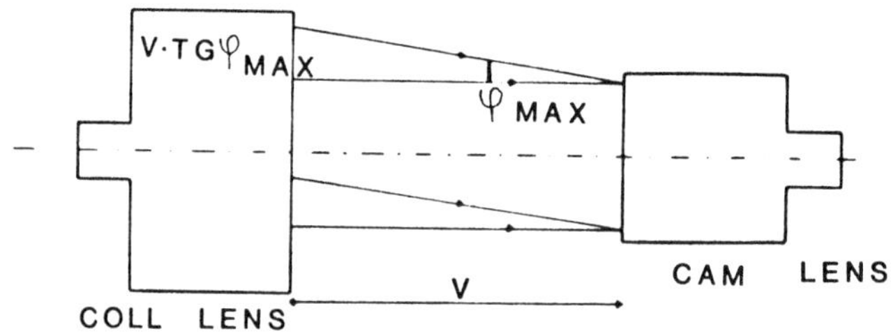

Fig.7. Elimination of geometric vignetting by increasing the lens diameter of the collimating lens.

D. <u>Intrinsic optical design:</u>

Once the focal lengths and apertures are fixed it is necessary to specify the intrinsic quality of the optics. Sufficient quality can be reached only when each lens is the compound of a number of lens elements. Lens optimization implies proper positioning of these lens elements within the compound lenses and optimal choice of glass types and curvatures.

There is risk if the optimization is done separately for the collimating lens and the camera lens, as the parallel light between these compound lenses is partially coherent. This makes it impossible to predict the performance of the combined compound lenses straight forwardly e.g. by multiplying MTF's.
Therefore a design for a combination of collimating and camera compound lenses can be applied in an optical image distributor only if their performance as a tandem combination is evaluated first, preferably by measurement.

Beyond the possible influence of the tandem partner there are two other external parameters that have to be taken into account continuously during the lens optimization procedure.
At first the spectral distribution of light, combined with the receptor spectral response, has to be specified as the refractive indices of all glass types are wavelength dependent.

Secondly the presence of glass elements in the optical path due to vacuum tube housings has to be taken into account. Ignoring these glass elements increases spherical aberration as illustrated in fig.8.

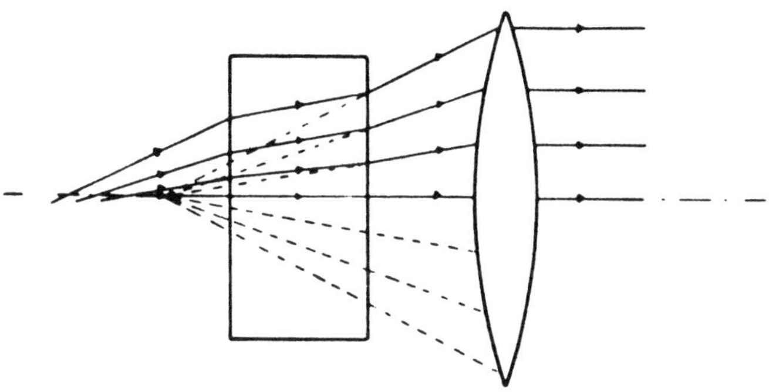

Fig.8 Spherical aberration if glass elements in the optical path are not compensated in the optical design.

The final design is determined by the design targets. Acceptable loss of luminous power reduces the number of useful glass types to types with sufficient transmission in the spectral region mentioned above.
Compactness necessitates a design like a telephoto lens i.e. the principal planes have to be located near the vertex of the compound lens at the parallel ray side, see fig.9.

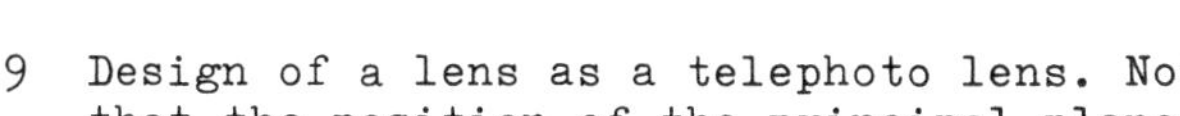

Fig.9 Design of a lens as a telephoto lens. Note that the position of the principal plane enables lens positioning very close to the image plane.

Additional compactness demands, especially for X-ray systems where the image intensifier is placed below the table, may make it necessary to fold the optical path. As shown in figure 10 this can be accomplished by designing elbow lenses, where a prism is inserted between the lens elements.

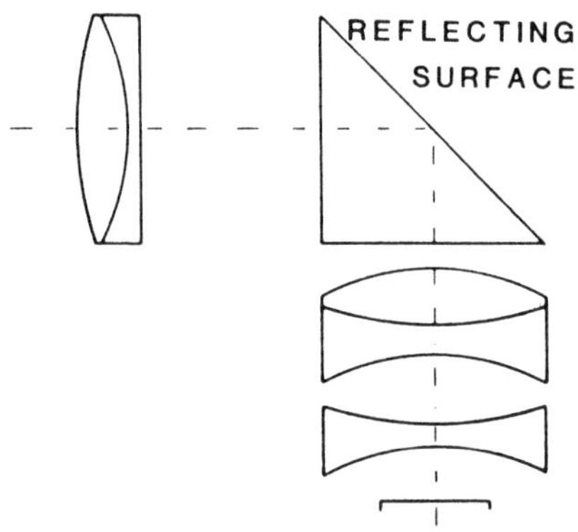

Fig.10. Design of an elbow lens.

Another solution is obtained by the application of total reflecting external mirrors, see fig. 11. In this way the vertex distance is increased. This leads to an increase of geometric vignetting which conflicts with the demand for maintenance of illuminance distribution.

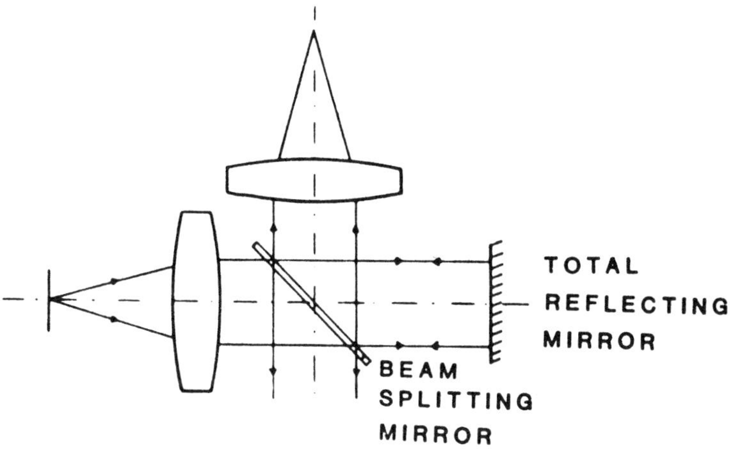

Fig.11 Folding of light path by external total reflecting mirror.

<u>Maintenance of illuminance distribution</u> also demands for lens elements with a diameter that is sufficiently large. The design must be such that the rim of only one lens element limits the light rays from both on axis and off axis points. Else the compound lens shows internal geometric vignetting, see fig.12.

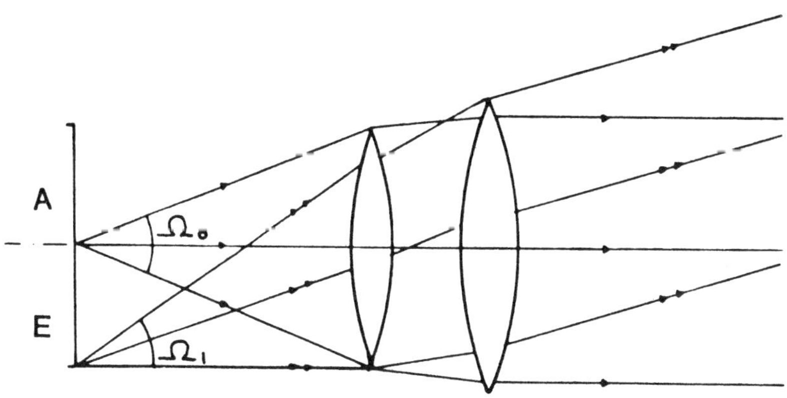

Fig.12. Illustration of geometric vignetting inside a compound lens. The size of the cone from E is limited by the lens element rim at T.

Minimum deterioration of object shape requires adequate compensation of distortion during the optics design.

Insensibility to temperature variations requires such a design that mechanical defocussing caused by thermal expansion of mounting support and lens housing is compensated by optical defocussing from thermal expansion of glass elements and thermal change of refractive indices.

Cost effective manufacturing implies the application of a minimum number of lens elements and small curvatures. This may conflict with the demand for minimum loss of diagnostic information, which requires optimal compensation of lens aberrations. For this purpose the optics designer needs the many degrees of freedom given by a large number of lens elements.

2. Beam splitting mirror

A. Glass substrate:

Compactness requires a design with minimum dimensions. Maintenance of illuminance distribution demands for a design that is sufficiently large to relay all parallel light beams that take part in the imaging. Else the system will suffer from geometric vignetting.
As the mirror is placed in one direction at an angle of 45° with respect of the incident beam, the above compromise leads to a mirror with an oblong shape.
Polishing oblong surfaces however can easily lead to aspherical surface curvatures, because the polishing disk is not supported uniformly by the mirror surface. Practical reasons require to accept some curvature of the mirror surface. By allowing some curvature the mirror will act as a lens which can sufficiently be compensated for by defocussing the optics.

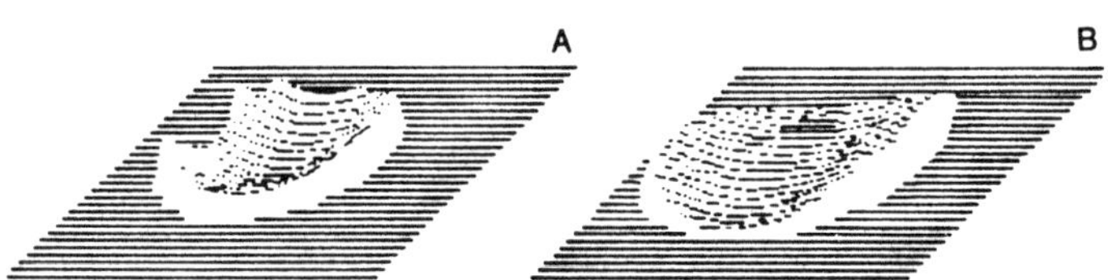

Fig.13 A. A typical effective curvature when positioning a mirror with spherical flatness deviation at 45°.

B. Idem after separating the spherical component by calculation.
Maximum flatness deviation approximates 1 μm.

Positioning of a spherical mirrir at 45° adds a cylindrical component to the lens action of the mirror, see fig.13. Combined with cylindrical curvature or even saddle shapes that may arise from the polishing process the optical focus can be smeared out, which cannot be compensated for by refocussing the optics. This conflicts with the demand for minimum loss of diagnostic information. Therefore the allowed flatness of an image distributor mirror is restricted to a few newton fringes.
For the same reason the mirror has to be designed sufficiently thick to prevent bending in its mounting frame. This in turn conflicts with the demand for fast switching between channels, which would benefit from a design with minimum mass.

B. Mechanical mounting:

Minimum loss of diagnostic information requires a mirror suspension that prevents vibration during camera operation, as vibration causes image blurring.
The mirror has to end up properly aligned after switching to an image distributor channel because tilt causes a shift of the image, mostly accompanied by asymmetric geometric vignetting.

C. Coating:

Acceptable loss of luminous power implies a coating with little absorption. Therefore a dielectric coating is applied. The reflection/transmission ratio has to be specified for the given spectral distribution of light and receptor spectral sensitivity. As most dielectric coatings are angle dependent with regard to their reflection/transmission ratio, this parameter has to be specified for the angular range : $45^\circ - \varphi_{max} \leq 45^\circ \leq 45^\circ + \varphi_{max}$.
Else the intensity distribution in the image plane will become slanting, which conflicts with the demand for maintenance of the illuminance distribution.
Good expertise requires an anti reflection coating at the rear side of the mirror to prevent multiple reflections.

3. Photosensor:

Contrary to the design of the previous components, where many trade offs have to be made, the design of the photosensor is straightforward.
In order to meet the demand for illuminance control and phototiming the available amount of light has to be measured.

As the X-ray shadow image contains darker and brighter parts one must be able to use a selected region of the image intensifier output image for phototiming.

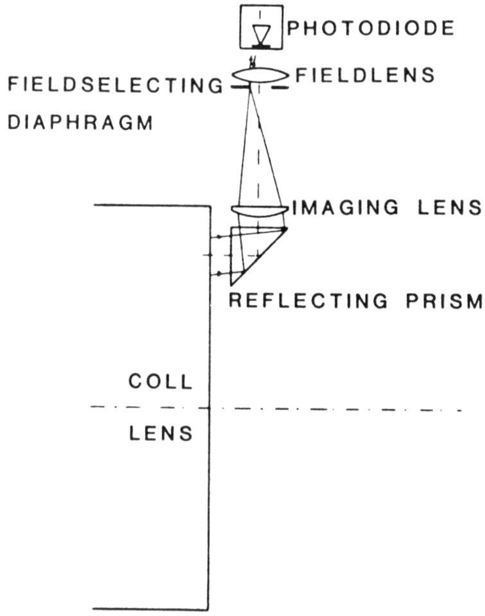

Fig.14. Schematic view of phototimer sensor.

This is accomplished by the unit shown in fig.14. A prism guides a fraction of the parallel light to an imaging lens with short focal length. A minified real image of the image intensifier output screen is shaped in the focal plane of this lens. A field selecting diaphragm in this focal plane transmits only the region of the X-ray image wanted for phototiming. A field lens concentrates the transmitted light upon a photodiode.

Adequate design implies proper alignment of the prism and vignetting free design of the lens system in the photo-sensor.

Trends in design

As stated in the introduction the design of optical image distributors continuously needs improvement. This need arises from several developments:

A. Increasing isoplanatism of image intensifiers:

With the improve in technology the performance of image intensifiers increases. This holds especially for the sharpness at the edge of the image, leading to an increasing demand for isoplanatism in the optics. Fig.15 shows the achievements for tandemcombinations with TV - camera objectives in the course of time which has resulted from tight cooperation between Philips and several manufacturers of optics.

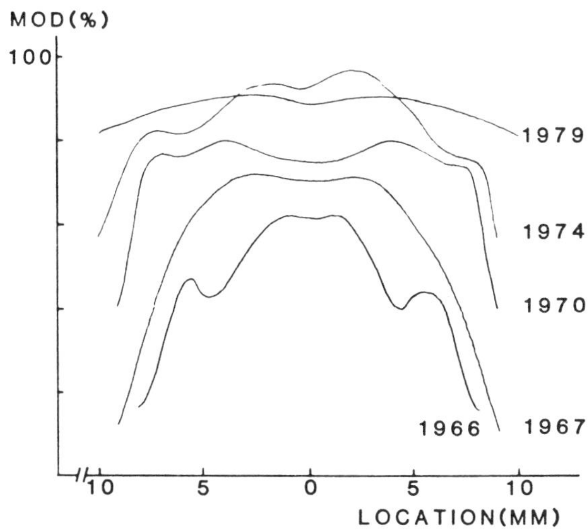

Fig.15. Increase of isoplanatism in the course of time for a tandem combination of collimating and TV-lens at maximum aperture.(Spatial frequency 10 lp/mm at image intensifier output plane).

B. Application of fibreplates in image intensifiers:

Among other factors the application of internal and external fibreplates in X-ray image intensifiers has resulted in a considerable decrease of the low frequency drop of these devices.Because of this decrease the low frequency drop in the optics has become of the same order of magnitude as the drop in the image intensifier. This sets increasing demands on paints for lens rims and housings and anti reflection coatings.

An additional advantage of the application of external fibreplates is the well defined object position for the optics, which enables the use of prefocussed collimating lenses.

C. Demands for optimum utilization of diagnostic information:

The increased isoplanatism and the associated increase of performance at the edge of the field has stimulated the tendency to image the complete entrance field of the X-ray image intensifier at the maximum size allowed by the receptor dimensions. This sets increasing demands for the alignment of mirrors and optics. Apart from vignetting caused by a tilting mirror there is a risk of light interception from the camera window.

D. Improvement of optical computing techniques:

Improved computing techniques enable manufacturers of optics to design more dedicated, e.g. by introducing barrel distortion in the collimating objective as a compensation for the inevitable pin cushion distortion in image intensifiers. These newer designs are not only made more cost effectively by reducing curvatures and numbers of lens elements but also the over all performance and isoplanatism appears to increase.

Concluding remarks.

Design of optical image distributors is an intricate activity. Manufacturing of the majority of the components requires the utmost of what can be reached at the present state of technology. This necessitates, apart from the design of the image distributor, a considerable research effort to develop new measuring techniques. These measuring techniques are used for :

- Specification adapted to image intensifier design.
- Incoming inspection for bought-in items and quality control during manufacturing.
- Adjustment of complete image intensifier systems after installation in the hospital or replacements by the service engineers.

For the next years it may be expected that advancing technology will still introduce many improvements in optical image distribution systems.

IMAGE INTENSIFIER CONTRAST RATIO MEASUREMENTS

Neil Bromberg
General Electric, Medical Systems Division
Milwaukee, Wisconsin 53201

Introduction

The contrast ratio measures the degree to which an imaging system maintains the grey scale presented by an object and, according to Schade[1], grey scale is the most important factor influencing image quality. Gould et al[2] report that the contrast ratio was a significant factor in predicting the clinical performance of image intensifiers. This paper uses the concept of Disc Transfer Function to explain the relationship between contrast ratio and image quality.

Contrast Ratio

Curve 1 in Figure 1 is the output of an image intensifier in which the left half of the input phosphor has been blocked by lead. Notice that scatter causes brightness to be recorded all the way to the edge of the tube behind the lead disc. Also observe, because scatter has been blocked on the left, the right side of curve 1 is not as high as curve 2, the brightness curve for the unobstructed tube.

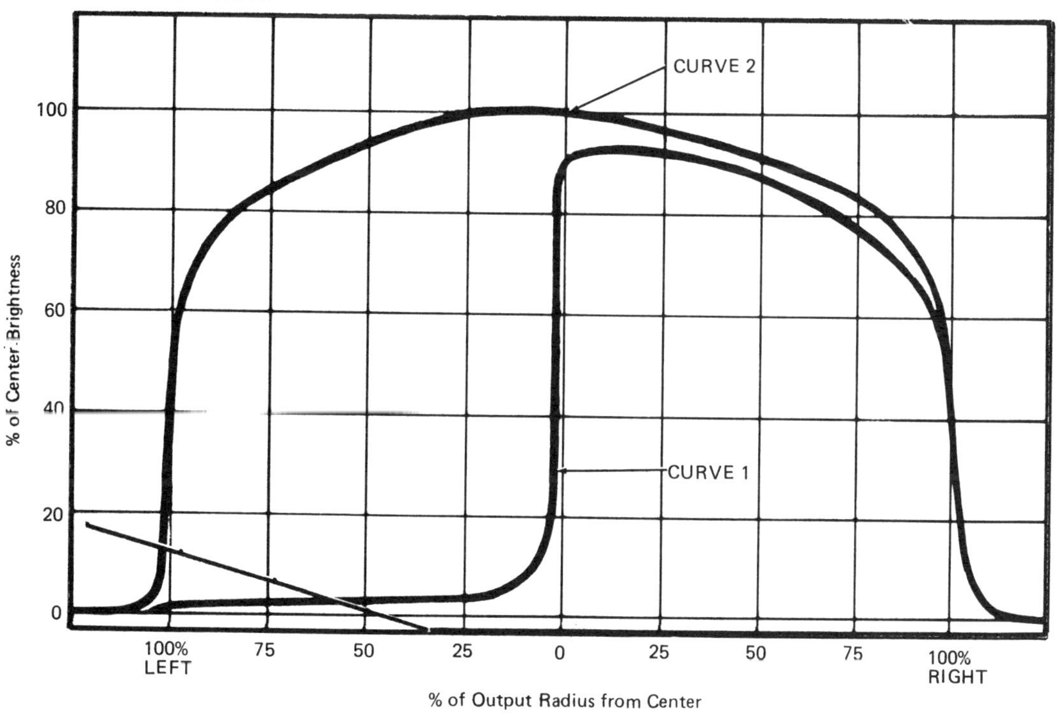

Figure 1. The Image Intensifier Tube Edge Response

Assume a lead disc of radius r is thick enough to block radiation. In order to quantify the amount of scatter, the following quantities are defined as shown in Figure 2.

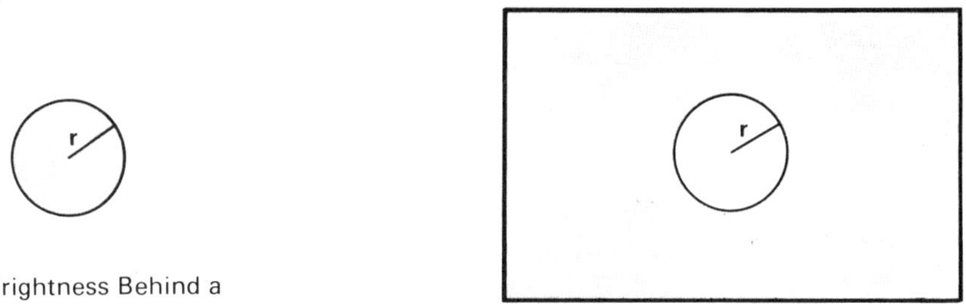

B_r = Brightness Behind a Lead Disc of Radius r.

$\bar{B}_r$ = Brightness Measured in Center of Hole of Radius r, Cut in Lead Sheet.

Contrast Ratio = $C_r = B_o/B_r$

Figure 2. Definition of Br and $\bar{B}$r

A useful relationship is: $B_r + \bar{B}_r = B_o$ B_o=Brightness with No Disc —(1)
One may find it convenient to work with K_r, the inverse of the contrast ratio.

Disc Transfer Function

If B_o is the brightness of the background we define the input and output signals determined by the lead disc as: Input Signal = B_o

Output Signal = $B_o - B_r$

and the Disc Transfer Function (DTF) is defined as shown in Figure 3.

Disc Transfer Function (DTF)

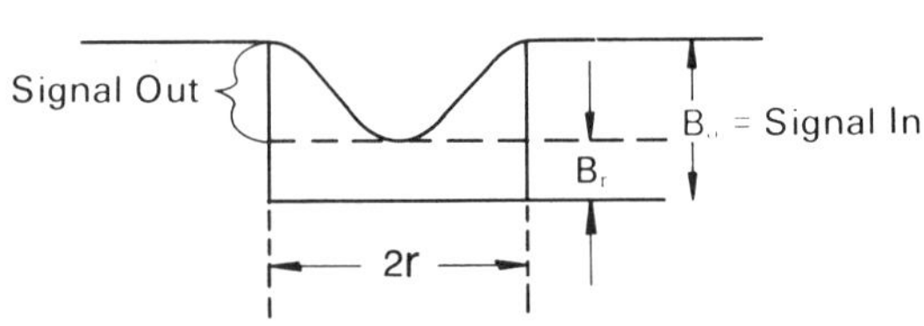

$$DTF = T_r = \frac{\text{Signal Out}}{\text{Signal In}} = \frac{B_o - B_r}{B_o} = \frac{\bar{B}_r}{B_o} = 1 - K_r \qquad \text{—(2)}$$

Figure 3. Disc Transfer Function.

T_r is a function of the radius r which increases to 1 as r increases (see Figure 4). From Equation (2) the DTF can be evaluated from the contrast ratio obtained from discs or more directly by measuring the brightness through holes. The latter method should be more accurate if the photometer used for the measurement tends to pick up ambient light.

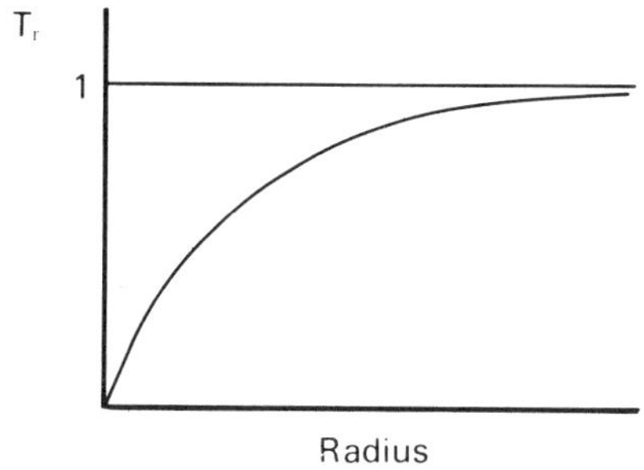

Figure 4. T_r vs. r

Scheid[3] has used expressions, heuristically derived, similar to Equation (2), in evaluating the MTF of a system. Here Equation (2) can be rigorously derived and shown valid when radiation penetrates the lead disc.

Scatter and Collimation

In measuring contrast ratios for a system it is important to separate out factors which influence contrast degradation that reside outside the system. In an x-ray system scatter and collimation modify the contrast of the object before the beam hits the input phosphor of the image intensifier.

Suppose we collimate a system down to a radius b. A variable with a superscript c denotes a system which has been collimated. The formulas for brightness behind a disc of radius r and for no disc are given below:

$$B_r^c = \bar{B}_b - \bar{B}_r + X_r \bar{B}_r + S_b B_b$$

$$B_o^c = \bar{B}_b + S_b B_b$$

B_b is the amount of light due to radiation hitting the input phosphor beyond a radius b when there is no collimation. The collimator does not eliminate all of this light and so S_b is the fraction of light seen because of scatter beyond the blades. Similarly X_r is the fraction of additional light seen because of x-ray scatter behind the lead disc.

Using Equation (2) yields the DTF with collimator:

$$K_r^c = B_r^c/B_o^c = 1 - (1-X_r)\, B_r/(\bar{B}_b + S_b B_b)$$

$$T_r^c = (1-X_r)\, T_r/(1-(1-S_b)K_b) \qquad \text{——(3)}$$

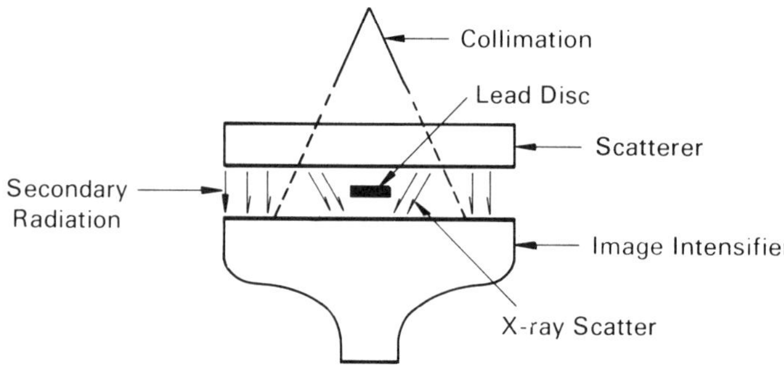

X_r = X-ray Scatter Fraction Behind a Disc of Radius r.

S_b = Secondary Radiation Fraction from Collimating to Radius b.

Figure 5. X-ray Scatter and Collimation

To test the result, the DTF was measured for a series of discs ranging from 4.8 to 63.5 millimeters in radius. An 80 kVp x-ray beam was hardened by a .0625 inch copper plate placed 28 inches from the front plane of the intensifier, far enough to eliminate any x-ray scatter behind the disc. The source to image distance was 40 inches and the beam was collimated down to a circle of radius 80 millimeters. Since $X_r = 0$, by Equation (3) T_r^c/T_r should be constant. K_b was measured to be .040. The results are recorded in Table 1.

TABLE 1

Radius (mm)	T_r	T_r^c	T_r^c/T_r
4.8	.839	.851	1.014
6.4	.845	.863	1.021
7.9	.858	.876	1.020
9.5	.868	.886	1.020
11.1	.876	.896	1.022
12.7	.880	.900	1.022
15.9	.889	.907	1.020
19.1	.897	.917	1.022
22.2	.905	.923	1.019
25.4	.907	.927	1.022
28.6	.918	.932	1.015
31.8	.920	.938	1.019
34.9	.922	.940	1.019
38.1	.928	.944	1.017
44.5	.936	.952	1.017
50.8	.940	.954	1.014
57.2	.942	.965	1.024
63.5	.953	.969	1.016

Taking the average from the last column in Table 1 we have: $T^c_r/T_r = 1.019$ and from Equation (3),: $S_b = .53$

If necessary one can simulate the x-ray scatter produced in a clinical environment by placing an acrylic phantom in the beam.

Parametric Model

The contrast ratio data on image tubes we have seen fits the parametric model shown in Figure 6.

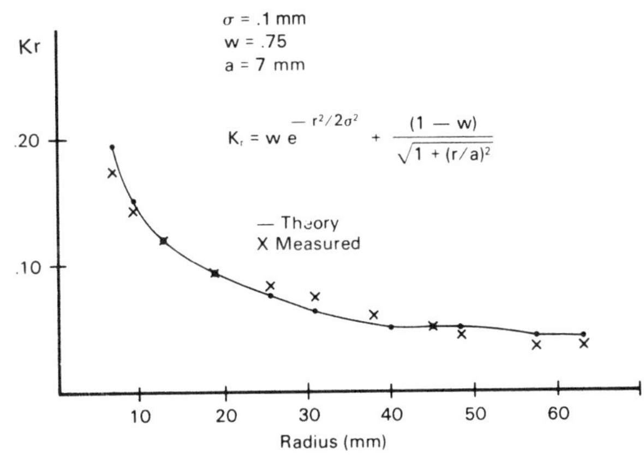

Figure 6. Parametric Model of Contrast Ratio

The model consists of two pieces derived from Gaussian and Cauchy point spread functions. The Gaussian part goes to 0 very rapidly yielding the interesting and useful result for T_r shown in Figure 7.

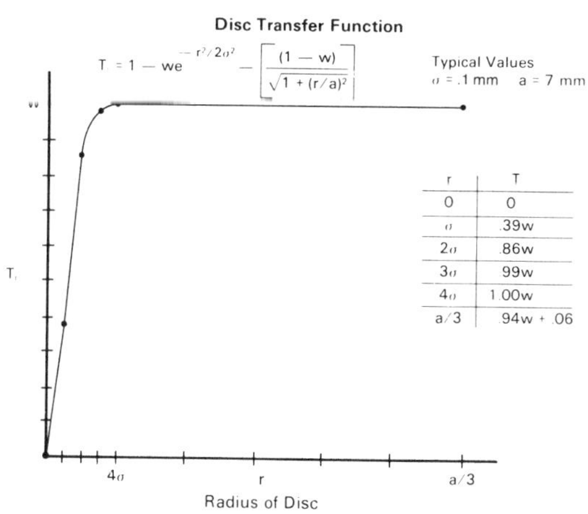

r	T
0	0
σ	.39w
2σ	.86w
3σ	.99w
4σ	1.00w
a/3	.94w + .06

Figure 7. Disc Transfer Function: T_r

As can be seen from the graph, for r between 2σ and a/3 the value of T_r is approximately w. Typically σ is about .1mm and a can run from 1 to 30 mm. Thus w largely determines the system's ability to see objects which are a size of clinical interest. The MTF formula for this model is shown in Figure 8 along with typical parameter values. Notice that w determines the drop in MTF at low frequency. This is usually referred to as the large area contrast loss.

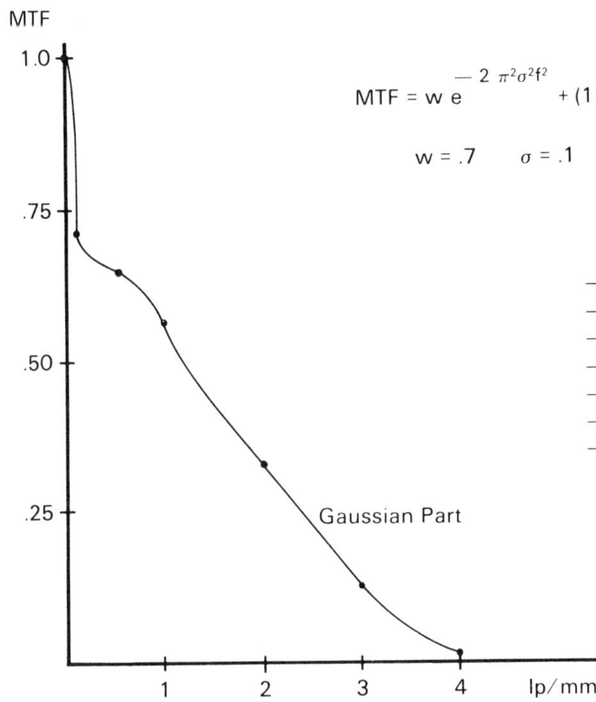

$$MTF = w\,e^{-2\pi^2\sigma^2 f^2} + (1 - w)\,e^{-2\pi a f}$$

w = .7 $\quad \sigma$ = .1 $\quad$ a = 7

f	MTF
0	1.0
.1	.70
.5	.67
1	.57
2	.32
3	.12
4	.03

Figure 8. The MTF Curve

The model parameters have varied with different scatterers in the beam implying a dependence on the energy spectrum of the radiation. Measurements taken with one inch of aluminum in the beam at 60, 80, and 120 kVp indicate as kVp increases a increases.

Experimental Setup

Figure 9 shows the experimental setup for measuring the contrast ratio of an image intensifier. It is important to make sure that x-ray scatter does not get behind the lead disc, therefore, the beam hardener should be placed as close to the x-ray tube as possible and should be as small as possible. You need also to take into account the distance the input phosphor is from the front plane and the aperture size of the light meter.

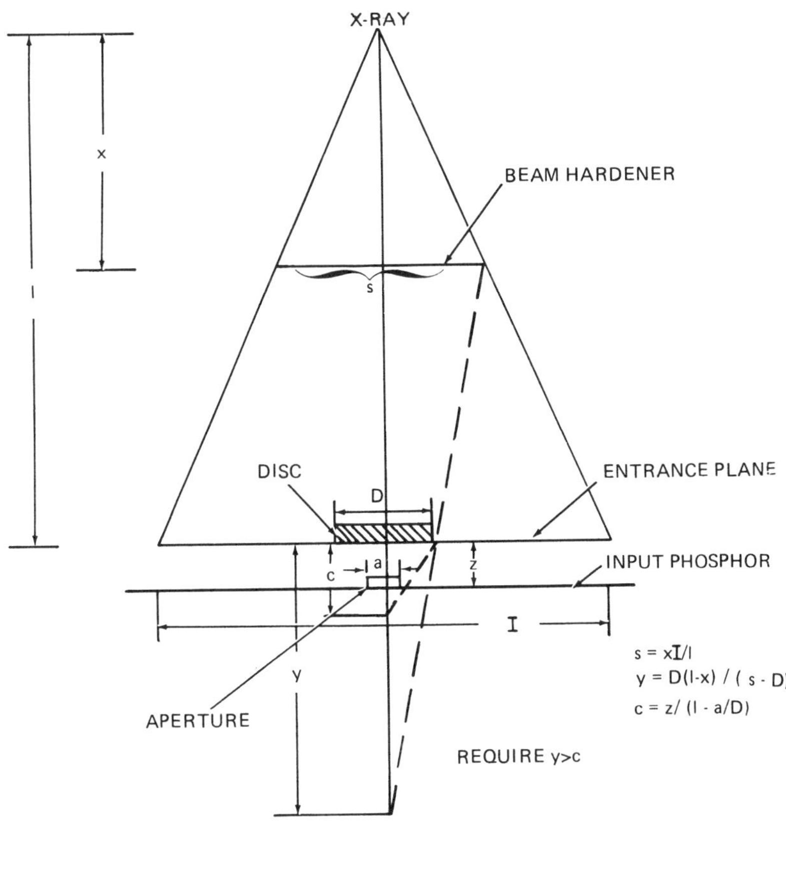

Figure 9. Experimental Arrangement

The contrast ratio should be measured for several size discs, two sizes at minimum to estimate the values of w and a. Although it is common practice to measure contrast ratio with a disc covering 10% of the input area, our analysis indicates that contrast ratio should be recorded by disc radius. Keeping constant the fraction of the input area covered is a poor standard because, without any change in the image intensifier characteristics, as the input area increases the contrast ratio will increase.

References

1. O.H. Schade, Sr., Image Quality. A comparison of photographic and television system. RCA Laboratories, Princeton, N.J., 1975.

2. R. Gould, B. Belanger, H. Goldberg, and A. Moss, "Objective Performance Measurements versus Perceived Image Quality in Intensified Fluoroscopic and Photospot Images". Radiology, Vol. 137, No. 3 pp. 783-788, December 1980.

3. C. Scheid, "Performance Measurement of Fluoroscopic Systems", Real-Time Radiological Imaging: Medical and Industrial Application. ASTM STP 713, D.A. Garrett and D.A. Bracher, Eds. American Society for Testing and Materials, 1980, pp. 168-179.

4. V.L. Davis and H.H. Glascock II, "Large Area Contrast of a Fiber-Optic Couples X-ray Image Intensifier". Medical Physics, Vol. 4, No. 3, pp. 211-213, May/June 1977.

DESIGN AND FUNCTIONAL DESCRIPTION OF 100 MM CAMERAS

Frank de Bruin
R & D and Marketing Department
N.V. Optische Industrie "De Oude Delft"
Oldelft P.O. Box 72 Delft, HOLLAND
and
Don E.C. Hall
Oldelft USA, 2735 Dorr Avenue
Fairfax, Virginia 22030

ABSTRACT

The design of photospotcameras in combination with Image Intensifier (I.I.) systems can be so that the limitations of the image quality of the total imaging chain is not influenced by the camera system, whereas the reliability of the mechanism should be such that the Mean Time Between Failures (MTBF) of the total installation should not drop considerably.

Another important factor in the camera design is the human engineering. Ease of manipulation without extensive instructions in operation and film handling should be strived for. Such extra features as data-recording, system function indicators on the camera, and optical adaptability are also discussed.

INTRODUCTION

Within the past decade the application of image intensifier photography for general diagnosis has proved its value. Developments in I.I. technology in the recent years support a wider acceptance of I.I. photography in more specialized applications, where resolution, short exposure times and contrast detection play a significant role.

Cost containment and radiation-consciousness are furthermore important factors leading to a broader acceptance and utilization. With the experience from the past decade and the trend towards better image quality in view, the design specifications for a second generation camera could be established.

DESIGN OF 100 MM CAMERAS

The basic function of the camera is to record the image from the image intensifier on to the emulsion of the film.

The heightened image as produced by the intensifier tube appears on the anode screen in a small format. The collimating lens collects this image to an infinitive beam. By way of a moving semitransparant or reflective mirror the different camera systems can be coupled.

Mechanical Interface of the Camera

The adaptation of the camera to the image distributor cubicle is different for most manufacturers, the camera front-end is for that purpose to be designed so that different adapter plates can be mounted. These adapter plates are (except for Philips) provided with adjustment aids to align the optical axis of the image from the distributor with the camera axis. This adjustment allows one to align the middle of the input screen of the I.I. to the middle of the film. With a Philips system this alignment is guaranteed through the positioning of the collimating lens on the image intensifier.

Optical Interface

The basic optical configuration used for image transmission from the output window of the image intensifier to the cameras is the tandem optics (Fig. 1) which allows image transmission with relatively low intensity loss and makes image distribution possible (Fig. 2). To obtain the correct image format on the film the camera lens has to have a specific focal length which depends upon the image format on the output phosphor and the focal length of the collimating lens.

The image format can be calculated with the Equation: $I = i \times \frac{fI}{fi}$

where I = image format in mm
fI = camera lens focal length in mm
fi = collemating lens focal length in mm
i = diameter of output image in mm

For example, with the output image size of 18 mm, collimator lens focal length 75 mm camera lens focal length: 400 mm will result in an image of 96 mm on the film.

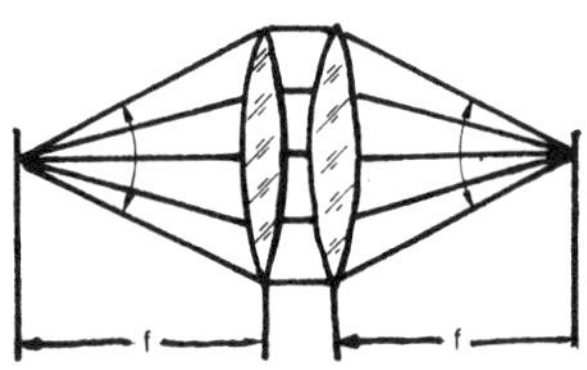

Figure 1. Tandem arrangement for image coupling

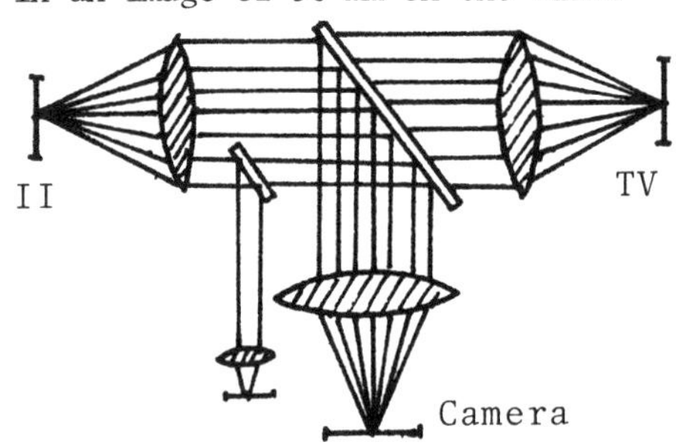

Figure 2. Image distributor with two output channels

Since x-ray manufacturers use different collimating optics, it is necessary to have a range of optics available to match these variations. For the Anodica there is a range of 322, 380, 400, 435, and 500 mm focal length available.

In order not to vary the outside dimensions of the camera, the back-focal-length of the different lenses should be kept constant while focal lengths actually differ.

Waterhouse diaphragm stops should be easily attached to optimize the input dose - film density relation, avoiding Quantum mottle. Minimum Modulation Transfer Function (MTF) values should be set so that the camera does not deteriorate the image quality obtained through the image intensifier. At 10 lp/mm in the film (approximately 4 lp/mm at a 9 inch input screen) the MTF in the whole field should not be below 70%. The contribution factor from stray-light, resulting in a low frequency drop, should be a low percent factor.

Camera Design Criteria

To finalize the image-related design criteria, the patient-ID photographic channel is included (Fig. 3). Both images from a written or typed patient ID card, and a sequential counter are photographed on the film.

Also the ease-of-use aspects of the camera should be considered. Easy loading and unloading of film supply and take-up magazines without film loss are definite concerns. Non-confusing and thus a minimal number of indications in a conspicuous viewing location on the camera, and ease of accessibility to the patient card slot should also be a major design consideration.

Of course the film transport mechanism should assure ultimate flatness and reproduciable positioning of the film.

Last but not least, the weight of the total camera is an important factor. Many overtable installations have required an additional ceiling suspension when the camera is in excess of 20 pounds. Installation, transport, and operation of the camera will vary considerably with increased weight.

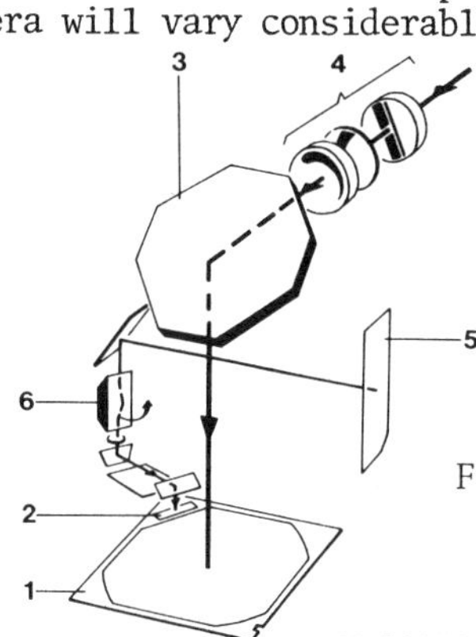

1. Film
2. Pat. I.D.
3. Mirror
4. Objective
5. Pat. card
6. Prism

Figure 3. Optical Elements In The Camera.

DESCRIPTION OF THE FUNCTIONS

The camera system should be compatible with the x-ray generator in the performance of its functions. Various safety features should be incorporated into the unit to avoid unnecessary radiation to the patient.

Camera operation indicators to inform the operator as to the number of film exposed, the number of unexposed film remaining and a ready command when initiation of exposure is ready. The interface with the x-ray generator, the power supply and all the logic functions must be grouped in the control unit. From their control unit, all electrical connections are made; to the x-ray generator, to the camera, and to the remote control and indication panel.

Camera Functions

Depending upon the camera's application, some of the functions, e.g. transport speed and magazine capacity, will vary. Gastrointestine, urology, and pediatric examinations generally require a limited number of exposures. The films should be available soon after the examination is completed. For these applications, the Anodica-2 was designed with limited speed of 2 frames/sec and equipped with a take-up magazine for 20 films.

A larger number of exposures and/or a higher frame rate is required for angiography and other specialized examinations. The Anodica HS, and its recent successor the Anodica-6 are designed for these applications.

Figures 4 and 5 depict the schematic diagram of the Anodica 2, and its relative location on a general purpose fluoroscopic image intensifier tower.

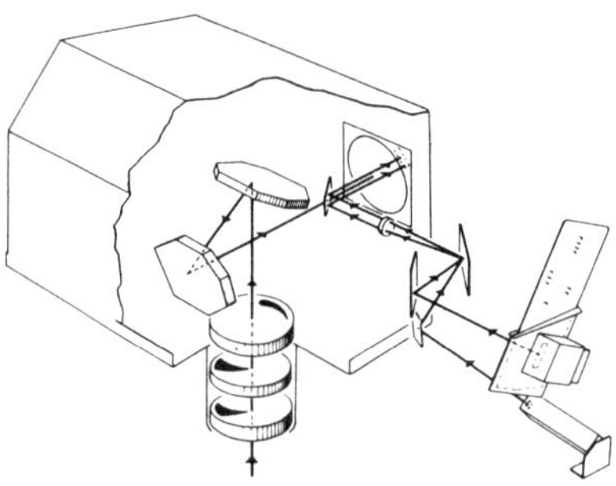

Figure 4. Schematic illustration of the Anodica 2 camera

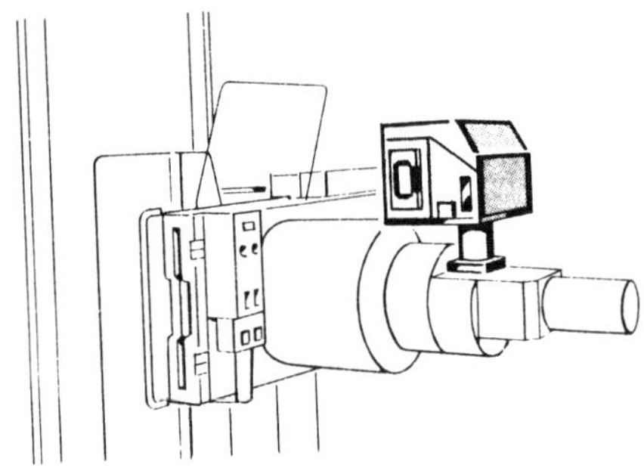

Figure 5. Anodica 2 in conjunction with examination table

Operation of Anodica 6 Camera

The camera mechanism has three operation modes: in transport of an unexposed film, out transport of an exposed film, and a combination of in and out transport. When the camera is set to the single shot operation, it performs an in transport function as soon as the preparation command is given to the generator. As the exposure is completed and the pushbutton has been released, the exposed film is transported into the takeup magazine. When the camera is in the serial exposure mode, e.g. several exposures in sequence, after the first exposure a simultaneous out and in transport takes place at the selected frequency of frame rate.

As shown in Figure 6 (see next page), the transport mechanism functions as follows: from the supply magazine (2) the next film is slowly shifted, with the transport shifter arm (3), into two fast turning rollers (1). As the film is loaded in place the pressure plate (5) is closed. Subsequently, the exposure signal is given by the logic system. After the exposure the excentric cam (7) opens the pressure plate and the film is pushed with the pawl arm (8) between the two lower transport rollers (1). It is then driven into the take-up magazine (4). The motor (6) directly drives the rollers (1). With each film transport, the low inertia motor is started and stopped, so that no camera vibrations are present during the exposure.

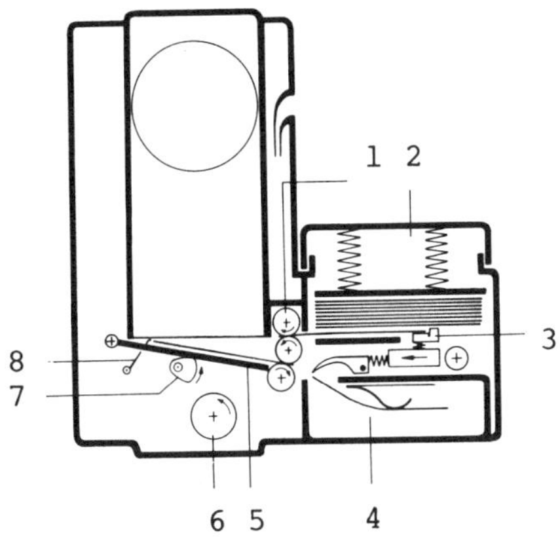

1. Transport roller
2. Supply magazine
3. Transport shifter
4. Take-up magazine
5. Pressure plate (open)
6. Motor
7. Exentric
8. Pawl

Figure 6. Schematic diagram of Anodica 6 camera

Magazines

The supply magazine has a maximum capacity of 130 films. As a design criterion the maximum successive number of exposures has been set at 80. The take-up magazine is, therefore, designed to receive a maximum of 80 exposed films. Within the camera logic an indicator is available to show that less than 80 films are in the supply magazine so that a new film pack of 50 films can be added.

The camera is also equipped with additional warning indicators for the following situations: (1) when there is less than 30 films in the supply magazine two packs of 50 film each can be loaded, (2) another warning is given as soon as the remaining capacity in the take-up magazine is less than 20 films.

An alternative to the darkroom loadable supply magazine is the daylight magazine. This supply magazine can be loaded with Delpack daylight film packs each containing 50 films at the site where the camera is in operation.

X-ray Synchronization

The film transport and the x-ray generator should work in synchronism with each other. The successive actions and interactions are as follows: suppose that the installation is switched on in the fluoroscopy mode, the filament of the x-ray tube is then low heated, and the x-ray tube potential is controlled by the fluoroscopic kVp-selector, the x-ray tube current is in the 0.5 to 5 mA range.

When a photofluorographic exposure is desired, the preparation push button is pressed first; this signal is sent to the camera control to check the readiness of the camera for exposure (the camera shows its green "ready" indication).

After this safety confirmation three actions follow simultaneously: (1) the x-ray generator is switched in preparation for exposure, so that fluoroscopy is switched off, the x-ray tube filament is heated-up, the rotating anode starts to speed up, and the kVp selector is switched to the photofluorographic kVp-control, (2) the image distributor is started to move from the fluoroscopic position (100% light to the T.V. pickup tube) to the camera-exposure position (90% to camera, 10% to T.V. pickup tube), and (3) the camera is put in preparation; the shutter is opened or a fresh film is transported in the film gate.

Contrary to the cine camera the shutter in a photospot camera is not used to block the light during film transport, it serves only as protection against stray light from the image distributor during fluoroscopic exposures or a cine run.

Heating up the filament, speeding up the anode and moving the image distributor requires in general 0.8 to 1.5 sec. A safety interlock is built-in the generator so that the exposure can only be initiated when all these conditions are attained.

For the next exposure in a serial run the whole installation remains in this "prep" position, and the camera controls the exposure of the generator in the following timing sequence: starting at the end of an exposure it takes 80 milliseconds (ms) to transport the previous and exposed film out of the gate and a new one in., 15 ms for pressure plate closure, the film is allowed another 10 ms settling time before the camera electronics send the exposure command to the x-ray generator.

Human Engineering

To facilitate an easy manipulation of the camera, all elements are grouped together on one side of the camera - such as: card slot, warning indicators, and magazines. The camera is designed in such a manner that this access side can be positioned toward the front of the table. The optical input can be turned 180° to accommodate varying requirements, inserting and removing of the film magazines can be performed easily with one hand.

There are function indicators on the camera that provide for "all clear", "supply low", and "take up nearly full". These indications are also duplicated on the remote control unit or at the generator control console.

DISCUSSION

The introduction of 100 mm photospot cameras in the daily routine of hospitals has been proven to give substantial savings in film costs. For efficient film handling accessories like the film feeder, film holder (Cadrix), framing material (Delcadro), enlargement viewers (Heliomat) can be utilized.

The integration of C.T., Ultrasound, and Nuclear Medicine modalities in the x-ray department places an emphasis on the recording aspect also. Several possibilities for recording are now in use: Polaroid photography, multiformat cameras, and 100 mm film transport camera back (Catodica). Because of the specific design characteristics a short description will be given regarding the 100 mm film transport camera back.

CRT Recording 100 mm Film Transport Back

Looking to an easy working method with instant (Polaroid) type films puts a burden on the designer to create an effective quality replacement. Any new unit should maintain the ease of operation and the benefit of a see-through exposure. It should incorporate a motorized transport system, as depicted in Figure 7.

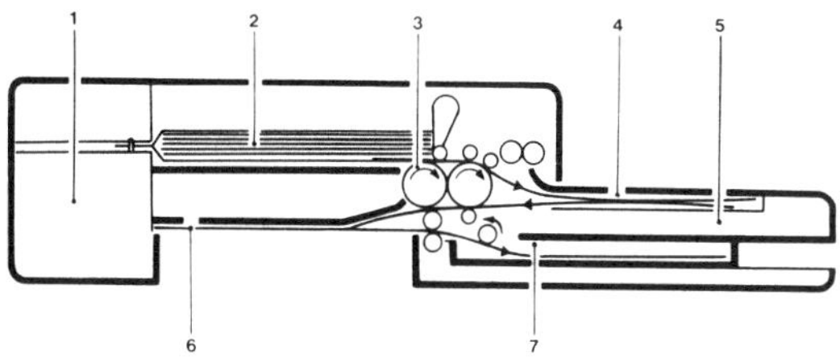

1. Motor compartment
2. Film supply Delpack
3. Transport rollers
4. Film parking position
5. Electronics
6. Film gate
7. Film take-up magazine

Figure 7. Schematic diagram of Catodica camera back

When the special Delpack daylight package is loaded, the lid is closed, and the wrapping material is transported out. The film pack (2) of 50-100 mm films is now ready to be transported by the rollers (3). From the parking positionin- (4), the film is transported into the active pressure plate (6), and after exposure to the take-up magazine. The electronic logic (5) controls the motor (1) in such a way that, after each exposure, the film is automatically carried out of the camera, and into the take-up magazine.

The Catodica film back is of an extremely adaptable design. The Polaroid back of most cameras can be replaced by this automatic 100 mm film back unit with the aid of dedicated adapter hardware shown in Figure 8. The use of the standard take-up magazine of the Anodica-2 guarantees the adaptation to the film-feeder. Implementation of this recording technique compliments and is compatible with existing systems and routines in the radiological department.

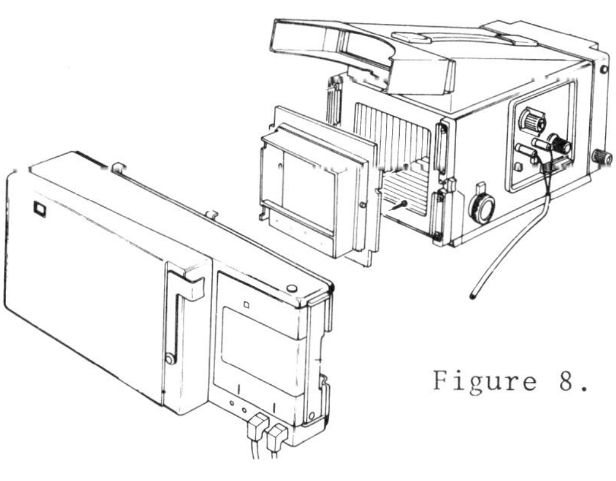

Figure 8. Schematic Illustration Of Catodica With Adapter And Camera.

CONCLUSION

Although the design of 100 mm cameras, which fulfill the needs of different disciplines and applications, is not a simple task, we have successfully constructed several types. Optical, precise mechanical, and electornic knowledge and skills are necessary to design, develop and produce 100 mm cameras for the medical market. Distribution and installation, together with the image intensifier is performed by the leading x-ray companies, assuring an optimal support to the end user.

AUTOMATIC BRIGHTNESS CONTROL OF IMAGE INTENSIFIER SYSTEMS

Gary F. Relihan
Imaging and Generator Control Design
General Electric Company Medical Systems Division
Waukesha, Wisconsin 53186

INTRODUCTION

Imaging is the prime purpose of the radiology department. The equipment developed around the image intensifier, which was introduced only 17 years ago, may represent today approximately 25% or more of the total x-ray equipment budget in many diagnostic radiology departments, not including computed tomography, nuclear medicine and ultrasound equipment.

The purpose of the image intensifier was to increase the light level of the fluoroscopic image to the point where better visual discrimination could occur without dark adaption.

Viewing the output of an image intensifier via a television chain introduced greater flexibility in the use and viewing of the radiologic image. Besides allowing the electronic recording of the image, it also provides a method by which contrast of the image could be electronically enchanced. Television viewing introduces flexibility in the use of fluoroscopic images.

Several recording methods have been incorporated in todays fluoroscopic systems. With the advent of the closed circuit television system use in fluoroscopy, the video tape recorder and video disc recorder has been used to electronically record the fluoroscopic image. In addition, spotfilm cameras and cine cameras have been added to the arsenal of recording media.

The physicist in today's modern x-ray departments are being over-whelmed with more and more sophisticated equipment. An understanding of equipment operation by the radiological physicist is important, sometimes required, for the carrying out of their duties. The following discussion should provide some insite into the operation and techniques used in automatic brightness control systems.

Automatic Brightness Control (ABC) Systems

Most image intensifier units in use today are equipped with automatic brightness control and the fluoroscopist no longer has to look at an image that gets light and dark with changes in patient anatomy. There are two major ways of controlling the brightness level. Controlling the electronics of the image system, in particular the gain of the television system; this type of control is called automatic gain control (AGC). The other technique is by controlling the radiation input to the image intensifier which is sometimes called automatic dose control (ADC).

Electronic Brightness Control

In an electronic brightness control system the gain of the television system is varied in order to maintain proper brightness on the monitor. The circuitry is relatively simple and inexpensive. The main disadvantage to this type of system is that as the gain of the television is increased, correspondingly the noise in the system is increased. Another disadvantage is that no consideration is given to the exposure rate to the patient. Adding a manual radiation control to this system would help this situation. However, this type of control does not provide brightness control when filming directly from the output phosphor. This type of system has not proven to be an adequate solution to the requirements of automatic brightness control.

Radiation Input Control

Controlling the input exposure rate to the image receptor has proven to be the best method of automatic brightness control. The obvious advantage of this type of control is that the output brightness of the intensifier is held constant. This allows filming of the output phosphor directly and obtaining proper density at all times. The quantum noise level can be controlled using this approach, and high quality films and fluoroscopic images can be obtained.

There are three basic ways to control the radiation input exposure rate. They are: (1) control of kilovoltage, (2) control of milliamperage, and (3) control of pulse width (exposure time). These three methods can be utilized separately or in tandem to provide various degrees of performance.

(a) "kVp" Brightness Control: In a kVp-control automatic brightness loop the kilovoltage is varied in order to control the radiation input rate to the patient and/or the input phosphor. This can be accomplished by regulating the secondary of the high voltage system using thyratrons, or controlling the primary voltage of the high voltage transformer. When utilizing this type of control it would be common for a manual mA selector to be available in order to keep the kVp in a range that is appropriate for a given procedure.

(b) "mA" Brightness Control: In a mA-control automatic brightness loop the x-ray tube filament current is varied to control the x-ray tube current. With this type of control the operator is provided a manual kVp-control to select the contrast of the image. The dynamic range of a mA-control loop can be as high as 200:1.

A simple voltage control filament drive can be used in this type of system, but the response time of this type of control will be sluggish. Using a current drive system, while more expensive, can decrease the response time to an acceptable level.

(c) Pulse Width Brightness Control: This type of system utilizes control of the exposure time in controlling brightness. The kVp and/or mA can be selected to optimize technique. Pulse width control is accomplished by using a grid controlled x-ray tube or by a constant potential generator with triode or tetrode switching. These systems are expensive, but provide the fastest response along with maximizing motion sharpness during recording modes. Three phase generators are generally used in this type of system to allow high speed cine recording and allow a wide dynamic range of pulse width, approximately 200 to 400:1.

(d) Combinations: Some systems on the market take advantage of more than one method of radiation control in order to obtain improved performance as well as make the system easier to operate. Such a system would be a combination of mA-control along with kVp-control. Another case might be pulse width control with kVp-control.

In this type of system one of the parameters controlled must be the primary brightness control and the other a subordinate to the first. Usually the mA-control or pulse width control will be the primary brightness control loop with the kVp-control as a minor control loop.

To further improve the performance a manual kVp-control can be added to allow operation at higher minimum kVp levels when a high-kVp/low-mA technique is more appropriate.

Automatic Brightness Control Systems

The following sections will detail some of the control systems used today along with some techniques used in their implementation.

One of the most general fluoroscopic systems is a mA-control system using peak video as the brightness control. Figure 1 shows a block diagram of such a system.

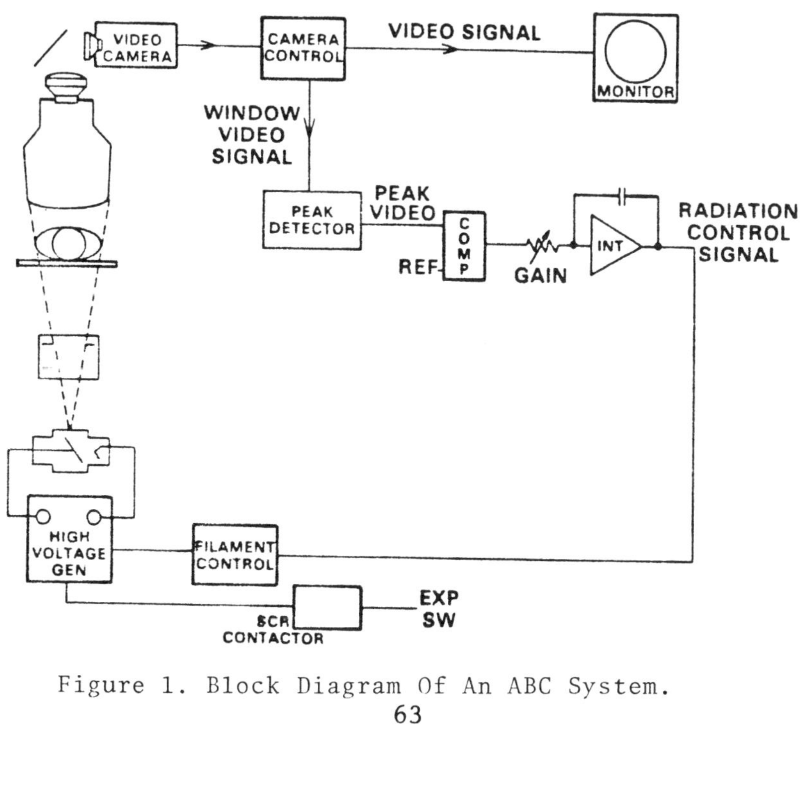

Figure 1. Block Diagram Of An ABC System.

In this system the video signal from the closed circuit television camera is split in two. One signal goes to the monitor and has the blanking and synchronization pulses needed to display the image. The other signal has no blanking or synchronization but rather contains only video information.

This signal can be further tailored by imposing a "window" to limit the area over which the brightness will be examined. Figure 2 shows the video waveforms as related to the video image.

The video waveform contains all the information displayed on the monitor screen. The dotted circle represents the "window" area. This signal is developed by combining the horizontal and vertical sweeps with a cosine function which produces the circular area on the image. Mixing the window signal and the video results in a window video signal which contains all video information inside the window area.

Figure 3 shows the selectivity of the sensing area when viewing an image using window video sensing.

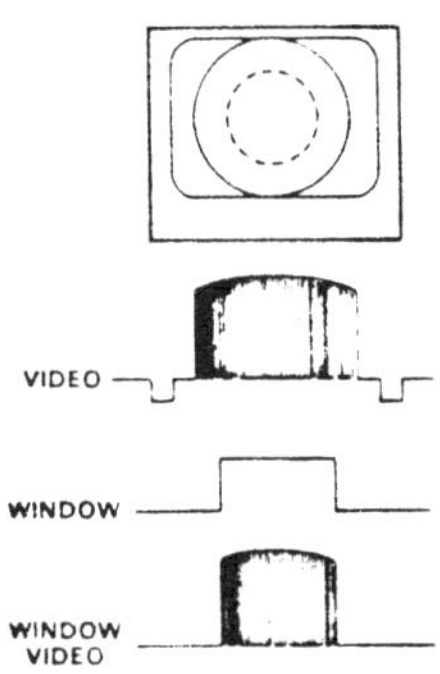

Figure 2. Video Waveforms.

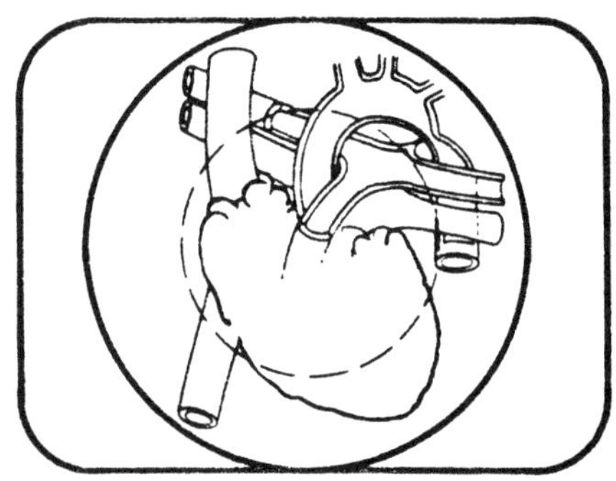

Figure 3. Selectivity of Sensing Areas.

The window video signal is applied to the peak detecting circuit of Figure 4 (see next page). The peak detector can determine the maximum brightness present during each video readout. The T.V. system is normally synchronized to the power line so 60 images can be analyzed each second. The timing diagram of Figure 4a (see next page) shows the sequence of events for each video sweep of the camera.

As can be seen by the timing of Figure 4a, the peak video stored on the C1 capacitor is transferred to the C2 capacitor during the S2 closure time. After the transfer, S1 closes and resets the peak detecting circuit for the next video input.

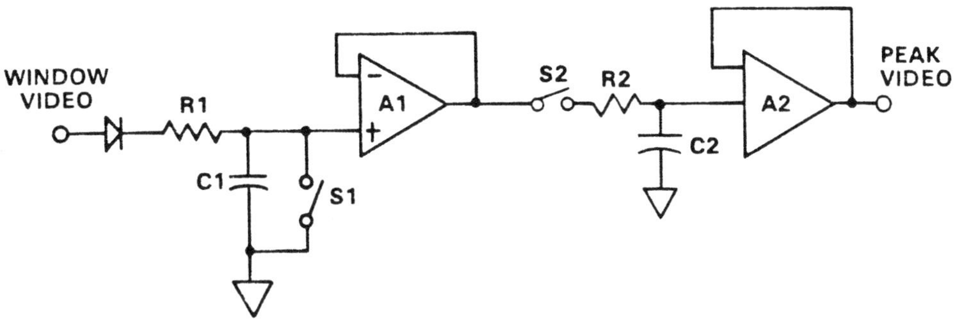

Figure 4. Peak Detecting Circuit.

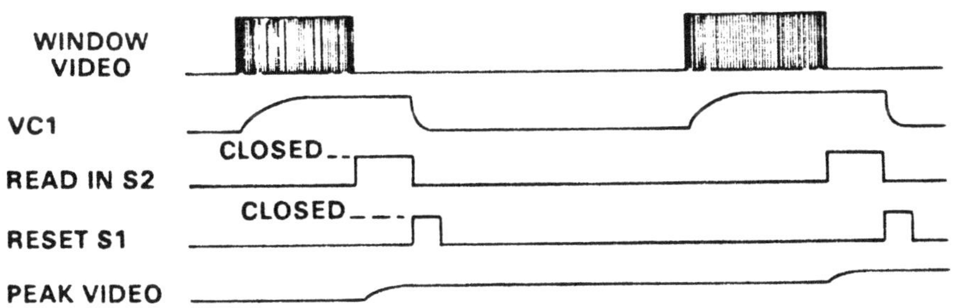

Figure 4a. Timing Diagram.

Diode D1, Resistor R1 and Capacitor C1 form the basic peak detection function. Diode D1 prevents discharging of the Capacitor C1 after a higher video level has charged the capacitor. The time constant $R_1.C_1$ can be adjusted to set the sensitivity of the peak detector to sharp rises in video level. If a long time constant is used the peak detector becomes "blind" to small areas of high brightness.

Amplifier A1 is used to read the peak value. Capacitor C2 is used as a sample and hold circuit. This circuit along with Amplifier A2 produce a dc voltage proportional to the peak video signal.

Referring back to the block diagram, Figure 1, it can be seen that the peak video signal is used to control an integrating circuit whose output controls the mA of the x-ray tube. Figure 5 shows more detail of this function (see next page).

A3 is a comparator circuit which determines if the peak video signal is greater than or less than a predetermined reference. The output of A3 will swing positive if the peak video signal is less than the reference and conversely negative if the peak video input is greater than the reference. The

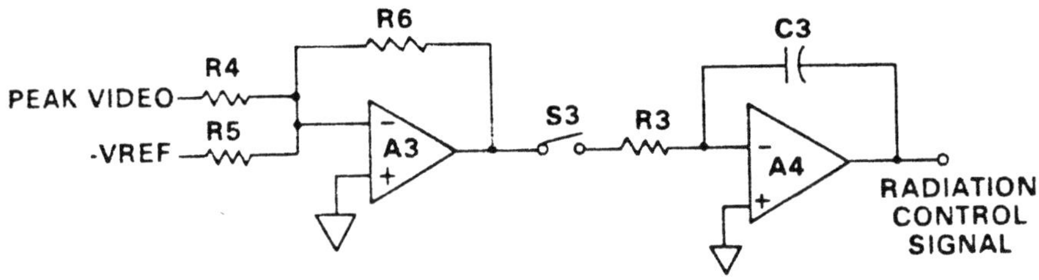

Figure 5. Detail Of A Typical Circuit Diagram. Peak video signal is employed to control the x-ray tube current.

output of A3 can then be used to drive the integrator more positive if brightness is low or reduce the radiation control signal if brightness exceeds the reference.

As can be seen from the timing diagram of Figure 5a, S3 closes after each update of the peak video signal. This is usually done during the vertical synchronization pulse so changes in radiation and consequently brightness will not cause shutter bars in the video presentation.

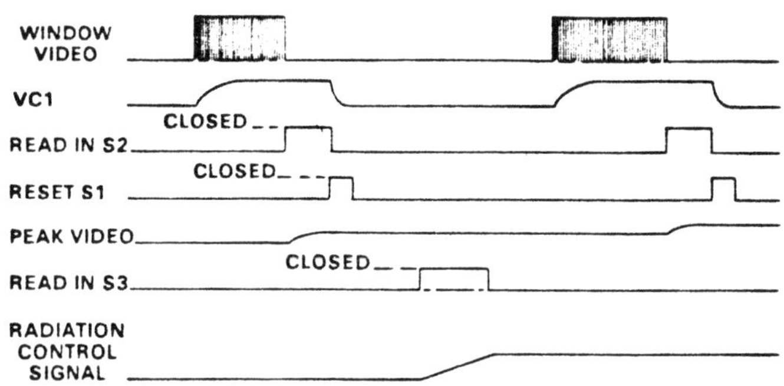

Figure 5a. Timing Diagram.

The last part of the control loop consists of the filament drive circuitry and a manual kVp-selector to control the radiation output of the x-ray tube. There are several techniques which are appropriate to control the filament temperature of an x-ray tube. Two popular techniques would be a filament current drive system as shwon in Figure 6 (see next page), or a pulse width modulation system as shown in Figure 7 (see next page).

The use of a current drive filament control provides relatively fast control of x-ray tube emission. The basic principle used is to provide a variable impedance in series with the primary of the filament isolation transformer. The variable impedance used in the circuit of Figure 6 is the control transformer.

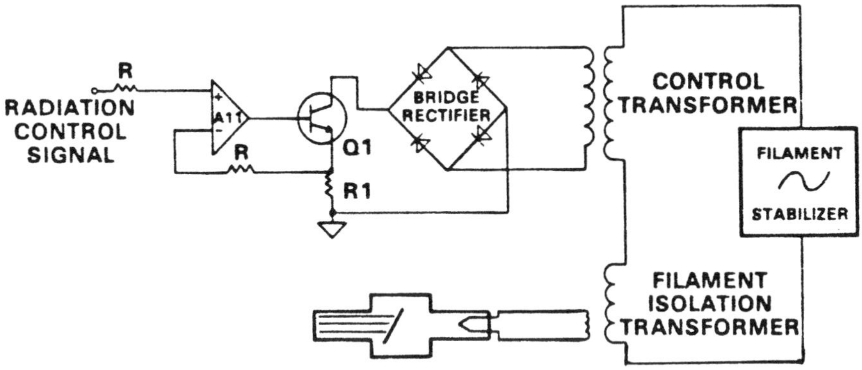

Figure 6. A Filament Current Drive System.

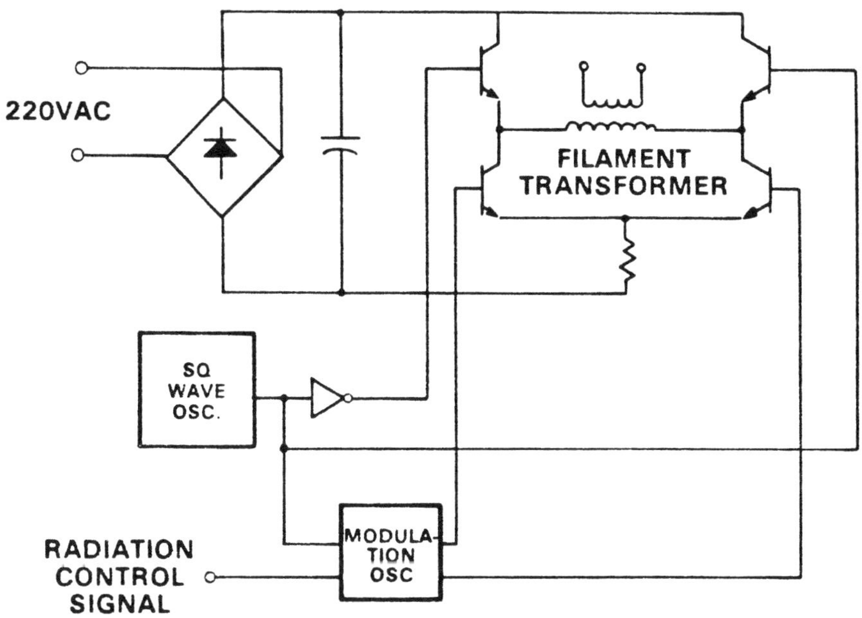

Figure 7. A Pulse Width Modulation System.

By using the current transfer characteristics of the transformer, a secondary load change will be refelected as a primary current change.

In Figure 6, the secondary is loaded with a bridge circuit to rectify the voltage so that a dc control can be used. The Transistor Q1 can be controlled by the Amplifier A1 in such a manner as to vary the voltage across Resistor R1 and consequently change the secondary load current. Any changes in the secondary load current will result in a corresponding current requirement in the primary circuit. The result of this change will increase or decrease the current in the filament transformer, and of course modify the filament temperature of the x-ray tube.

A slightly more sophisticated technique is the pulse width modulation circuit of Figure 7. This design incorporates a special filament transformer that will operate at higher frequencies than the normal 50 or 60 Hz systems. The filament stabilizer voltage, in this case the 220 VAC input, is rectified and filtered. This dc voltage is then placed across a transistor bridge circuit as shown. An oscillator is used to develop a square wave across the filament transformer by turning on Trnasistors Q1 and Q4 for one half the period of the oscillator and then turning on Q2 and Q3 on the other half cycle. This process will provide the alternating voltages needed to drive the filament transformer.

A second oscillator running at a frequency much higher than the square wave oscillator is now used to vary the duty cycle and as a result the current output of the filament transformer will vary accordingly.

The waveform of Figure 7a shows the voltage waveshape across the primary of the transformer. A major advantage of this type of control is the large control range available. The higher frequencies used allow smaller transformers to be used saving space in the high voltage system. Another type brightness control system is the system shown in Figure 8 (see next page). This approach uses the output brightness of the image intensifier as the brightness control. This type of control is ideally suited for brightness control of a cine fluoroscopic system. For that reason a pulse width control system is shown as the radiation control technique.

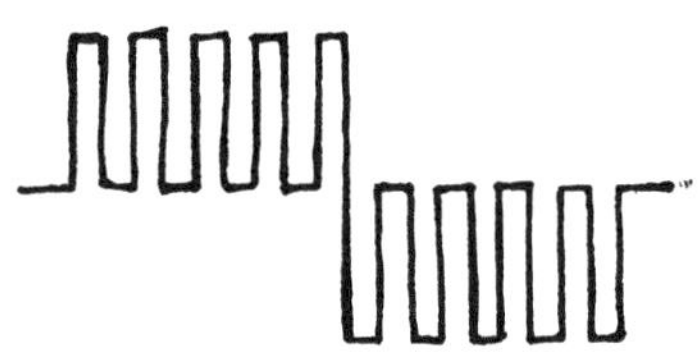

Figure 7a. Voltage Waveshape.

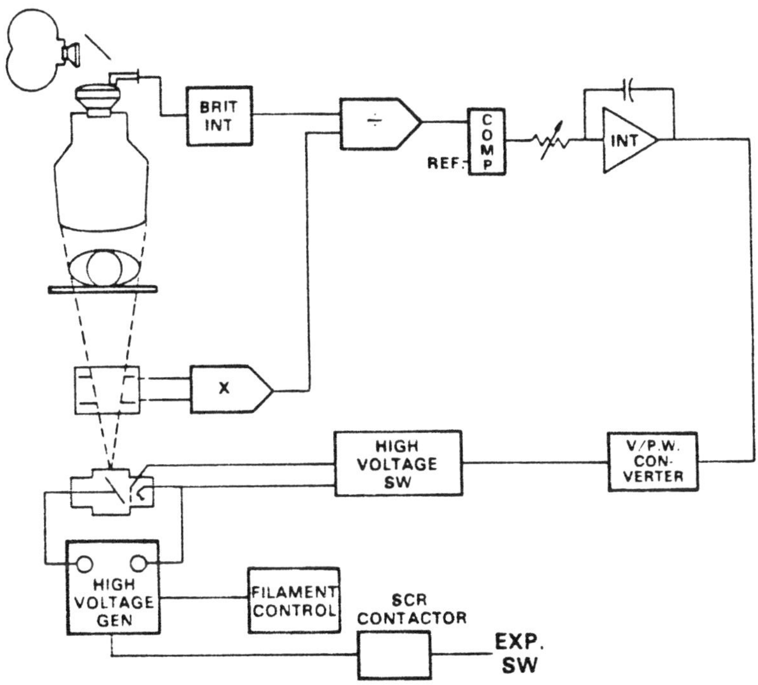

Figure 8. Automatic Brightness Control For Cine System.

By using output brightness sensing, the control system can accurately control the image brightness as seen by the cine camera film. The pulse width control of radiation provides the pulsed radiation required by the cine camera and also allows extremely fast brightness control.

Figure 9 (see next page) shows the circuit used to develop the brightness signal from the output brightness sensor. In the past, photo multipliers were used to detect the output brightness. In modern systems solid state photo voltaic cells, (solar cells) are used. The solid state devices required no high voltage supplies as do photo multipliers and reliability is greatly increased.

In operation the current from the detector is integrated during the exposure time. Figure 9a (see next page) shows the timing of the integrator circuit. By calibrating the maximum output voltage reached by the integrator during an exposure which properly exposes the cine film the control circuits can determine when an exposure is correct.

This type of integrator is the same technique used in phototiming systems.

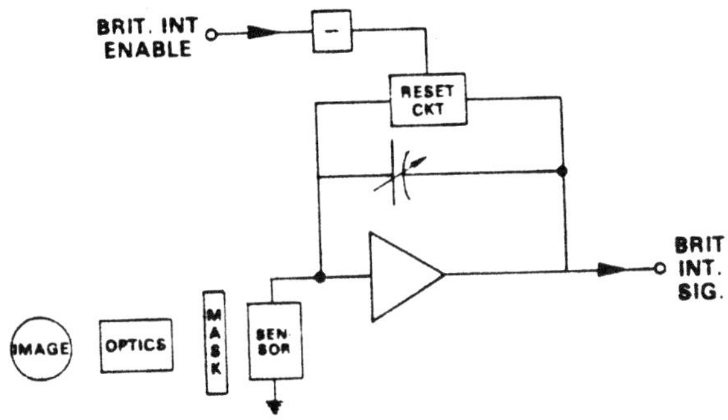

Figure 9. Typical Arrangement Of Output Brightness Sensor.

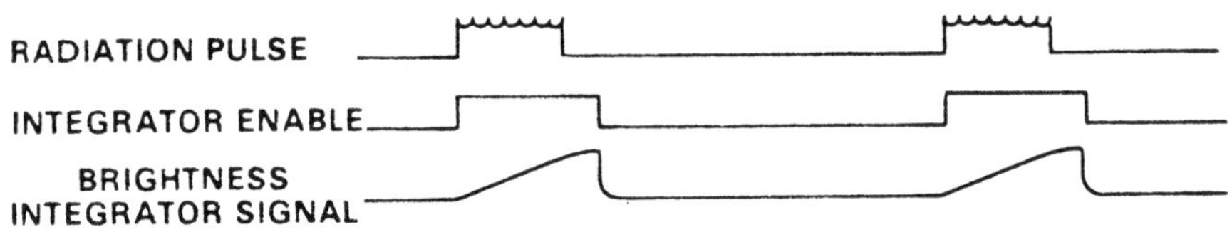

Figure 9a. The Timing Of The Integrator Circuit.

However, when using a pulse width control system the exposure times are too short for direct phototiming. The output of the brightness integrator is sampled after an exposure and a correction in pulse width is made to the following exposure.

Figure 10 shows the same comparator/integrator circuit used to convert brightness error to a radiation control voltage. The technique is similar to the system previously described. In Figure 8, a shutter compensation circuit

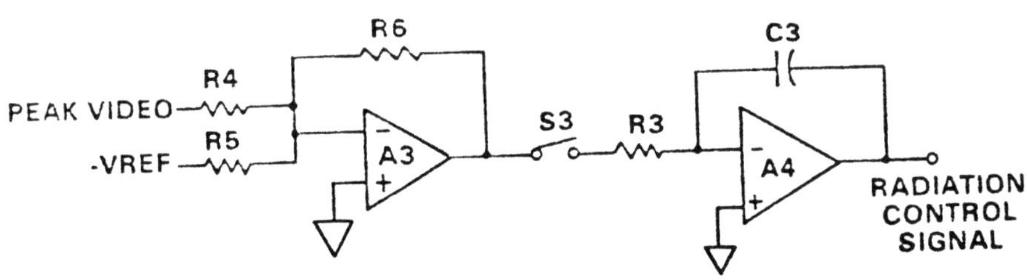

Figure 10. Comparator/Integrator Circuit.

inserted between the brightness signal and the comparator circuit. This circuit is used to compensate for changes in the sensor output current as the x-ray shutters are closed, reducing the area of the image. Without such a circuit, film density would increase as the viewed image is reduced in size. By monitoring the shutter position the brightness signal is modified by the expression:

$$B^1 = B/(X \cdot Y)$$

where B=Peak Brightness input

X = X dimension of collimator

Y = Y dimension of collimator

X and Y dimensions are equal to 1 when collimator is full open.

In order to drive the pulse width or x-ray tube, the radiation control signal must be converted to a variable pulse width. Figure 11 shows one technique used in implementing a voltage to pulse width converter.

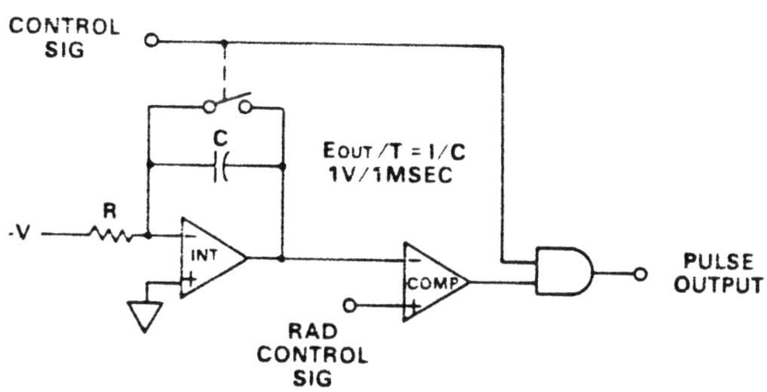

Figure 11. A Voltage-to-Pulse Width Cinverter.

The timing diagram of Figure 11a shows the operation of this circuit. The control signal input provides the frame rate timing which is obtained by processing the shutter position of the camera. When the camera shutter opens the ramp generator starts. It's output is fed to a comparator which detects when the voltage of the ramp generator equals the radiation control signal. By calibrating the slope of the ramp, a pulse is created using the "and" gate which is proportional in length to the radiation control signal.

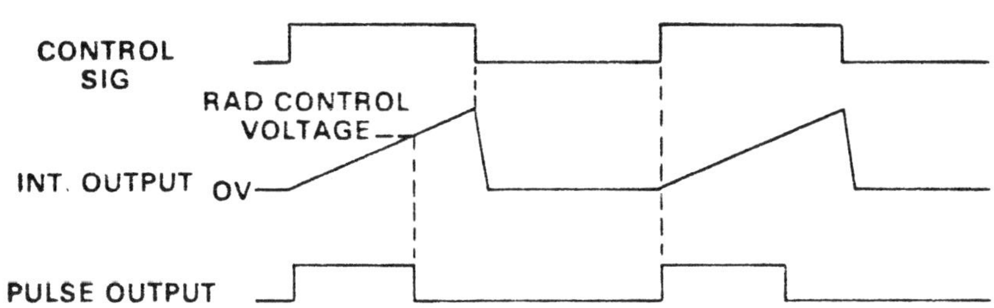

Figure 11a. Timing Diagram.

The pulse output is then used to turn on and off a solid state high voltage switch which controls the grid of the x-ray tube.

With either type of ABC control system the kVp can also be controlled automatically to provide additional automation and extended dynamic range of the Automatic Brightness Control System.

Figure 12 shows such a kVp-override system. When used with a mA or pulse width control, the kVp can be increased when the primary brightness control reaches an upper limit. When the brightness signal reduces, the kVp will first return to the preset minimum level and then the mA or pulse width will be reduced to correct the brightness level. In Figure 12, the brightness and radiation control signals are used to determine when additional kVp is required. By examining the equation for kVp error it can be seen that no error will result as long as the actual pulse width can correct for brightness error.

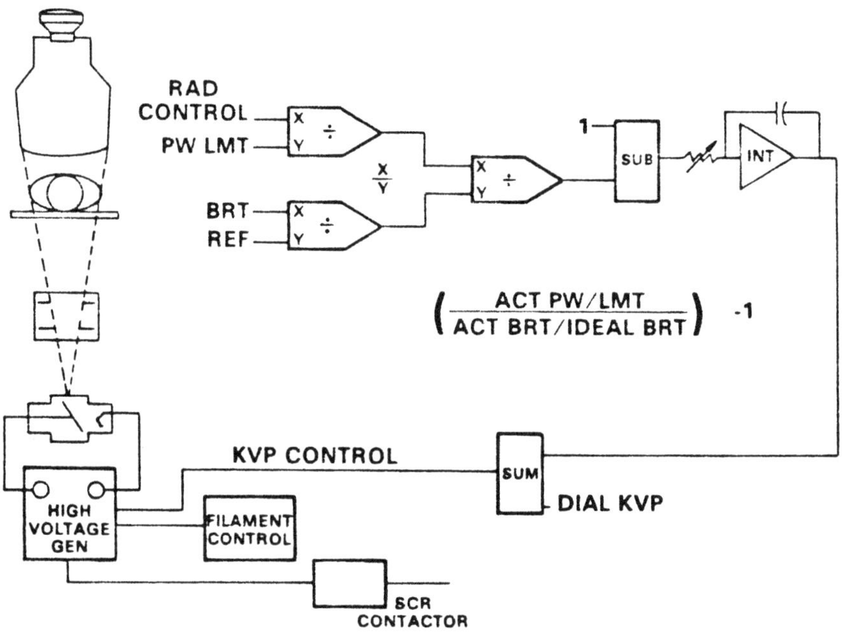

Figure 12. A Typical kVp-override System. The kVp can be increased when the system is utilized with a mA, or Pulse Width control as the primary brightness control reaches an upper limit.

CONCLUSION

We have explored two types of automatic brightness control systems in this article. There are many more possible control techniques which could be and are used. The techniques explained here represent the basic concepts used in most ABC circuits in use today.

Figure 13 shows the block diagram of a traditional control system using feedback. This control system is the basis for all closed loop systems such as the ABC circuits just reviewed.

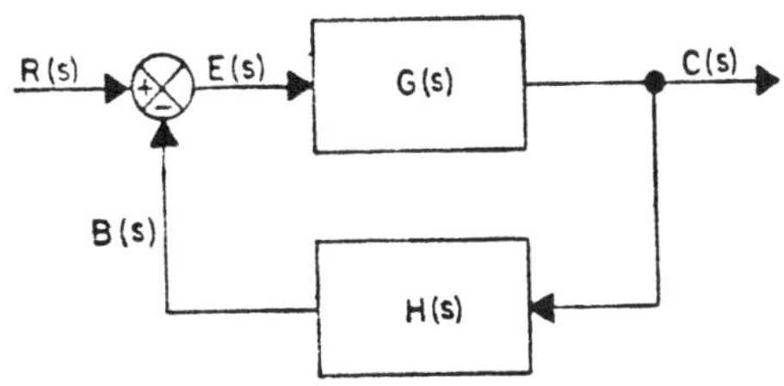

Figure 13. Feedback Circuit.

All the components required by this block diagram are present in the ABC circuits presented here. R(s) is the brightness reference voltage. The circle is the comparator circuit. G(s) is comprised of the radiation control circuits with C(s) being the radiation output of the x-ray tube. These circuits contained in the G(s) block, determine the transfer function of the circuit. This is where stability and speed of response are controlled. The H(s) block represents the feedback in our system. Included here is the patient absorption, intensifier response and brightness sensing circuits. This completes the control loop. Hopefully, some insite has been gained into the workings of automatic brightness control systems. As technology moves forward, I am sure more and more requirements will be placed on brightness control systems, and at the same time, more and more modifications to the above techniques will be available when implementing these control systems.

REFERENCES

1. Radiologic Physics, Equipment and Qaulity Control, Yearbook Medical Publishers Inc., Chicago.

2. A Practical Approach to Modern X-ray Equipment, Little, Brown and Company, Boston.

EVALUATION OF X-RAY TUBE FOCAL SPOTS IN THE FACTORY

Viktor W. Pleil
Picker-Dunlee Corporation
Bellwood, Illinois 60104

Introduction

The focal spot size is widely used as the determining factor of the imaging quality of diagnostic X-Ray tubes. Even though other methods have been considered in recent years, it is still the pinhole method which is used in the industry to determine focal spot shape and energy distribution as the basic factors of the imaging quality of a particular X-Ray tube.

It is the purpose of this presentation to describe this method, its possibilities and limitations and also to address the problems inherent to this method.

Specifications and Tests

I will start with the NEMA-Specifications, documented in the Standards Publication/No. XR5-1974. This standard specifies the material and dimensions of the pinhole (Figure 1),

Pinhole Diagraphm Dimensions in Millimeters

Nominal Dimensions of Focal Spot, f, Millimeters	Nominal Diameter	D	L
$0.3 < f \leqslant 1.2$	0.030	0.030 ± 0.005	0.075 ±.0.010
$1.2 < f \leqslant 2.5$	0.075	0.075 ± 0.005	0.350 ± 0.010
$2.5 < f$	0.100	0.100 ± 0.005	0.500 ± 0.010

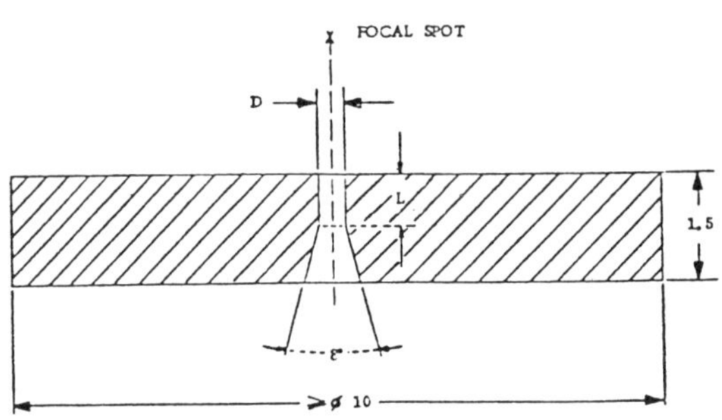

Dimensions in millimeters

Cross-sectional View of Pinhole Diaphragm Indicating Essential Dimensions

Figure 1

the enlargement factors to be used for various focal spot sizes (Table I),

Table I

Enlargement Factor

Nominal Dimensions of Focal Spot, f, Millimeters	Enlargement Factor
0.3 < f ≤ 1.2	2
1.2 < f ≤ 2.5	2
2.5 < f	1

NEMA Standard 9-11-1974.

the specifications for tube voltage and current which are to be used for the subject X-Ray tube type (Table II),

Table II

Test Voltage and Current *

Maximum Rated Voltage, U, kV	Test Voltage	Test Current
U ≤ 75	Maximum rated voltage	50 percent of maximum permissible current at the test voltage for 0.1 second at the highest rotational speed for which the tube is rated.
75 < U ≤ 150	75 kV	
U > 150	50 percent of maximum rated voltage	

* For tubes intended for fluoroscopy only, the test conditions shall be at maximum rated voltage and maximum rated current.

the maximum dimensional tolerances permissible for the individual focal spots, with the 0.7 multiplier for the length (Table III),

Table III

Nominal Size of Focal Spot, Millimeters	Tolerance, Percent Minus	Tolerance, Percent Plus
< 0.8	0	50
0.8 through 1.5	0	40
> 1.5	0	30

In addition, for line focal spots only (to allow for the peaked distribution of energy in this direction), a multiplying factor of 0.7 shall be applied to the dimension corresponding to the direction "a" dimension (see Fig. 2-2).

and a specification as how to measure the pinhole image of the focal spot and to derive the proper geometrical size (Figure 2).

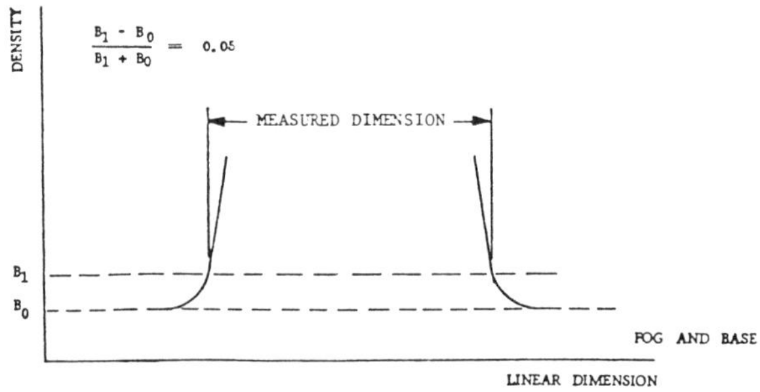

Figure 2

Figure 2 brings to mind the question: How do we determine the borders of the pinhole image? I would like to show one sketch of a focal spot as an example (Figure 3):

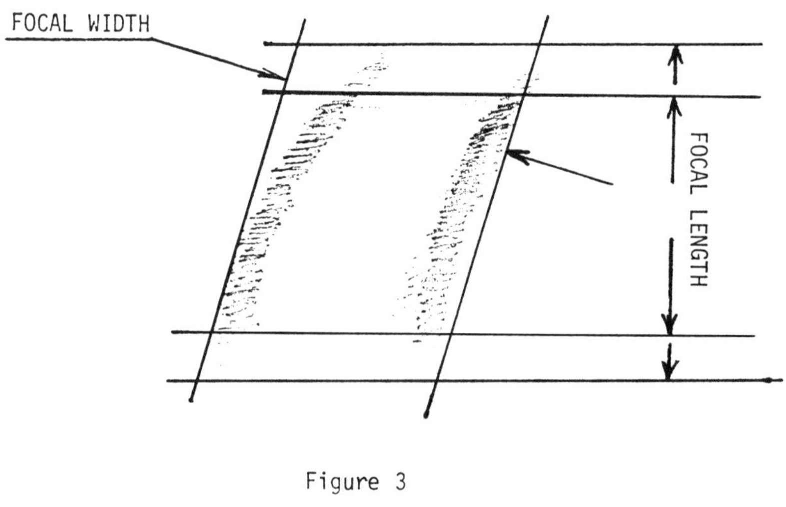

Figure 3

While a fairly accurate determination of the focal width is not a major problem, it may be difficult to measure the focal length. The problem becomes apparent if more than one examiner try to compare results. We even observed that the same individual examiner obtains different results if he or she measures the same focal spots again. The error is random, and can be as much as ±5%.

Problems

This leads to the communication problem about focal spot sizes between the user and the X-Ray tube manufacturer. Besides the aforementioned measurement inaccuracies, other parameters contribute to the problem. From personal experience I know the following:

1. Other exposure values than those specified are used.
2. Film-screen combinations instead of single coated dental films are used as image receptors.
3. The largest contributor to the communication problem, is the controversial interpretation of tolerance factors. Most of our customers are of the opinion that nominal focal spots are obtainable on request. It is not common knowledge that the manufacturers have utilized the maximum tolerances to achieve optimum ratings and emission current values.

I would like to present a slide from my last presentation on this subject, which clearly shows that the optimum compromise between the unsharpness of motion, determined by the load capacity, and the geometrical unsharpness caused by the focal spot size has to be met in order to achieve an optimum radiograph (Figure 4).

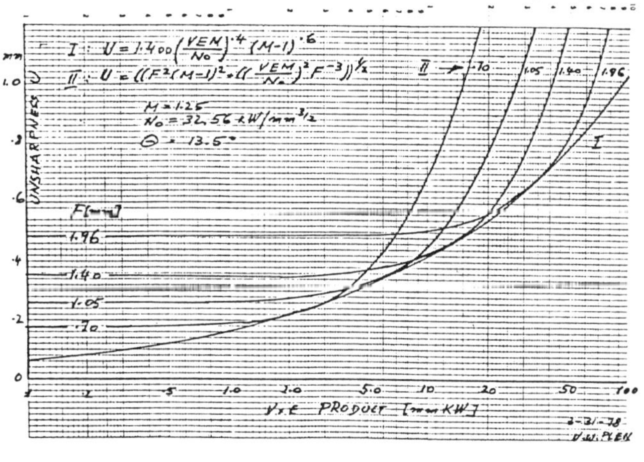

Figure 4

The abszissa represents the product of object motion and load requirements while the ordinate represents the unsharpness. For larger V x E products, it is evident that the utilization of the maximum permissible focal spot size produces higher quality radiographs than the use of the nominal focal spot size. This is the actual cause for the discrepancy between "nominal" and "actual" focal spots.

There is another factor to consider: with the introduction of generators of higher power output, the larger focal spots produce sufficient images and cause less focal track wear than the smaller "nominal" focal spots. For this reason, the larger focal spot tolerances have been utilized in X-Ray tube production but the focal spot nomenclature has not been changed. This fact was not considered objectionable at that time.

The situation has changed in recent years: with the upcoming sophistication of imaging systems, the radiological community would prefer to see a more realistic focal spot size determination by using the actual focal spot sizes for the proper definition.

To change over to this new nomenclature would cause much confusion for the radiological community if some tube manufacturers do not comply. Therefore, all major tube manufacturers have agreed to maintain the existing tolerance factors, relying on the understanding of the user that oversized focal spots are customary.

Search for Better Solutions

In order to overcome the above dilemma, several attempts have been made to derive a better definition of the imaging quality of a focal spot than its size determination by a pinhole picture. Since the resolving power of a particular focal spot is under question, it is obvious to use a method which measures just that. Bar patterns and star patterns have been in use for quite some time. These objects are used to determine the limitations of the resolving power of a focal spot by the use of the zero contrast method which involves measuring the line frequency at diminishing contrast (Slide). This method is much appreciated since it is easy to use in the field for there is no need to remove the collimator. This is the obvious reason why the acceptance test criterion of the A.C.R. radiographic formulary uses the star pattern for focal spot size measurements (Tube crane and tube-item 13). A simple nomographic method (Figure 5) can be used to derive the focal spot size. This is done by determining the magnification and the diameters of the disappearance frequency arches (D, right scale). The magnification is determined by measuring the diameter K or the distance H of the star radiograph (Left scale). The intersection of the line connecting K (or H) and D with the center scale gives the focal spot dimension.

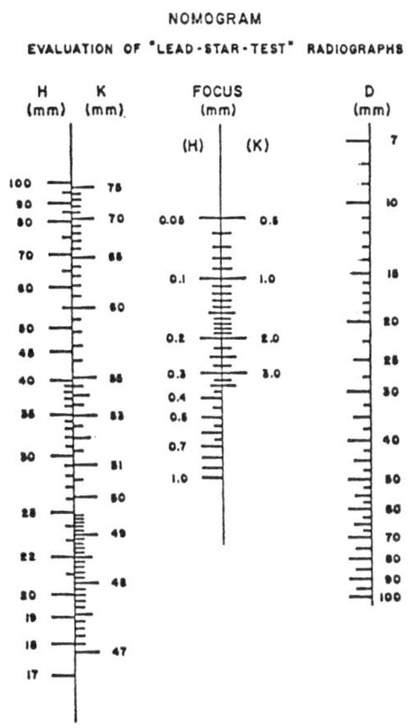

Figure 5

One can not expect that the focal spot dimensions determined by the star method will agree with the actual focal spot dimensions derived by the pinhole method. The resulting focal width and length can be up to 20% larger and 15% shorter respectively when compared to the dimensions of the pinhole picture. The reason for the discrepancy in the focal width are the so called "edge bands", the reason for the discrepancy in the focal length is due to the diminishing density towards its borders. This is illustrated on a 0.6 focus of maximum NEMA-tolerance in Table IV:

Table IV

Nominal		Pinhole	Star
.6	Width	.9 mm	1.1 mm
	Length	1.2 mm	1.05 mm

At present, the NEMA - X-Ray tube subcommittee is preparing a resolution method which identifies the existing nominal focal spot size as a dimensionless number and brings it into relation with the minimum resolving power (lp/mm at zero contrast) which has to be achieved at a specified magnification factor.

The existing draft uses a 1 1/2° Vertex angle star pattern and lists focal spots from 0.05 to 2.0.

The Table V below depicts some examples:

Table V

Nominal Focal Spot (F)	Magnification	Minimum resolution lp/mm at state M
.3	2.5	2.78
.6	2.0	1.54
1.0	1.5	1.50
1.5	1.30	1.44
2.0	1.25	1.35

This new procedure has not yet been perfected, however, I personally feel that the resolution method is headed in the right direction so that a better way of communicating the X-Ray tube imaging qualities between the radiological community and the tube manufacturers can be established.

REQUIREMENTS OF MODERN AUTOMATIC EXPOSURE CONTROL SYSTEMS

Walter Weigl
Product Department
Siemens Corporation
Iselin, New Jersey 08830

Abstract

A complete system of automatic exposure controls (AEC) is discussed. Included are applications in general radiography, tomography, mammography, fluoroscopy and film recording from an image intensifier.

Introduction

Variations in size, density and positioning of the patient, as well as variations in imaging geometry and a number of other factors such as type of examination, type of grid, collimation etc. influence the radiation exposure factors tremendously.

Efforts have been continuously made to design an automatic exposure control to achieve consistent film densities regardless of variations in exposure conditions. In the meantime, early crude systems have been refined. Today, the AEC is an appreciated component in the search for optimum imaging results.

Prerequisites

Optimum Image Quality

Optimum image quality is not necessarily the most beautiful image, but rather is an image which supplies the radiologist with the most information for proper diagnosis. To achieve this, the factors affecting film density, contrast and resolution, must be at an optimum for the particular examination (correct exposure, correct kV and optimum relation of kW output, geometry and film screen combination). Naturally, positioning and collimation are of decisive influence. Depicted in Figure 1 are the factors influencing the image quality of a radiograph.

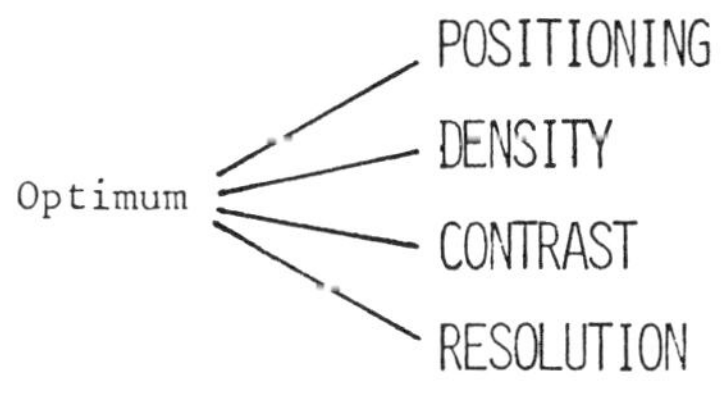

Figure 1. Factors influencing image quality.

The classic AEC controls only the film density, and all other factors are controlled by the radiographer.

Influence of Various Parameters on Film Density

As depicted in Figure 2, there are four major factors that affect the film density. They are: (1) the patient (subject) with a certain size, shape, density (tissue, bone, etc) and scattered radiation, (2) the examination unit (geometry) with or without anti-scatter grid, fixed or adjustable collimation, fixed or adjustable SID, and the amount of filtration, (3) the speed or gain of the imaging system fixed or variable over the range of exposure factors and the processing conditions, and (4) the x-ray tube/generator system with a certain waveshape of the high voltage, variable exposure factors (kV, mA and time), and the radiation characteristics and distribution (tube performance).

All these factors require particular design elements and performance characteristics of an AEC. For the following discussion, an ideal processing condition will ab assumed.

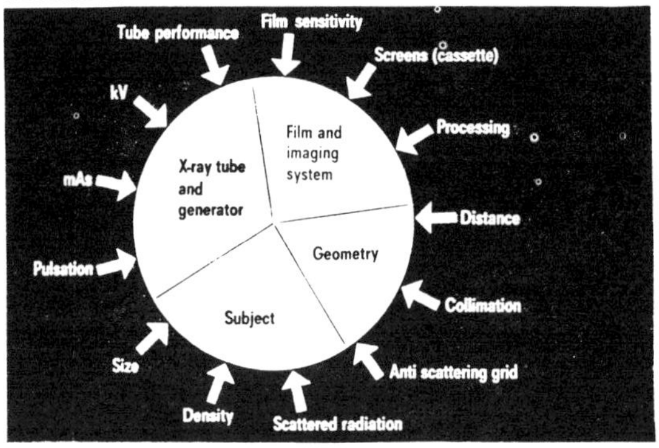

Figure 2. Influence of various parameters on film density.

Basic Design of An AEC

In order to control the film density, basically two components are required: a radiation sensor, and a control unit to interrupt the radiation when the correct film density is reached.

Location of the Radiation Sensor

There are three possible locations for the radiation sensor as shown in the block diagrams depicted in Figure 3. The simplest approach is to mount the sensor directly in the port of the x-ray tube. Naturally, this arrangement cannot detect variations of the subject or geometry, but is capable of maintaining the output at a constant level. Its application is limited to similar subjects and standard geometry such as that of dental examinations.

The most common location for the sensor is behind the subject, but in front of the film cassette. Here it can detect variations of the subject and geometry. It can even detect different grid types (or no grid), if the grid is mounted between the subject and the radiation sensor. A radiation sensor placed in this location must be very low in absorption. It must also be shadow free, even with large film sizes. To keep the patient-film distance small, this sensor must be as thin as possible.

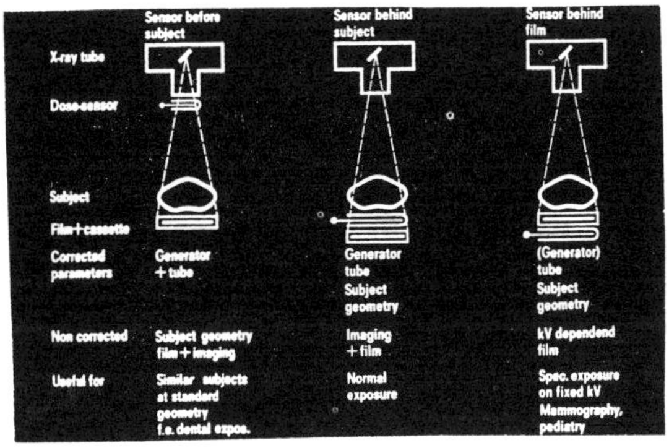

Figure 3. Various locations of the radiation sensor.

The third possible location is to mount the sensor behind the imaging system (which means behind the film cassette). Here a lot of space is available, the sensor does not need to be shadow free, and its absorption can (and should) be high. Minimum patient film distance (least geometric unsharpness) is possible. However, there is a disadvantage. The screen system is designed to absorb as much radiation as possible (to transfer it into visible light). Hence, the intensity of radiation behind the cassette is very low, and it even varies considerably depending on the potential and the patient because of varying absorption in the screens. However, there is still a number of applications such as operation within a constant tube potential range (like pediatrics of mammography), or operation with a preprogrammed AEC technique where the effect of the tube potential dependence can be counteracted by corresponding programming of the gain in the control unit.

The output of the radiation sensor carries only very small signals (in the nA to uA range), either because of the required low absorption (if located in front of the cassette) or because only a very low radiation intensity is available (if located behind the cassette). For this reason, either a preamplifier is used at the radiation sensor, or the measuring signal is connected directly to the central amplifier (control unit) by a highly insulated shielded cable (See Figure 4). In the central amplifier, parameters can be corrected, which cannot be measured by the radiation sensor (e.g. the properties of the screen film combination).

The amplified signal is connected to a discriminator where it is compared to a nominal value. When the signal reaches the nominal value, the exposure is terminated.

Radiation Sensors

Basically, there are three types of radiation sensors available: (1) ion chamber type, (2) photo tube type, and (3) photo diode type, as shown in Figure 5.

An ionization chamber can be designed to be relatively shadow free, therefore, it is ideal for application in front of the film cassette. Unfortunately it delivers only a very low signal with high impedance.

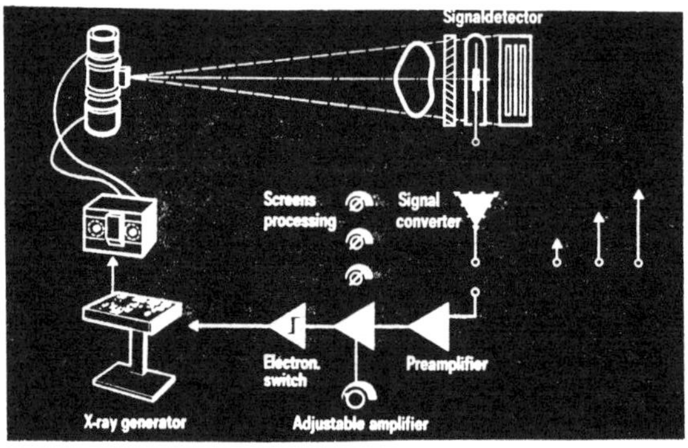

Figure 4. Block diagram of AEC.

To achieve a higher signal, an ionization chamber for AEC application is designed with thin metal walls in contrast to "air equivalent" walls used in chambers for application in dosimetry. This design increases the ionization current by approximately a factor of 100 compared to a typical air equivalent chamber. However, the chamber becomes kV-dependent. A compensation for this kV-dependency is necessary in the control unit (central amplifier).

Instead of an ionization chamber, a very thin fluoroscent screen can be located in front of the cassette (it must be thin, to fulfill the requirement of low absorption). This fluoroscent screen is tilted slightly, therefore, its emitted light can be directed to a photo multiplier or a photo tube.

A semi-conductor radiation receiver can also be used, its signal level is relatively high. It is not shadowfree, therefore it can only be used for measurements behind the cassette. As indicated before, a compensation for tube potential dependence in the control unit is necessary.

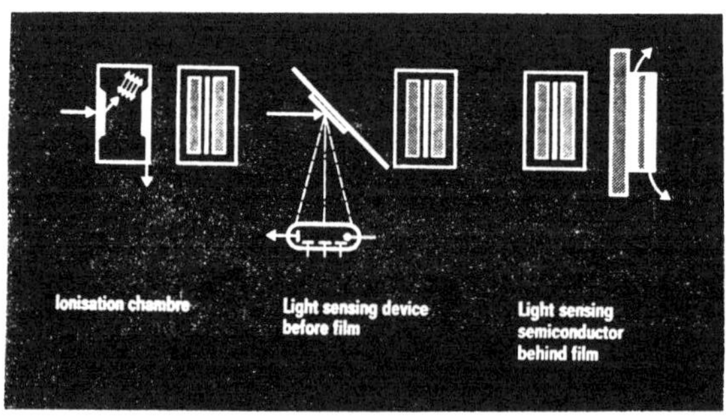

Figure 5. Three types of radiation sensors.

Today, ionization chambers are used as radiation sensors quite frequently. The discussion hereafter will be focused on the ionization chamber type AEC devices. Different shapes and positions of small measuring fields are possible and can be selected individually or in parallel in any combination in accordance to its application requirements. Figure 6 shows a typical threefield AEC device. For the p.a. view of a lung, the side fields (II and/or III) are used. For the lateral view, the center field (I) is used. In this manner, the x-ray transmission at the primary important location is measured.

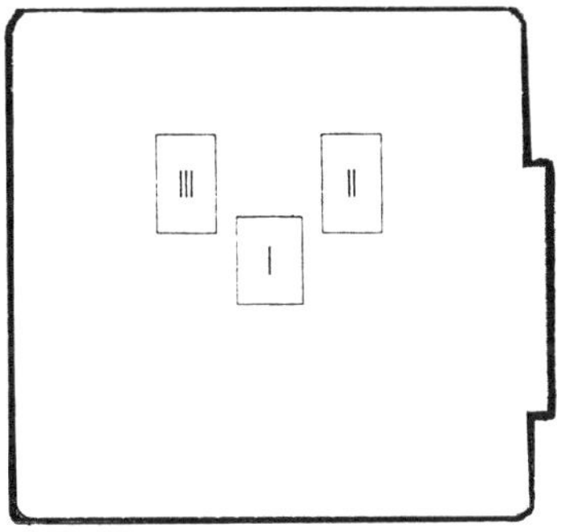

Figure 6. Threefield ionization chamber.

Exposure and Exposure Rate Regulation

The film density increases in direct proportion to the exposure received by the film. In other words, at a certain exposure rate for a given exposure time, the processed film will show a certain film density. The exposure rate is influenced by many parameters (as has been shown at the beginning of this paper). But it can be measured by the AEC continuously until the exposure for the required film density is reached, and subsequently the radiation can be switched off. This means, the AEC controls the time to reach a certain exposure for the desired film density, with an unknown exposure rate at the film (Figure 7). Because of the continuous measurement, small variations of the exposure factors (tube potential and current) are also compensated for by the AEC. This type of time regulation is very applicable for a table bucky or a spotfilm device as shown in Figure 8.

It is quite useless, however, for some other applications, such as tomography, where the exposure time is a fixed parameter; there the AEC needs to be an actual regulation circuit. It then varies tube potential and current to achieve the regulation of the radiation output.

For tomography, this controlled radiation output then results in the correct exposure, and therefore, the film density for the desired exposure time as depicted in Figure 9.

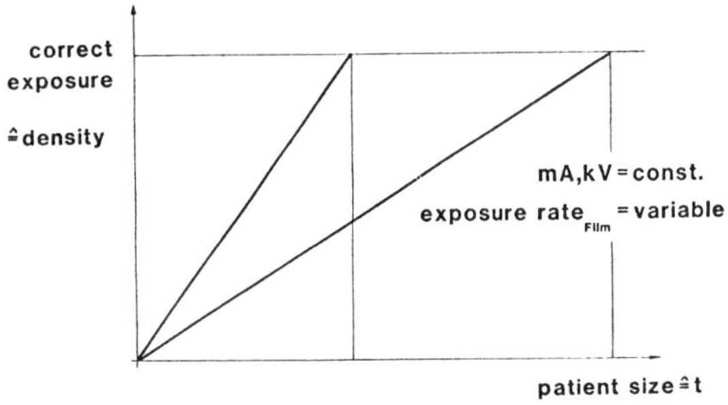

Figure 7. AEC employing time regulation.

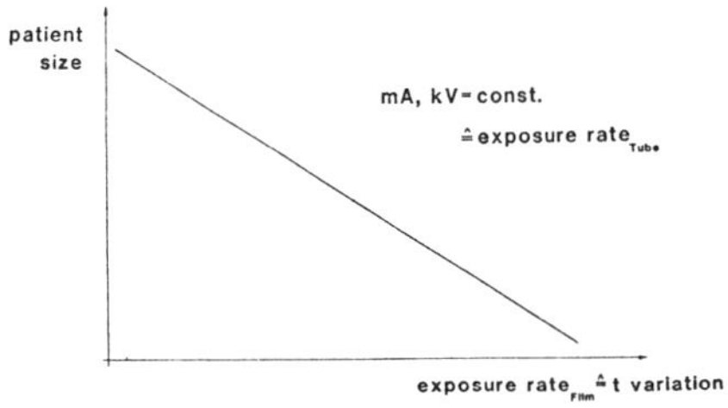

Figure 8. Patient size and time regulation.

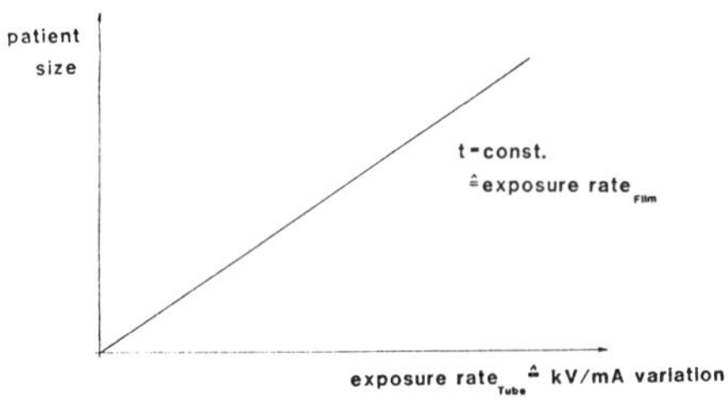

Figure 9. Patient size and exposure rate regulation.

Integration of the AEC into the X-ray Generator

Be designing the AEC as a part of the X-ray generator system, a number of problems, pertaining to (1) tubeloading, (2) voltage dependency of the radiation detector and film screen combination, (3) regulation for photospot and cine application and (4) organ programmed exposure technique can be solved.

Tubeloading

The method of using constant tube potential and current and hence only regulating the exposure time, results in the following disadvantages: The required exposure time is not known at the start of the exposure. To avoid potential destruction of the x-ray tube due to overloading, the loading factors have to be adjusted for the longest expected exposure time as illustrated in Figure 10.

This prohibits in many cases the maximum loading of the x-ray tube, therefore, resulting in longer reexposure time with increased motion unsharpness. An additional timer is necessary to terminate the exposure in case the maximum exposure time is reached (back-up timer). This problem may be solved by integration of the AEC into the design of the generator.

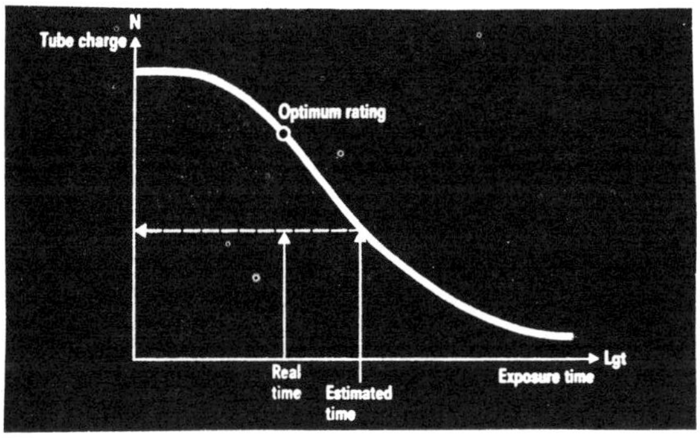

Figure 10. AEC and tube loading.

Optimum tube loading and much simpler operation have been achieved with the introduction of falling load generators as shown in Figure 11.

First, the static tube loading curve found in the documents with each x-ray tube is converted into a dynamic loading curve. This curve now still represents a 100% tube loading for all exposure times. To extend tube life, a loading of approximately 85% of this curve can be used for small and medium sized patients, and 100% is used for heavy patients.

This continuous falling load curve is incorporated into the filament circuit of the generator. It allows all exposures to be started with the highest possible tube loading which then is reduced according to the dynamic loading curve. As a result of this design, shortest exposure times, and simple operation in setting up the radiographic techniques can be realized.

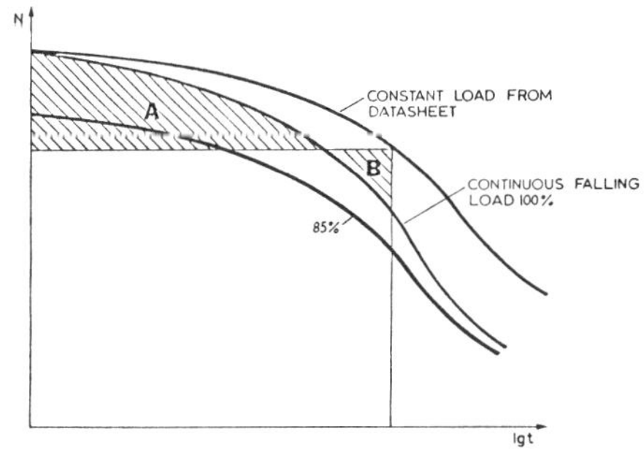

Figure 11. Continuously falling load.

Voltage dependency

Until recently AEC was reuqired to provide approximately constant exposure at the film plane, regardless of variations in tube potential, object size, geometry, grid system etc., based on a relatively linear tube potential dependence of the then exvlusively used calcium tungstate screens. Today a number of different rare earth screens is widely used. Their light output varies considerably over the tube potential range of 50 to 150 kVp, and the variations differ depending on screen type as illustrated in Figure 12.

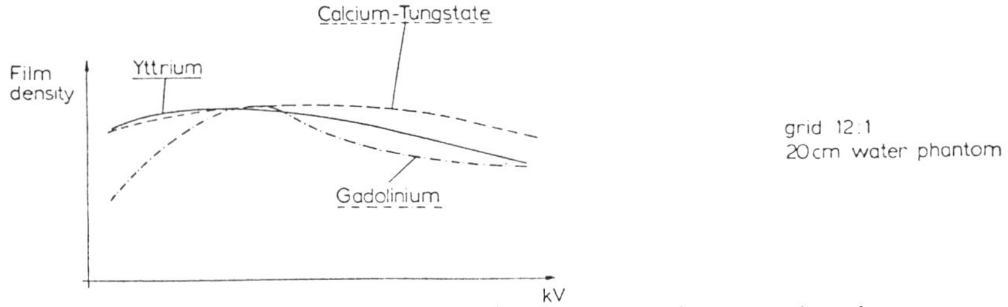

Figure 12. Tube potential dependence of various screen film combinations.

Now, in addition to the tube potential compensation for the radiation sensor, a tube potential compensation for the particular screen film combination is necessary. Depending on the design of the AEC, these compensations can be achieved with various methods. For example, in the case of a microprocessor based AEC, different preprogrammed proms for the various film screen combinations.

With these fast screen film combinations, forced extinction is absolutely essential. Additionally, it might be necessary to reduce generator output to avoid reaching the shortest possible exposure time of the generator/AEC combination (e.g. using the 0.6 mm focal spot rated at 40 kW for high voltage p.a. chest exposures).

Regulation for Photospot and Cine Application

An automatic exposure rate regulation is used for fluoroscopy. This is commonly achieved by using a photo-multiplier at the output of the image intensifier. Comparison of the output from the photo multiplier with the reference signal is made and the difference is signals is used to regulate the exposure rate to achieve constant medium brightness at the I.I. output window. The necessary regulation range fro the exposure rate of 1000:1 (object thickness from infant to obese adult, addition of barium, and SID changes of the spotfilm device from fully compressed to fully extended), can be easily achieved with a combined "mA-kV" regulation.

Naturally, this fluoroscopy regulation can also be used as a base for controlling photospot and 35 mm cine films. The tube potential value arrived at during fluoroscopy (which is a measurement for correct penetration) can be used as automatic adjustment of x-ray tube potential for photospot and cine: kVph. Depending on the type of examination, a correction to higher or lower kVph can be programmed. By using this method, optimum contrast of photospot and cine films is assured. The photo-multiplier at the image intensifier output is then used as a radiation sensor for phototiming of each individual frame as demonstrated in Figure 13.

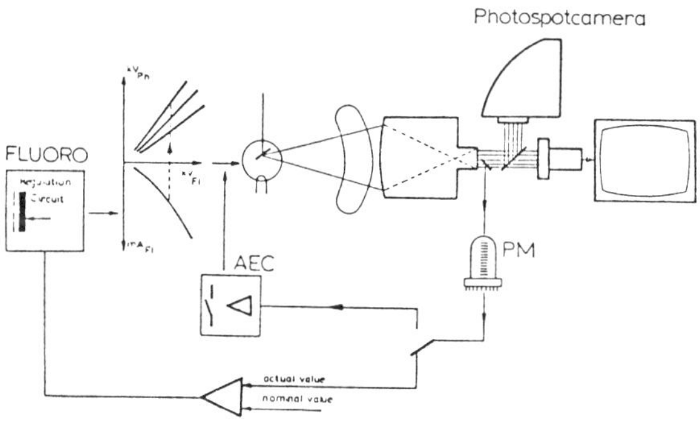

kV_{ph} . . . .Tube Potential for Photospot Camera

kV_{fl} . . . .Tube Potential for Fluoroscopy

mA_{fl} . . . .Tube Current for Fluoroscopy

Figure 13. AEC for photospot and cine.

Organ programmed exposure technique

Organ programmed generators are aimed at simplified operation, and at standartization of the exposure factors within a department. Selecting a particular program adjusts a pre-determined set of factors for a particular examination. It not only adjusts the necessary AEC factors (such as field selection and sensitivity of the radiation sensor for a particular screen film combination), but can also select the focal spot, tube potential and the particular examination unit. Naturally, for examinations requiring fixed exposure factors, these can be programmed instead of the AEC factors. This method of operation allows the user to spend more of ones valuable time on the patient. The required manual operational steps are: correct positioning (including correct positioning of the important anatomy for the radiation sensor), and correct collimation.

Performance Testing

As mentioned previously, an AEC is designed to deliver consistent exposure in order to obtain a constant film density. The exposure at the image receptor plane depends on a number of factors with wide variations, such as scatter radiation from the patient and the examination table, among others. Therefore, for performance testing, only exposure measurements in the image receptor plane should be taken into account; see Figure 14.

Patient Size

Tube Potential (kVp)

Saturation

Back-up Timer

Reproducibility

Measurements taken at the imaging plane,

Most suitable phantom: Water

Figure 14. Performance Testing

Phantoms should represent the human body in absorption and beam hardening characteristics as close as possible. For this reason, water is probably the best material, but is cumbersome to handle. Aluminum is used widely today, but one might expect slightly different beam hardening characteristics.

Summary and Outlook

Today, the AEC is used worldwide for routine and special examinations. Organ programmed technique for routine examinations and falling load are widely accepted. In the near future, an adaptation of the AEC to extreme short exposure times with corresponding low "mAs" values as required by the continuous increase of modern film screen system speeds can be expected. One will also see more developments towards combined exposure-exposure rate regulation systems, to be able to achieve not only optimum film density, but also optimum contrast even for complex examinations, such as peripheral runoffs on a stepping table. These developments of modern technology will bring us closer to the goal of the radiology departments: optimum diagnostic image quality.

This goal can be achieved by a radiographer concentrating on medical details, and by fully automatic equipment concentrating on technical details.

References

1. Aichinger, H: Fortschritte in der Technik der Belichtungsautomaten in der Roentgendiagnostik. Der Radiologe, August 1968, 233-238.

2. Franke, K: Aspects for Designing Modern X-ray Generators. ICR Madrid, October 1973.

3. Hunold, M: Automatic Exposure Systems, Benefits and Limitations. ACR Miami, March 1977.

4. Lin, P: Phototimers, User's Point of View. SPIE Vol. 70 (1975) Medicine IV, 196-200.

5. Siemens AG: Automatic Exposure Control in X-ray Technology. Slide presentation 1977.

6. Weigl, W: Design of Falling Load Generators, Proc. Midwest Chapter AAPM, Symposium "Workshop in purchase specifications and performance evaluation of radiological imaging equipment" , pp 163-172. Obtainable from National Technical Information Services NTIS Publication Number 287-985.

THE DESIGN AND FUNCTION OF AUTOMATIC X-RAY FILM PROCESSORS

Robert L. Shanebrook
Photographic Technology Division, Eastman Kodak Company
Rochester, New York 14650

Abstract

Today's automatic x-ray film processor consists of many complex components designed to process silver halide x-ray films uniformly and reliably. This paper describes the design and function of these component parts including film transport, chemical replenishment, chemical reactions, solution agitation, temperature control, film washing, film drying, and energy efficiency. Care in the design and control of the interaction of process components is essential to obtain quality radiographic images.

Introduction

Prior to 1956, x-ray film processing was done manually by suspending the film in tanks of chemicals. This process was slow, requiring as much as one hour to obtain a properly processed, dry radiograph. This required a skilled operator who was depended on to properly handle the films, monitor and maintain temperature control, agitate the film in the baths, move the film from tank to tank, and maintain the solutions to provide a constant chemical activity. This was an expensive and difficult process to operate properly.

In 1956, the introduction of the first Eastman Kodak Company roller-transport x-ray film processor, the Kodak X-Omat processor, model M, led to a revolution in medical film processing. While the model M was a major improvement over manual processing, it was quite different from today's processors. It was very large, approximately 10 ft long, 5 ft high, and 3 ft wide. It cost $25,000 and required 6 1/4 minutes to fully process and dry a sheet of x-ray film.

Today's processors have evolved from the model M. The basic function remains the same; that is, to process photographic film uniformly and reliably in the chemicals necessary to develop the latent image, then remove any chemicals from the film that would interfere with future evaluation of the image.

The automatic film processor serves as the mechanical link between the x-ray film and the processing chemicals. To satisfy the needs of radiologists, the films, chemicals, and processors must be designed to optimize the performance of the entire system. Design of any portion of the system must consider its influence on the whole system to produce reliable images for diagnosis.

Processing Cycle

Before processor mechanics can be discussed, it is necessary to consider the chemical reactions that take place in an automatic x-ray film processor.

The developer's primary function is to convert exposed silver halide to silver. This is done using a Phenidone-hydroquinone developer with a pH of approximately 10. During development, the solution causes the gelatin to swell. The degree of swelling is controlled by hardening the gelatin with an aldehyde. The hardener reduces the gelatin's susceptibility to physical damage. The degree of swell also influences the time necessary for chemicals to diffuse in and out of the film. Therefore, the hardening will influence all chemical reactions that take place in the film.

The fixer must stop development and desensitize the film to light by converting the unexposed silver halide to a soluble complex that can be washed from the film. Converting the silver halide to soluble complexes takes place in a series of reactions. Inadequately fixed films cannot be properly washed. An old rule-of-thumb still holds true; fixation time should be twice as long as it takes the film to lose its milky appearance.

An additional requirement of the fixer is to further control gelatin swell. Because fixer composition has a great influence on the amount of liquid the film will absorb, it influences the drying of the film and may adversely affect the resulting film density. All fixers are not equal. Poorly formulated fixers can reduce film density and contrast. This is largely caused by the degree of film swell, which in turn affects the covering power of the developed silver.

Film washing is basically the removal of chemicals, particularly the thiosulfate complex ion, from the swollen gelatin emulsion by allowing clean water to replace the chemicals. The water is then evaporated from the film in the dryer.

In order to produce satisfactory images, these reactions must take place uniformly and at controlled rates. The medical profession has generally accepted 1 1/2 to 2 1/2 minutes total processing time. The times available for each processing step in a 90-second 35.6°C Kodak X-Omat processor are:

Developer:	22 seconds
Fixer:	16.5 seconds
Wash:	9.8 seconds
Dry:	25.3 seconds

The remainder of the 90 seconds is consumed in transporting the film between the processing steps.

The processor must allow each chemical and physical action to take place uniformly and repeatably. To do this, processors have become complex; but the result is increased confidence that the films are properly processed.

The Processor

The major sections of the processor are:

roller-transport system
processing solution sections

(developer)
(fixer)
(wash)
dryer
chemical replenishment system
solution recirculation system
solution temperature-control system
operational support, i.e., standby control, electrical power, plumbing, and mechanical and electrical safety features

Roller-Transport System

The roller-transport system consists of: (1) feed tray, (2) area accumulator, (3) entrance assembly, (4) developer rack, (5) developer-fixer crossover, (6) fixer rack, (7) fixer-wash crossover, (8) wash rack, (9) wash squeegee, (10) dryer, and (11) receiving bin (See Figure 1).

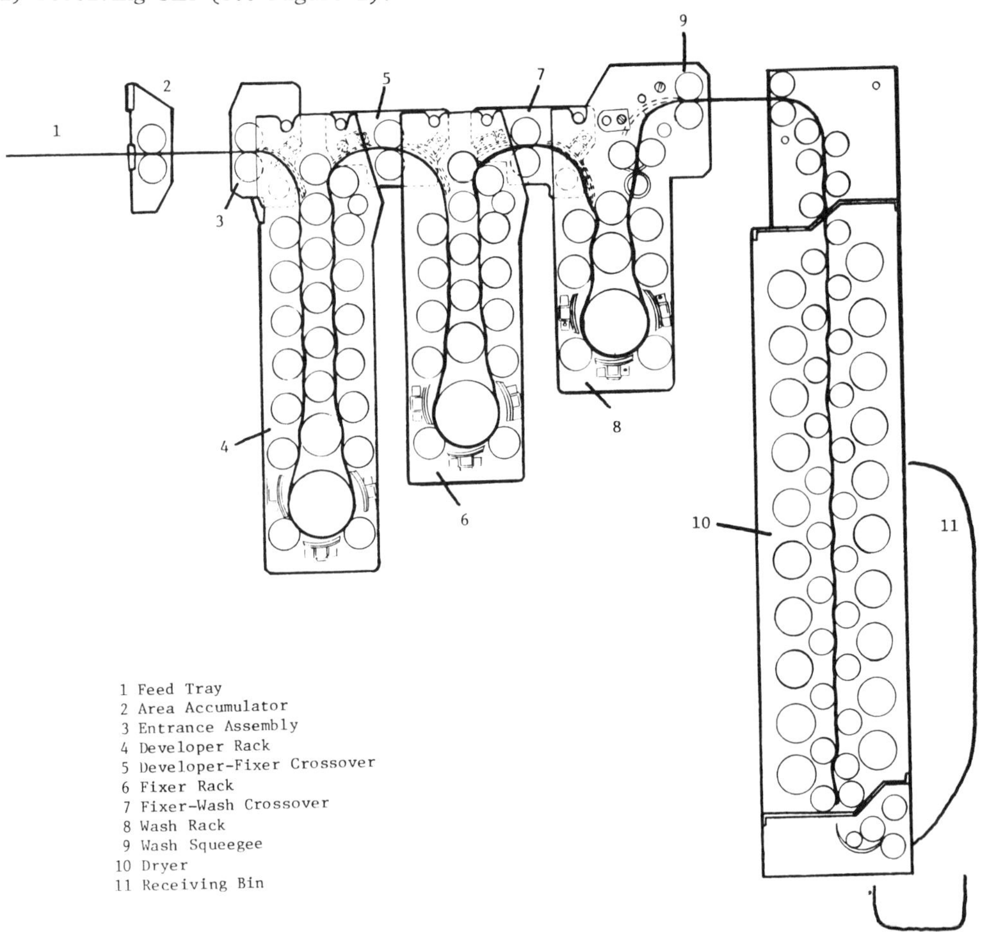

Figure 1. Principal Components Of The Roller-Transport System.

Feed Tray

The feed tray must be smooth to prevent surface abrasion marks, have an edge-guide to provide proper alignment of the roll or sheet of film, and be large enough to support the sheet and prevent buckling.

Area Accumulator/Entrance Assembly

The film is pulled from the feed tray by the driven rollers of the entrance assembly. This assembly properly orients the film for entering the developer. Detectors located in the entrance assembly area are also used to approximate the area of the film processed. Film-area measurement in the Kodak RP X-Omat processor, model M8, is used to determine the volume of replenisher solution necessary to maintain proper chemical activity. The film-area measurement also can be used to estimate the silver concentration in the fixer. The area of film processed is useful for accounting purposes or can be utilized to control a silver recovery cell.

Processor Rack Design

The function of the roller racks is to transport the film through the processor and to provide adequate agitation for the chemical reactions to take place.

All of the processor's racks have basically the same design, but the developer rack is the longest and generally the most critical. A staggered path, three-roller developer rack is shown in Figure 2. The major components

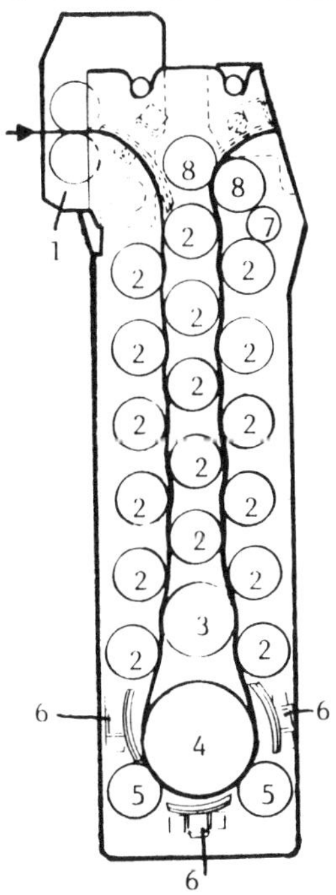

1 Entrance Assembly
2 Transport Rollers (1-inch diameter)
3 1.22-inch Roller
4 Master Roller
5 Cluster Roller } Turnaround Assembly
6 Guide Shoe
7 Rewet Roller
8 Squeegee Roller

Figure 2. A Typical Rack.

of the rack are the entrance rollers, transport rollers, rewet rollers, squeegee rollers, turnaround assembly, drive mechanism, and structural parts.

The rollers used in the racks have a great influence on processing uniformity and reliability. To achieve high quality, specific roller materials are selected from a wide range to produce the desired characteristics. Examples of roller materials are: ABS, synthane, stainless steel, EPT (ethylene propylene rubber), phenolic, and silicone.

These materials provide different hardness and surface characteristics. If the rollers are too smooth, there will not be sufficient friction to transport the film. If they are too rough, the rollers can emboss the gelatin and produce a processing artifact.

Improper roller hardness can also cause processing imperfections. If the leading edge of a film squarely strikes a hard roller, it will stop momentarily causing an artifact on the film. In the same situation, a soft roller will allow the film to continue to be transported.

In determining the configuration of the rollers, great care is given to the position of each roller. The film must be transported at a uniform speed so that all portions of the film are processed equally. In doing so, the various sizes of flexible sheets must be held so that they don't slap against rollers, causing a processing mark. In designing a processor, it is common to eliminate a processing imperfection by relocating a roller by as little as an eighth of an inch.

As the film is processed, its sensitivity to types of forces can change. Therefore, each time the film touches a surface, the contact must be considered on an individual basis. The same rollers cannot be used for all purposes.

The entrance rollers of each rack must redirect the film and begin driving the film. The rollers are soft so that the lead edge of the film isn't damaged as it strikes the roller and also to prevent hesitation of transport. Entering the fixer rack is particularly sensitive to hesitation because uneven termination of development can result.

The rollers in the middle portion of the rack are in a staggered three-roller configuration to minimize squeezing the film, since squeezing can cause damage to the soft, swollen film emulsion. The three-roller configuration allows the center roller to be used both for the descending and ascending film paths.

The bottom portion of the rack is the turnaround. The turnaround assembly redirects the film path 180°. This area is critical because the film has to be forced around a radius. It is very easy to damage film in this area.

Successful turnaround is accomplished by transporting the film past a 1.22" diameter roller then around a 2" diameter master roller. In so doing, the film is directed by metal guide shoes and held in place by spring-loaded cluster rollers. Since considerable force is exerted on the film while it is in this area, care must be taken to assure that the assembly is carefully maintained or film damage will result.

The film then ascends through the staggered rollers to the squeegee rollers. The squeegee rollers remove surface liquid from the film to reduce carryover of chemicals into the next tank and improve process uniformity.

Since the developer and fixer squeegees are laden with chemicals and exposed to air, chemical deposits can accumulate on them. To reduce this, a rewet roller is used to keep the rollers wet at all times.

The film then enters a crossover assembly. The crossover assembly turns the film path 180° and directs the film downward into the next tank. Two major problems can occur in the crossover area: chemical deposits and solution "runback". Air flow from the fixer area can carry fixer fumes to the crossover resulting in concentrated fixer deposits on the crossover rollers. These deposits will nonuniformly stop development resulting in a processing imperfection. Developer solution can also evaporate on the rollers leaving crystals that will mark the film. Crystallization is reduced by properly venting the processor, rotating the rollers periodically (by means of a stand-by control), and cleaning the crossovers daily.

"Runback" imperfections are streaks on the trailing edge of the film. It is caused by the bead of developer on the trailing edge of the film "running back" onto the film as the film begins its downward path into the fixer.

Runback can be reduced by squeegeeing the film as it exits the developer rack and the crossover. Squeegeeing eliminates most of the liquid bead. For single-coated films, processing the film with emulsion facing up will also reduce runback.

Fixer-Rack, Fixer-Wash Crossover, and Wash Rack

The fixer rack and wash rack are similar to the developer rack although they are usually shorter. The exit squeegees in the fixer rack are very important because they can improve silver recovery by minimizing fixer carryout.

The fixer-wash crossover is similar to the developer-fixer crossover. However, to prevent contamination they should not be interchanged.

Dryer Transport

Transporting the film while it is passing from wet-to-tacky-to-dry stages can be difficult. The slightest hesitation in transport will cause a localized surface imperfection. A "two-roller" staggered path works well. Many alternative approaches to film drying have been studied over the years, such as infrared, microwave, and radio frequency. None of these has proved to be as effective as hot-air impingement drying.

After drying, the film is transported to the receiving bin.

Receiving Bin

The receiving bin is designed to accept a variety of film sizes including 10 x 10 cm films and roll films.

Temperature-Control System

The temperature of the developer and fixer must be controlled in order to ensure desired sensitometric performance from films and chemicals. Processors should be designed to maintain the film manufacturer's recommended developer temperature within ±0.2°C over a wide range of environmental conditions. Sensitometric variation that can occur over a ±2°C range is shown in Figure 3 (see next page). Based on the ±2°C data, the variation over ±0.2°C should not be perceptible.

The fixer temperature is also important. A maximum range of ±3°C from the developer temperature is adequate. If the temperature is too low, proper fixing and washing will not take place. If the fixer is too warm, evaporation will concentrate the fixer, reducing its effectiveness, and crystal build up may occur.

Precise temperature control of the developer can be accomplished by many means. A method favored by Kodak is the alternate heating and cooling of the solutions. This allows precise control of solution temperature. The fixer temperature can be controlled by its own system or by heat exchange through the developer-fixer tank wall or a developer heat exchanger in the fixer tank. Heating is accomplished using electrical heaters. The size of the heater used

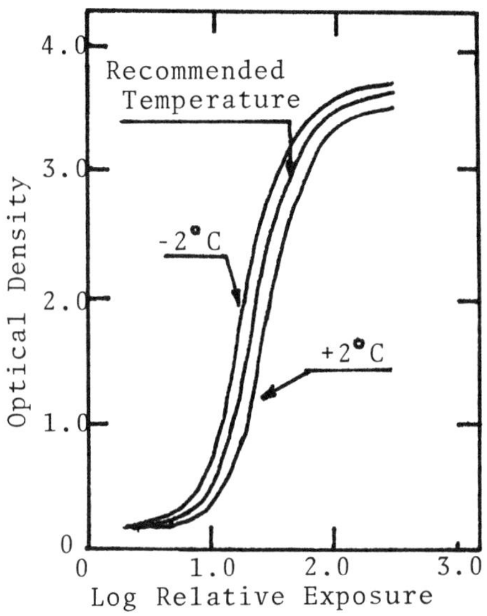

Figure 3. Sensitometric Curves Of An X-ray Film Processed At The Recommended Temperature And ±2°C.

is largely determined by how long it takes to bring the chemicals to processing temperature from ambient temperature.

Cooling is generally accomplished by passing the developer through a heat exchanger using room air or water to cool the developer.

Solution Recirculation

The recirculation of the developer solution is critical to maintain temperature control, solution cleanliness, and process uniformity. Rapid recirculation allows simple temperature measurement and temperature control by heating and cooling. Cleanliness is accomplished by filtering and by keeping the solution homogeneous; thus the displacement of liquid by replenishment gradually washes away unwanted material that would normally settle out. The recirculation also provides film/chemical agitation. This removes processing by-products from the surface of the film and introduces fresh chemicals to the film. Failure to do this uniformly across the film will result in uneven development that will manifest itself in streaks on the film. By directing the solution into the center of the rack, uniform processing can result. If the recirculation is too great, it can cause mistracking of the film that could result in a film-transport failure.

Film Washing

The function of the wash section is to simply replace unwanted chemicals in the emulsion with water that can be evaporated away during drying. The most popular way to do this in x-ray processors has been to immerse the film in a tank of warm water that is changed many times per hour. Energy and water conservation has led to a re-examination of washing techniques. As a result, many processors currently can use less water and at ambient water temperatures. This has resulted in large savings in energy and water. However, problems have been encountered with low water-flow rates. The water supplied in many geographic areas contains micro-organisms that thrive in water with low concentrations of gelatin and fixer.

Low-flow rates do not stir up the wash tank, thus allowing the organisms to grow in the tanks. Such problems can be controlled by using biocides.

Another contributing factor to wash tank design is that many national and municipal codes require a 20 mm free air gap between the water supply and liquids such as those found in processors. The purpose of this is to prevent water contamination that could occur if there is a water-supply pressure loss followed by a pressure reversal.

From a chemical engineering point of view, the most effective way to reduce the concentration of fixer in the film is to pass it through a series of continuously replenished countercurrent wash baths. With each successive bath the concentration of fixer is lower in the bath and in the film. Doing this with multiple tanks in an x-ray processor is obviously cumbersome and costly.

However, the basic concept can be used by applying fresh water to the film as it ascends in a rack that is in an empty tank. The water that is applied runs down the film. In doing so the film goes through a continuous countercurrent wash in which the concentration of fixer in the film decreases as it ascends in the wash rack.

This scheme uses less water but will produce washing equal to that of a conventional submerged-wash processor. In addition, this scheme will provide an air gap and effectively eliminates biological growth in the wash tank.

An alternative method for using the same equipment is to fill the wash tank and use a final wash. With high-quality processing chemicals, either method will provide adequately washed films.

Wash Squeegee

The final stage in the wash section is a squeegee to remove as much water from the surface as possible. Soft, resilient, silicone squeegees greatly reduce the amount of heat and air necessary to dry film particularly with cold wash water. Good squeegees also will provide the film with a better surface after drying because only a uniformly wet film will dry uniformly.

Drying

The final processing stage is drying. Drying is more complex than is normally recognized. The task sounds simple: evaporate the water out of the film. However, to do this quickly a large volume of heated air is uniformly blown onto the film as it passes through the dryer.

The air in the dryer must strike the film uniformly. If the gelatin is dried differentially, streaks will result on the film that can interfere with the image. Uniform drying can be aided by blowing the heated air on both sides of the film through slots in tapered tubes.

The temperature of the air impinging on the film is governed by a thermostat that controls the dryer heaters. The actual temperature of the air blowing on the film should not vary more than ±5°C from a set point. The set point is determined so that the films are just warm to the touch as they exit the dryer. This will provide the best surface quality. Too fast drying will result in drying imperfections.

To conserve power much of the air is reused. The remainder of the heated air is vented from the processor. Careful attention must be given to the installation of the vent. Too much or too little venting will upset the balance of new and reused air, resulting in insufficient drying. Improper venting can also change the airflow over the racks, allowing fixer fumes to deposit on the entrance assembly and the developer-fixer crossover. The resulting chemical contamination will interfere with development.

Tanks

The processing tanks isolate the solutions from the dry portions of the processor. The volume of solution in the tanks is important in maintaining stable chemical activity and controlled temperature. A large tank will provide a "fly wheel" for maintaining uniform process control. The volume of chemicals used must be weighed against the time and electrical power needed to initially bring the chemicals to the proper operating temperature. Also with larger tank volumes, the time necessary to reach a chemical steady state is longer. Popular tank volumes are 2.5 to 5 gallons.

Newer processors use plastic rather than stainless steel for tanks. This offers a number of benefits. Plastics can be molded to a complex form with all three tanks combined in a single piece. The filter housing and mounting fittings are also formed into the tank. This one-piece concept allows the exposed wet section to be totally isolated from the rest of the processor. This makes cleaning easier and reduces the probability of chemicals contacting the dry portion of the processor.

Plastic materials also offer improved insulation characteristics over stainless steel. This thermally isolates the developer and fixer from the internal thermal environment of the processor as well as cold water in the wash tank. These materials also minimize condensation on the wash tank walls.

The strength of the plastic aids in the structural design of the processor can support the racks.

Material Compatibility

Photographic film and processing solutions are very sensitive to contamination. Care has to be taken to assure that materials that are used in the vicinity of the film and solutions are not photographically active. Photographic inactivity has to be verified by long-term testing of the processor construction materials with photographic film and chemicals.

It is also important in material selection to consider that the chemicals can rapidly deteriorate some materials. Any materials that will be subjected to chemicals or chemical fumes must be tested to see if they will survive this application.

Chemical Replenishment

To maintain chemical stability and to remove unwanted or spent materials from the solution by replacement, it is necessary to replenish the processing solutions.

Approximating the film area processed by measuring the length of film processed is satisfactory for many applications. Film measurement can be done by using rollers that are displaced when film is fed. The displaced rollers activate a switch that turns on the pumps. Errors in measurement occur when small films or rolls of narrow film are processed.

A more accurate way of measuring the film area is to use a photoelectric array of detectors that actually measures the area of film processed. The approach used in the Kodak RP X-Omat processor, model M8 is to transport the film at a constant speed past an array of 22 scanning infrared light-emitting diodes. When film is present, the IR signal is not received by the IR detector. The area accumulator determines the time and the number of signals missed. When the equivalent of a 35x43 cm sheet of film has been processed, the replenisher pumps are activated for a predetermined period of time.

Knowing the area of film processed is only part of obtaining proper replenishment. Care must be taken to assure that the volume of replenisher delivered to the processing tanks is correct. The most accurate way to measure this is to collect replenisher that will enter the solution tank. Inserting a "J"-shaped tube in the replenisher inlet in the developer tank will allow collection of the solution and take into account variables in the replenisher system. Some processors have a measuring graduate built into the processor. Other processors have only ball floats showing the approximate volume of replenishment or indicate only if replenishment is taking place.

A major consideration in replenisher pump selection is sensitivity to head pressure on the input side of the pump. Since the processor tanks are always filled to the same level, the head on the output side of the pump is constant. However, as solution levels in the replenisher barrels are lowered, the head pressure also decreases. With some pumps this can cause large differences in replenishment. Earlier processors provided ball floats for estimating replenishment and control valves that required daily attention to compensate for these differences. The replenishment pumps in current processors are of a special bellows type. These pumps are relatively insensitive to input head pressure and will provide very repeatable replenishment regardless of input head pressure. With these pumps, daily monitoring of replenishment is not required. As a convenience to the user, ball floats and built-in vials are included in some processors to aid periodic monitoring of replenishment.

It has become common to replenish the solutions each time a piece of film is processed. The process doesn't require this, but frequent replenishment simplified processor design. Actually, 15 chest films 35x43 cm size, can be processed without significantly lowering chemical activity in 2.5 gallons of developer. This characteristic allows the use of very simple processors that can be manually replenished after a given number of sheets are processed. The disadvantage of this type of replenishment is that it is somewhat inconvenient, requiring constant attention.

Processor Monitoring

To be certain that critical processor functions are performing properly a monitoring system is included in the M8 processor. This gives added assurance that films will be properly processed. The critical functions that are monitored are:

- Developer and fixer solutions (total of eight functions)
  - Temperature limits required to maintain process control
  - Temperature in thermowell
  - Presence of solution in thermowell
  - Solution recirculation
- Water
  - Correct mode
- Dryer
  - Overtemperature

The thermal functions are monitored by various temperature sensitive devices. The presence of solution is determined by the electrical conductivity of the solution. Solution recirculation and water mode are monitored using pressure sensitive switches. If a critical function fails, an audible and a visual alarm are activated. This indicates to the operator that the processor is not operating correctly. The particular LED that lights indicates which system is operating incorrectly.

The monitoring of these functions plus a convenient means of checking the temperature of the solutions, testing the area accumulator, and measuring the replenishment rates decreases the chances of film being processed improperly.

Cabinetry

The cabinetry must be light tight but also serves other functions. As the shroud of the processor, it must protect operators from the potential mechanical and electrical hazards in the processors. Another function is related to thermal control. It provides insulation from the environment and when necessary also serves as a heat shielded from unwanted heat. For interfacing with other equipment or for through-the-wall installation, the cabinet provides a mounting surface. The cabinetry also provides noise control.

Other Important Factors

In addition to designing for image quality, other factors such as energy conservation, safety, and serviceability must be considered. Environmental-chamber testing and computer modeling are utilized in designing processors with low energy and water requirements without sacrificing processor performance. Safety aspects must be considered in order to meet worldwide governmental requirements as well as the requirements of various testing laboratories such as UL and CSA. To assure long-term performance, the processor must be designed to be reliable and simple to maintain.

Manufacturer's Recommendations

The interactions among the processor components and the rest of the imaging system are considered during design and testing of processors. The intention of the manufacturer is to provide a product that performs satisfactorily under specified conditions. In many cases subtle changes in the processor, film, or processing chemicals can adversely influence image quality or other performance characteristics. These components are designed to function as a system when used in a predetermined fashion. Any departure from the manufacturer's products and specific instructions should be very carefully evaluated before they are implemented.

Conclusion

The successful design of an automatic x-ray film processor requires that careful attention be given to each of the processor's systems such as film transport, temperature control, chemical replenishment, and drying. In designing each of these parts, it is essential to consider their influence on the final film image. It is the quality of the film image that determines the value of the processor.

Acknowledgment

The writer would like to thank Ronald Lillie and John Widmer for their advice.

INITIAL START-UP OF AUTOMATIC FILM PROCESSORS

William E.J. McKinney
E. I. DuPont de Nemours & Co., Inc.
Photo Products Department
X-Ray Technical Service
Processing Training Center
Wilmington, Delaware 19898

Abstract

Acceptance testing of radiological imaging equipment should include all components. The processor is responsible for visible image production through development. This paper presents an approach to the installation and start-up of automatic processors from the electromechanical, chemical, and sensitometric viewpoints.

Introduction

Many processor manufacturers provide installation manuals, start-up and operation instructions. Some even provide pre-installation instructions. Unfortunately, this literature is often overlooked, read by the wrong individual or not understood. Often there is a lack of communications between installers and users, with the result being a problem installation and a processor that doesn't work properly.

Understanding the importance and function of the processing operation is quite easy. According to the Gurney-Mott Theory of the Formation of the Latent Image[1], an exposed crystal contains a development site of 5 atoms of silver. This silver halide crystal is reduced by the developer to black metallic silver made up of 10^9 atoms of silver. This is an oxidation/reduction reaction. It is a chemical reaction controlled by elements of time and temperature which are provided by the electromechanical systems of the processor.

Processing converts the latent image into the useful visible image. The processor controls the chemical reaction. Should the processor or processing in general be anything other than optimal, the result is an alteration in radiographic film quality (sensitometric or physical).

Thus, it can be seen that the overall radiographic department is as good as any one component, be that screens, generators, viewboxes, or processors. The following discussion is based on assumption of the importance of processing and that the processor is a technical device. Some discussion is devoted to site preparation and installation since prior planning will reduce post installation problems.

From the moment the decision is made to purchase a processor for an area, this event and all further work (site preparation, installation, acceptance testing, quality control, etc.) should be documented. A log of activity should be begun at this point.

Testing should follow two paths: does the equipment meet its advertised specifications and what is the overall "quality" (usefulness) regardless of specifications? Both parts should be investigated and answered. The second part may show problem areas (or potential problems) that might be corrected immediately.

Installation Site

Once the processor brand and type has been chosen and the method of installation chosen (through the wall, daylight, etc.) location of the unit is considered. Any processor should be located so that there is at least 24 inches of access space on all sides. If this is not possible then the unit should be placed on a skid or rollers to allow easy removal for service. Working with pre-installation instructions, have the wall opening cut and framed out. Make certain utilities are supplied exactly as specified. There is no valid reason to permit fluctuating or low electrical power or water pressure. It is recommended that a water mixing valve (to blend hot and cold water) be installed even if the process is of the cold water type. Install filters to protect the valve. Follow the National Electric Code and the National Plumbing Code. In the latter code a reverse flow preventer (vacuum breaker) is required on the water line. A pressure gauge should be included as an inexpensive quality control instrument.

The floor drain must be able to handle 2 1/2 times the maximum effluent possible. Centralized or located at-site chemical replenishment and recovery lines should be run. The exhaust hose should be installed with few bends and run to an open ended exhaust duct.

The floor material must be moisture resistant and there should be an emergency floor drain on both sides of the darkroom wall.

Check doorway openings for size. Locate an unpacking and staging area and walk the route for delivery.

Installation

All work should be performed by the dealer personnel <u>and</u> all institution personnel responsible for routine operation and maintenance.

Inspect the unopened package for damage. Look for instructions for unpacking or for locating the instructions. As each unit is uncrated it must be inspected. This inspection is two-fold:

- Shipping Damage
- Manufacturing Quality

In the second case there are a variety of steps to be done to determine quality and to help understand how the components function. Some components may be judged of poor quality but are not included in the warranty so that you would want to correct the problem before installation. The most obvious example is sharp edges or sharp burrs on the metal frame or parts.

All components should be inspected when first observed or unpacked. A written record of this activity is important for subsequent referral. The main drive motor and blower motor, the main drive shaft, the pumps, etc., will be factory installed but should be visually inspected.

Install the frame in its prepared site. Level the frame using built-in adjustment screws but do so near their upper limit. This raises the unit off the floor about one inch to provide air space and easier cleaning under the unit. Connect utilities. Install seals and/or finishing trim.

Initial Start-Up for Cleaning and Inspection

This is for systems checking rather than film production.

Fill each processor wet tank one at a time with water looking for leaks. Wipe up all spills. Turn on the water and look for leaks especially at the water solenoid, if included. If there is a pressure regulator included, adjust it to the specifications. Observe that it works.

Disconnect all electrical devices except the circulation pumps and turn on. Inspect for leaks and observe operation. Pumps should start immediately and be very quiet. Allow to run for five minutes, drain and refill a second time.

Operate just the main drive motor. Notice start-up and general operation. Look at meshing (or slippage) of gear teeth (or sprocket teeth and chain) and/or drive shaft roundness. Turn off power. Rinse all racks in water, inspect for tightness and squareness. Look at all rollers and shafts. Install each rack separately; activate the drive for each rack in sequence. Run the drive motor with all racks, except dryer, in place.

Turn on just the dryer system. Work the thermostat, observe the indicator light and blower sounds. Run for five minutes. Turn off and install dryer tubes, rollers or rack. Hook up replenishment tank lines and put water in the tanks.

Using a level recheck the frame. Check the racks and feed tray. In each case check all four sides and diagonally. Adjust if necessary.

Turn on the processor completely. Observe sounds. Measure water flow rate. Have an electrician measure start-up amperage and run amperage later when heaters turn off. Inspect all components visually, feel for heat rise and vibration, listen for abnormal sounds. If satisfied put on all sides and covers. Allow to run for five minutes while area is cleaned up. In the darkroom (if attached) inspect for light leaks.

Shut down processor, disassemble and drain out water. Inspect all components. Rinse racks and tanks.

Initial Start-Up for Sensitometry

Obtain fresh chemicals. If mixed in-house, use a calibrated bucket to calibrate the mixing/replenishment tanks. If the tank has markings verify them. Follow mixing instructions exactly. Avoid excess agitation of the developer. If pre-mixed is used, have it prepared the day prior.

Fill the processor with the correct volume. Make certain replenishment lines are filled. Add the correct amount of starter solution of the same make as the developer to insure the correct sensitometry. Install racks and turn on the processor.

Set the thermostats at a low level. As the indicator light turns off at a temperature (verified with a separate thermometer), this will allow calibration of the thermostat. To verify or calibrate thermometers requires at least five minutes for response time to be achieved. Suggested operation: at 33°C (92°F) for developer, 20°C (68°F) for water and the dryer at 50°C (120°F) or as recommended by the manufacturer for 30 minutes. Check for leaks. Set the replenishment detection system and inspect the operation of the replenishment pump. Set rates and verify. Process 35 sheets of green 14 x 17 film from the film bin. These will clean away debris missed in the rinsing step. Continue feeding with inspection until the films are free of debris. These tests are made with the processor open to observe transport.

Close the processor and feed three unexposed 14 x 17 films. Inspect for fog, clarity, debris, guide shoe scratches, rundown, stain, wetness, etc. If acceptable, proceed with sensitometric testing.

Sensitometric Testing

The processor should be set-up to the film manufacturer's recommended specifications for a given chemistry brand with satisfactory results. The user may, however, desire slightly different results and that is permitted. Changes in excess of 15% equivalent exposure should not be taken lightly. Ideally, a new processor would be matched sensitometrically to the previous unit or to other existing units.

Each manufacturer of film has performed tests commonly called "Gamma (meaning contrast) Series" in which they determine the contrast response of the film to a variety of times, temperatures and activities (different brands of developers). This data produces the film manufacturer's recommendations for optimal processing. It is very important to follow the film manufacturer's recommendations for processing and not those of the processor manufacturer or the chemicals manufacturer. The Du Pont Company using its Bit System[2] has determined that 15 Bits of processing (any reasonable combination of time, temperature, and activity) will produce acceptable quality for any combination of film, chemicals and processor. It allows the matching of quality between different processing sites or conditions. Bit values are available for all films/screen combinations, brands of chemicals, and both time and temperature ranges.

Preferably, an institution should set-up the processor by performing their own Gamma-versus-Temperature series. Process a film with a sensitometric image at 2°F (1°C) intervals from 86°F to 96°F (30°C to 35°C). Allow five minutes at each temperature for stability before processing. Read the densities, plot H&D curves and calculate speed, contrast, maximum density and base plus fog density. Plot the film's response against temperature to determine the level that suits your needs. Verify this setting of optimal quality by processing first phantom and then several routine patient exposures. If these are judged acceptable by the radiologist(s) or quality review committee, then relative quality has been identified and a "master" film has been prepared for reference in Quality Control Work.

Further Testing

Establish the minimum amount of heat for the dryer to dry eight 14 x 17 films fed in a series. This should not exceed 60°C (140°F). Measure the blower and exhaust air velocity (use a velometer) and compare to the specifications. Measure the air velocity of the exhaust where it connects to the in-house ventilator. If possible measure the noise. Record all data for reference in future troubleshooting.

Process all types of film and sizes according to the processor manufacturer's recommendations. Next feed a variety of films slightly askew or rotated 90° to present a different dimension. If normally used, process roll (serial) and cine film. Misalignment of rack components will cause long strips of film to "walk" or migrate from side-to-side. Monitor for this and re-align turnaround assemblies or recheck level if found. Also, a long strip of film may cause a load on the main drive motor resulting in a slower rate of transport. Monitor and record transport rates in inches or feet per minute (IPM, FPM).

In acceptance testing, the processor will probably work fine initially. If possible, the testing period should extend over a one or two week period with 30 days being ideal. Additional data should be obtained as to immersion times in each solution, temperature stability, thermostat cycle times, and the influence of ambient conditions. All measurements and records will be necessary for your quality control program and helpful should some problems arise after acceptance but during the warranty period.

It is important to assign priorities of usefulness. Does the equipment perform as advertised? Does it match its specifications? If it does not, is the error acceptable or not? How good is the quality or performance of unspecified items?

The Last Step

With careful planning for installing, starting up and testing the processor the operator will be able to inspect the majority of components. Also, the result of this work will be the documentation of the processing conditions and the production of the foundation of a Quality Control Program. The Quality Control Program, by monitoring and controlling quality, is in itself a continuation of the original testing program. It will be very useful, along with routine cleaning and maintenance, in keeping the processor operating optimally electromechanically and sensitometrically.

Usually a processor will function properly at start-up. The operator makes adjustments to produce a level of quality. Most problems can be easily overcome. All problems are easier to overcome with adequate knowledge of the forces involved. The quality of the processor is relatively consistent. The quality derived from the processing carried out in the processor set-up by the operator will dictate the ultimate quality and its consistent production. What is done to a processor and how processing is controlled is perhaps more important than the finite quality of the assembled electromechanical components.

References

1. R.W. Gurney and N.F. Mott, Proc. Roy. Soc. 164 (138).

2. Robert J. Trinkle, Bit System of Technique Conversion, 7th Edition, E.I. DuPont de Nemours & Co., Inc., Wilmington, DE 19898

Bibliography

R.E. Wayrynen, The Photographic Process, The Physics of Medical Imaging: Recording System Measurements and Techniques, Arthur, G. Haus, Editor, AAPM, 1979, American Institute of Physics, NY 10017.

W.E.J. McKinney, Radiographic Processing: Processing and Chemistry, The Physics of Medical Imaging, AAPM.

ACCEPTANCE TESTING OF RADIOGRAPHIC X-RAY GENERATORS

Raymond P. Rossi, M.S.
Department of Radiology
University of Colorado Health Sciences Center
Denver, Colorado 80262

Abstract

Methods and instrumentation requirements for the evaluation of the performance characteristics of radiographic x-ray generators during initial acceptance tests are described. Parameters to be evaluated include exposure timer function and kVp and mA calibration. Guidelines for acceptable performance are presented along with a discussion of common problems and pitfalls.

Introduction

The evaluation of the performance characteristics of radiographic x-ray generators during initial acceptance testing should fulfill three distinct purposes: 1) to verify compliance with agreed upon purchase specifications; 2) to determine potential system problems prior to clinical use; and 3) to establish a set of baseline data for future quality assurance activities and system life performance testing. In order to fulfill these purposes it is necessary that testing be carried out using suitable test instrumentation, proper procedures for conducting the tests and careful documentation of the results. For purposes of discussion in this paper, it will be assumed that the testing is to be carried out on a general purpose radiographic/fluoroscopic unit equipped with a three-phase twelve-pulse high voltage generator. Procedures for other types of generators (e.g., single-phase, capacitive discharge, grid bias controlled) are similar for the most part and only important differences will be noted.

Required Test Instrumentation

In order to assess properly the performance of a radiographic x-ray generator, it is essential that the proper test instrumentation be available. A listing of typical test instrumentation required is given in Table I. While no reference is given to a particular type of test instrument, several points should be noted. The high voltage divider ratio should be voltage and frequency independent over the range of voltages and frequencies for which it will be used and its calibration should be well known (1). The National Bureau of Standards now provides calibration services for these devices (2). The calibration of the storage oscilloscope should be well established by use of a precision DC voltage supply for the range of voltages to be measured. If scope probes are to be used, the oscilloscope/probes should be calibrated as a system. If a mAs meter is to be used, its calibration should be established using a precision current source and accurate timer. Some high voltage dividers require the use of high impedance probes (i.e., 10 MΩ) for connection to the oscilloscope as their measuring resistors are of such a value that direct connection to the 1 MΩ input impedance of the typical oscilloscope will result in as much as a 20% change in their calibration due to loading. One should always refer to the vendor's instructor

Table I
Instrumentation Required for the Performance Assessment of X-Ray Generators

Device	Application
High voltage divider	kVp measurement, timer measurement
Dual trace storage oscilliscope	kVp, timer and mA waveform measurement
10 Ω Precision Resistor	mA measurement
mAs meter	mA measurement
Oscilliscope Camera	Waveform documentation

manual to determine proper use (3, 4). Jumper cables should be as short as practical as the impedance introduced into the filament heating circuit by the cathode cable will perturb the mA calibration due to a change in the voltage applied to the filament.

Instrumentation Hookup

Two general methods may be employed for connecting the high voltage divider to the x-ray generator and are illustrated in Figure 1. In the first of these, the high voltage divider is connected directly at the x-ray tube such that the jumper cables are in series with the high voltage cables from the generator. Alternately, the high voltage divider can be connected at the high voltage transformer in a similar fashion. This approach must be utilized on x-ray generators which employ rotary high voltage switching for multiple tube selection. This configuration has several disadvantages. First, it is frequently difficult to obtain ready access to under table fluoroscopic tubes to facilitate easy connection of the high voltage divider. Second, the presence of the jumper cable on the cathode side of the tube will result in some perturbation of the mA. Third, when dealing with grid bias tubes (e.g., cine) a special adapter box (5) is required to interface the normal three conductor high voltage cables to the four conductor cables required for grid bias operation.

The second installation configuration overcomes these difficulties and may be used on high voltage transformers which employ parallel high voltage switching and which have an unused set of high voltage receptacles available. Using this approach on a typical tri-port transformer, the high voltage divider is installed in the auxillary parts of the high voltage transformer. The auxillary ports are then placed in parallel with the x-ray tube under evaluation and the high voltage measured in this fashion. The advantage of this technique is in ease of installation and the elimination of an adaptor box for grid bias systems. Regardless of the approach taken, great care whould be taken to ensure that connections are made properly, that suitable insulating oil or grease is used, and that grounding rings are present if required.

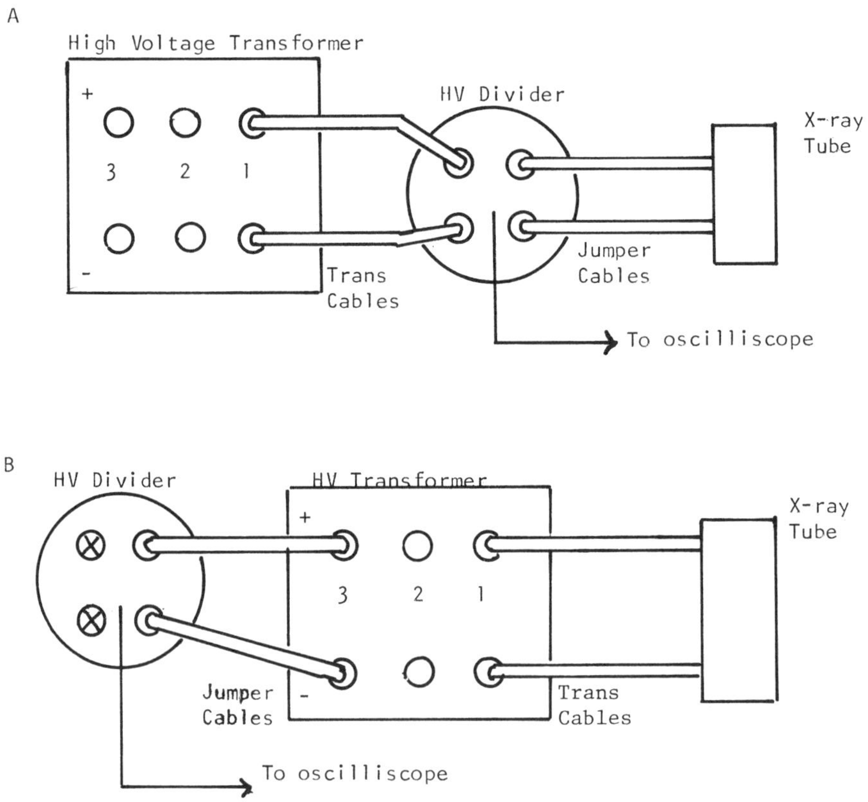

Figure 1: Arrangement of high voltage divider for kVp measurement. A) Divider inserted between high voltage transformer at tube or transformer (required for rotary switched generators); B) Divider inserted in auxillary ports of high voltage transformer and placed in parallel with tube under test.

Measurement of tube current can be carried out by insertion of a mAs meter or 10Ω precision resistor in the mid secondary of the high voltage transformer in series with the mA metering circuit. Access to the proper connections is normally available within the operator's control. When using a mAs meter, the mA is inferred from the measured mAs and the time of exposure. While good results may be achieved with this technique, the inferred mA may be influenced by timer inaccuracy, mA spiking, and capacitive tails on the high voltage waveforms. Our preferred technique is to utilize a precision 10Ω resistor and measure the resultant voltage drop across this resistor to evaluate mA where the resultant voltage will be 1V/100 mA. As mA waveforms are typically noisy and as it is the average mA which is of interest, the technique illustrated in Figure 2 is used (6). The voltage signal developed across the resistor is applied directly to channel 1 of the

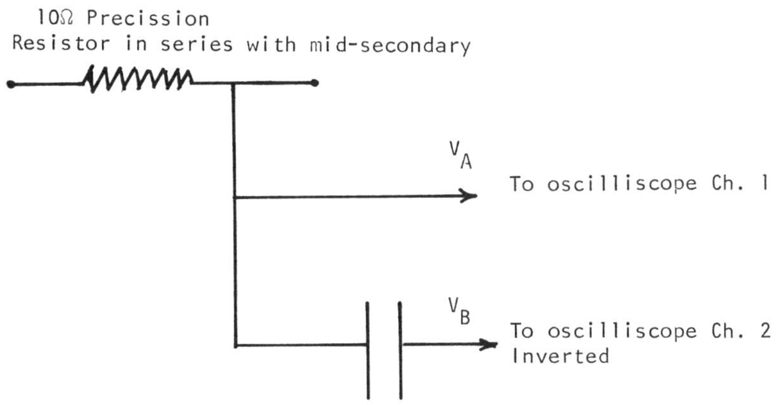

Figure 2: Use of precision resistor for measurement of tube current.

oscilloscope. It is also applied to channel 2 of the oscilloscope after being capacitively filtered. Channel 2 is inverted and algebraically added to channel 1 and the resultant waveform interpreted to determine mA. Illustrations of typical waveforms obtained with and without the capacitive filter are shown in Figure 3. As can be seen, the capacitively filtered waveform is far easier to interpret. Many generators have such a mA calibration resistor built in which may be utilized during testing if desired. Some commercially available high voltage dividers will provide direct readout of mAs or mA as well as other parameters when connected, using the installation configuration of Figure 1A and may be used to advantage. However, caution must be noted not to rely solely on digital displays and there is no substitute to viewing the actual waveforms during testing. Caution is also given to carefully read the manufacturer's instruction manual for any test device to be used.

When making either high voltage or mA measurements, the oscilloscope should always be floated with respect to ground. Failure to observe this caution may result in serious equipment or operator damage.

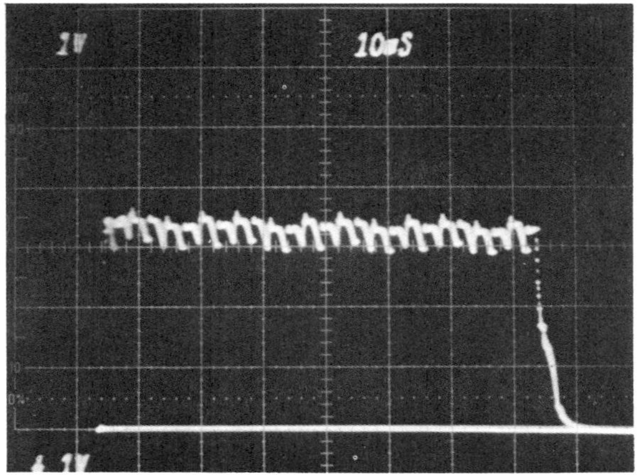

(A)

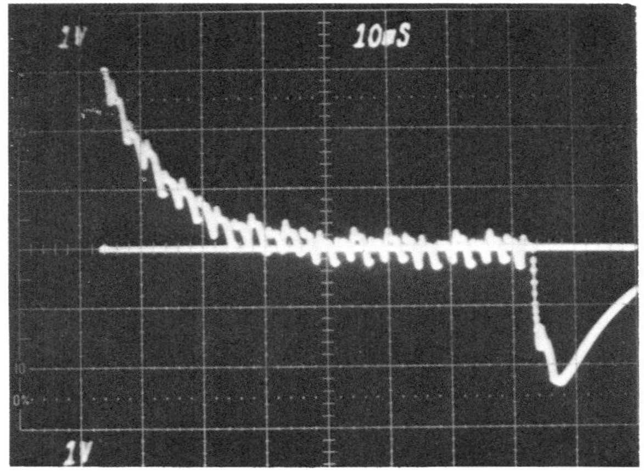

(B)

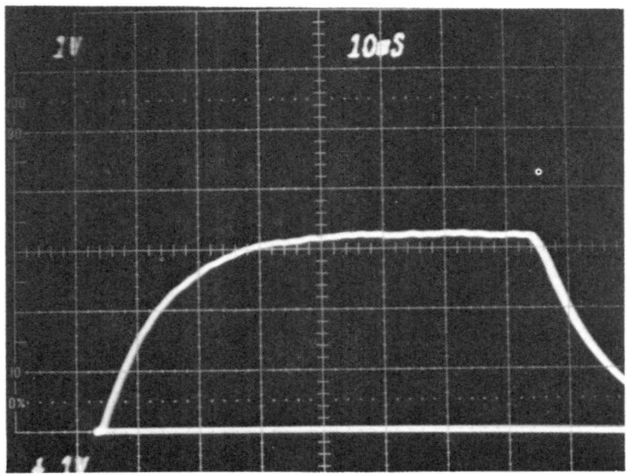

(C)

Figure 3: Illustrations of tube current waveforms. A) Voltage signal developed across resistor, V_A; B) Voltage signal developed across resistor and capacitively filtered, V_B; C) Voltage signal resulting from algebraic addition of V_A and inverted V_B.

Testing Protocol

While the order in which the generator evaluation is carried out may be selected by personal preference, we have found the following sequence to be most efficient.

Line Regulation

The regulation of the incoming power line is normally checked by the vendor during installation and need not be measured unless a problem is suspected. Poor regulation of the incoming power line will adversely affect the performance of the generator and most vendors will not complete an installation unless it is adequate. If desired, it may be checked by measurement of the load and no-load line voltages under the conditions specified by the vendor.

Tube Protector

The function of the tube protection circuitry should be verified by selecting various kVp, mA, focal spot size and time combination to verify that the tube is neither under nor overrated with respect to the appropriate rating charts.

Exposure Timer

The exposure timer should be evaluated to determine accuracy with respect to indicated times and reproducibility. For three-phase equipment, exposure time is normally defined as a given percentage of the width of the high voltage plus, typically 75-80%. To carry out these measurements, the nominal kVp is adjusted to obtain a high voltage waveform that is four divisions high on the oscilloscope and at a mA which is sufficiently high (e.g., 300) to negate effects of cable capacitance. This point is illustrated in Figure 4 where the two high voltage waveforms were obtained at the same exposure time but at different mA's. The importance of using a specified percentage of high voltage pulse peak is demonstrated in Figure 5. For longer exposure time, the high voltage pulse is approximately square wave and thus its width approximately independent of the percent of peak at which it is measured. For very short exposure times, however, the pulse shape deviates greatly from a square wave and thus the exposure time is highly dependent on the percent of peak at which the width is measured. Actual determination of measured exposure times is made from oscillograph recordings of the displayed waveform.

For single-phase equipment, the exposure time is simply the number of measured pulses times 1/120th of a second.

For cinefluorographic systems, exposure time is taken as the time between removal of grid bias potential to reapplication of grid bias potential as illustrated in Figure 6. Such measurements are normally carried out at 80 kVp, low frame rate and the maximum allowed mA for the pulse width, mA, frame rate combination at 80 kVp.

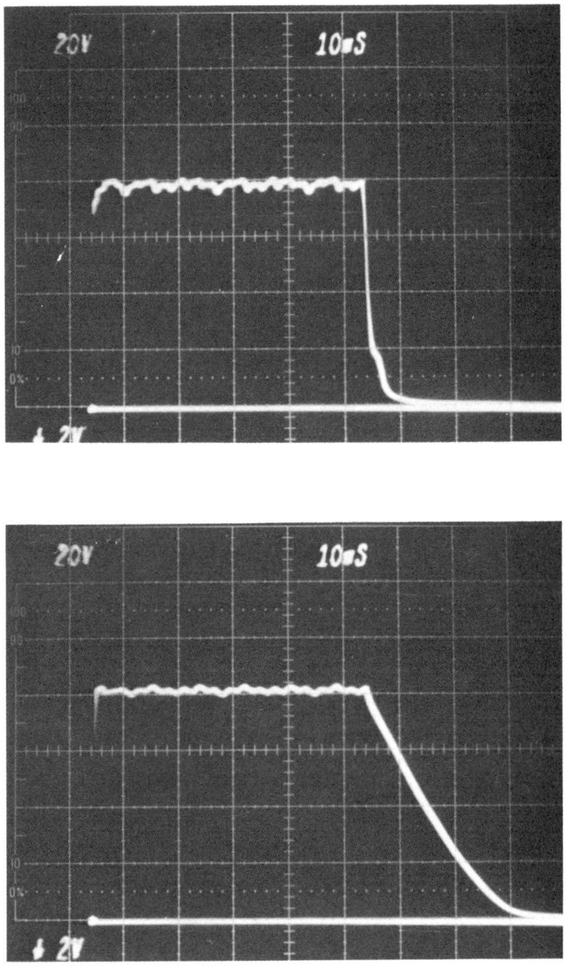

(A)

(B)

Figure 4: Effect of cable capacitance on measured exposure time. A) 300 mA; B) 10 mA. Nominal exposure time of 50 msec.

(A)

(B)

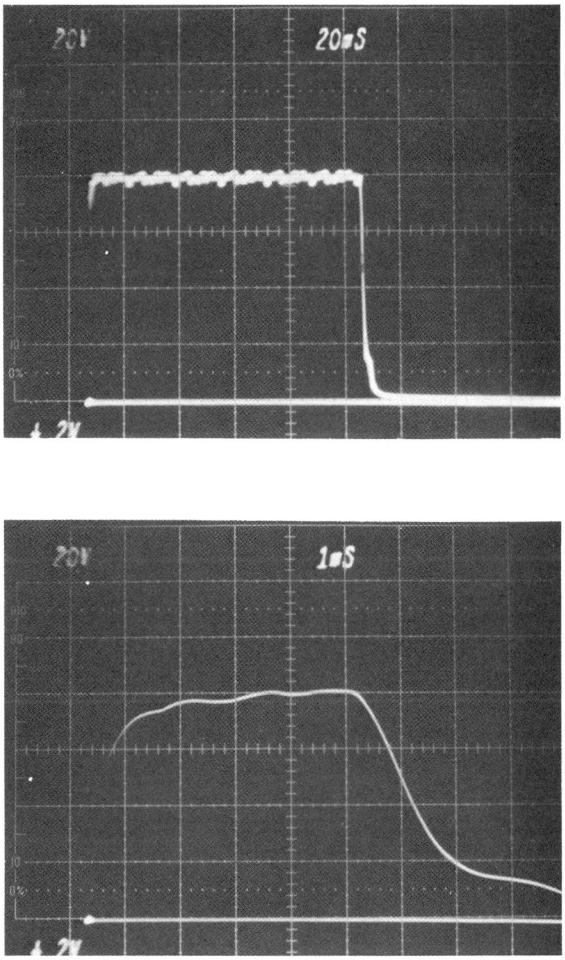

Figure 5: Influence of specified percent of peak on exposure time. A) 100 msec exposure; B) 5 msec exposure, both at 80 kVp, 300 mA.

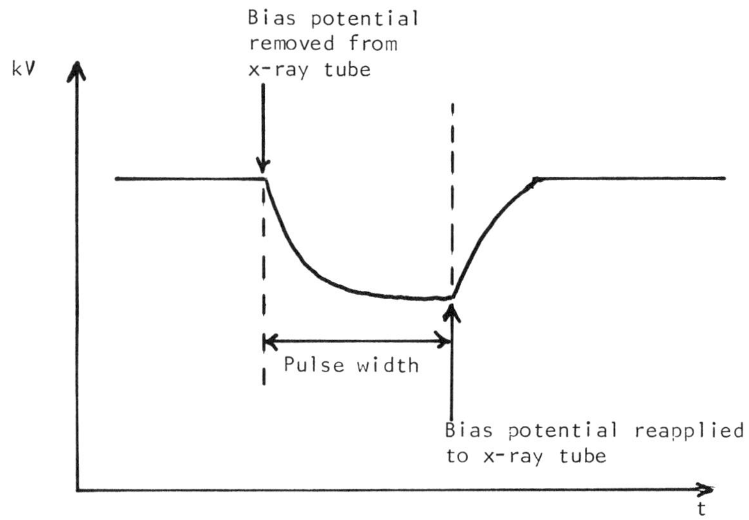

Figure 6: Diagram illustrating meaning of exposure time is grid bias (cine) systems.

<u>kVp Calibration</u>

The kVp calibration of the x-ray generator should be evaluated for accuracy with respect to indicated settings, reproducibility, and independence of mA for all connected x-ray tubes and operating modes.

As a preliminary measurement, the x-ray generator should be checked for symmetry of high voltage applied to the x-ray tube from anode to ground and cathode to ground at low, medium, and high mA at 80 kVp. Resultant waveforms of such a test are illustrated in Figure 7.

Next, the kVp calibration of the generator is evaluated over the typical working kVp range (e.g., 50-150 kVp) at all mA stations consistent with the tube ratings. The typical exposure time employed is 100 msec and measurements are typically made in ten to twenty kVp increments. The oscilloscope is used in the algebraic addition mode and oscillograph recordings made for interpretation. Typical results are illustrated in Figure 8. In analyzing the waveforms, one should pay particular attention to kVp overshoots at the beginning of the waveform, unusual appearance in the high voltage ripple, and uniformity of the kVp over the time exposure.

There is little difference in interpretation of kVp waveforms between three-phase and single-phase systems but cinefluorographic systems present a different problem. One method of assessing kVp for cinefluorographic systems is illustrated in Figure 9 and yields acceptable results.

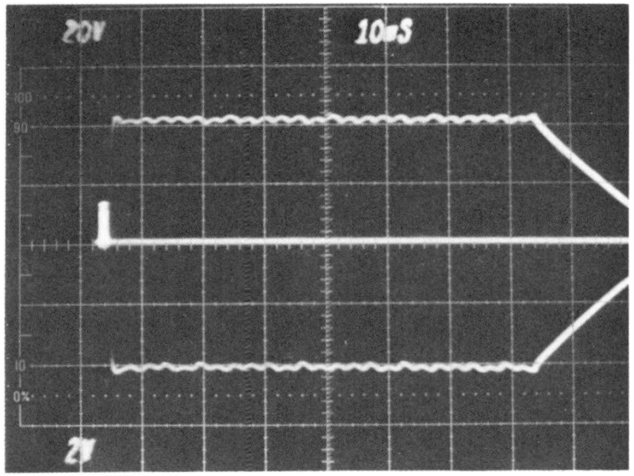

(A)

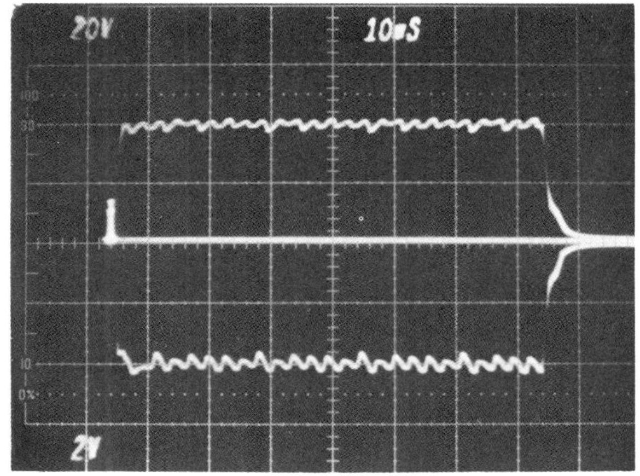

(B)

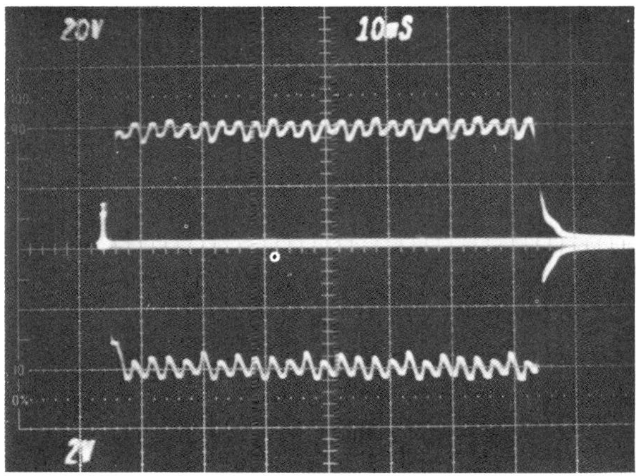

(C

Figure 7: Waveforms for determining symmetry for high voltage applied to x-ray tube. A) Low; B) Medium; and C) High mA. All measurements at 80 kVp.

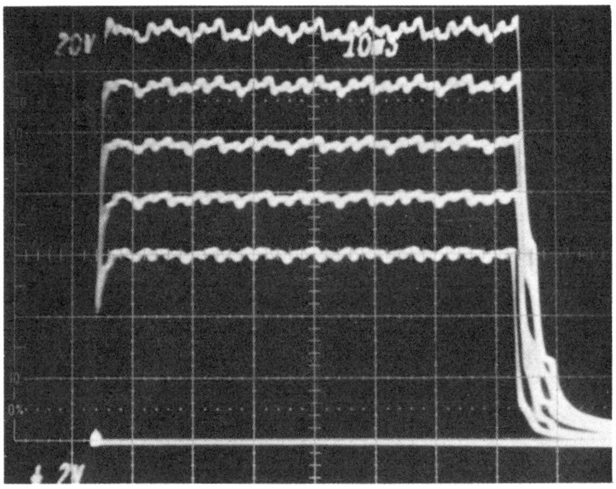

Figure 8: Typical waveforms used to determine kVp calibration, 60, 80, 100, 120, 140 kVp at 300 mA, 70 msec exposure.

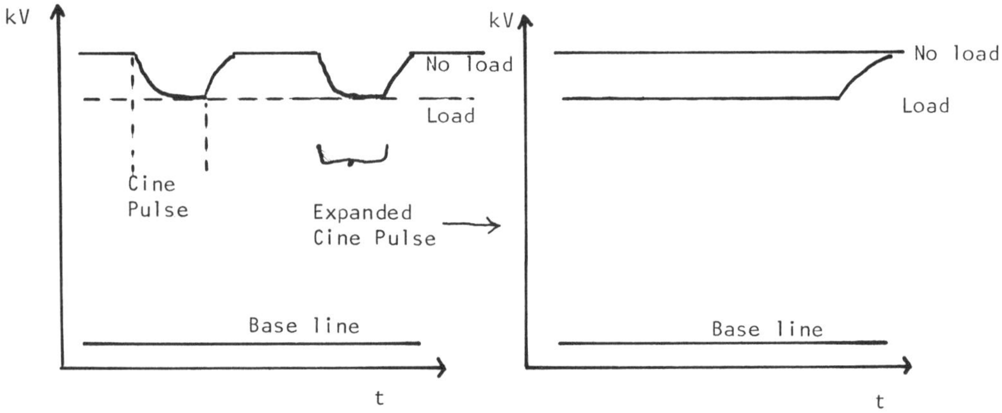

Figure 9: Method for determining cine kVp. The kVp is given by: kVp =

$$kVp = kV_{Load} + .33\ (kVp_{No\ Load} - kVp_{Load})$$

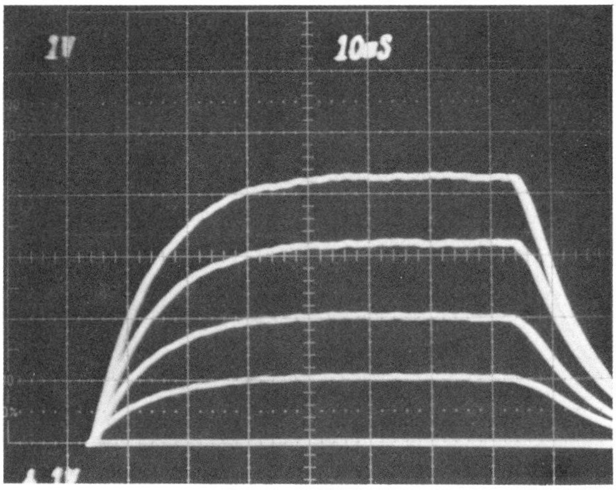

Figure 10: Examples of waveforms from which mA calibration is determined. 80 kVp, 100, 200, 300 and 400 mA.

mA Calibration

The mA calibration of the x-ray generator should be evaluated for accuracy, reproducibility, and independence of kVp for all supplied x-ray tubes and operational modes. Evaluation is normally carried out at each mA station and focal spot size and over the working kVp range of the generator (e.g., 50-150 kVp) in ten to twenty kVp increments consistent with tube ratings. For three-phase equipment, oscillographs of the voltage pulse developed across the 10Ω precision resistor are taken for subsequent interpretation as illustrated in Figure 10. Typical exposure times used are 70 to 100 msec. For equipment operated in the fluoroscopic mode, direct measurement of tube current is made utilizing a digital multimeter operated in the DC and mA mode in the mid-secondary of the high voltage transformer. For single-phase equipment, oscillograph recordings of the voltage developed across the 10Ω precision resistor and subsequently interpreted with the mA being given as

$$\text{mA} = \text{Recorded Peak Voltage} \times 100 \text{ mA/V} \times .637$$

Measurement of mA during cinefluorographic exposures may be accomplished using the precision resistor technique.

In analyzing the resultant mA waveforms, one should look for unusual spiking at the beginning of exposure and inconsistency of the mA during exposure which may be due to an improperly adjusted mA stabilizer, and decrease of mA with exposure time for long exposures which may be the result of electron cooling in the x-ray tube (7).

Data Recording and Documentation

The results of all measurements taken during the initial acceptance testing of a radiographic generator should be carefully documented. If desired, the actual oscillographs may be retained for future reference but we have found this is only necessary if some unusual behavior is noted. All data taken is summarized in a fashion similar to that shown in Table II which proves useful for future comparative purposes. It is also good procedure to document the test instrumentation employed along with a brief description of the procedure by which measurements were made.

Table II
Summarized kVp/mA Acceptance Testing Results

Nominal mA / Nominal kVp	25(S)	50(S)	100(S)	200(S)	300(L)	400(L)	500(L)	600(L)
60	60 25	58 50	60 100	57 210	60 320	61 410	61 500	62 600
80	80 24	81 49	80 105	78 210	80 315	81 405	82 500	80 600
100	101 24	101 49	100 100	98 207	99 305	100 410	101 500	99 610
120	122 25	121 49	121 100	118 210	118 315	120 415	119 510	118 620
140	143 24	141 49	140 100	138 210	138 315	140 420	139 515	138 630

Criteria of Acceptable Performance

For a given radiological imaging system, the criteria of acceptable performance should have been established as part of the original purchase contract. If this is not the case, the specifications of the manufacturer should be utilized. Reasonable criteria for performance of radiographic generators is given in Table III. Should the results of testing indicate noncompliance with established performance criteria, the vendor should be required to correct the deficiencies found prior to acceptance of the equipment.

Table III
Criteria for Acceptable Performance
of X-Ray Generators

Parameter	Performance Level
Exposure time	The measured exposure time shall be accurate to within ±5% of the indicated exposure time for times of 10 milliseconds or greater and shall be accurate to within manufacturers specifications for times of less than 10 milliseconds. The exposure time shall be reproducible to within ±5%.
kVp	The measured kVp shall be accurate to within ±5% of the indicated kVp and reproducible to within ±5%. Precontacting overshoots shall be less than 5 kVp. The rms value of the high voltage shall not change by more than 5% during an exposure.
mA/mAs	The measured mA/mAs shall be accurate to within ±5% of the indicated mA/mAs. The mA/mAs shall be reproducible to within ±5%. For constant load systems the DC level of mA shall remain constant to within ±5% during an exposure.
Tube protection	The x-ray tube protective circuitry shall prevent exposure techniques in excess of the allowed exposure techniques for the given x-ray tube but shall not under-rate the x-ray tube by greater than 30% with respect to the maximum load for single radiographic exposures.

Problems and Pitfalls

The procedures outlined above are reasonably straightforward and if carefully followed little difficulty should be experienced in carrying out the initial acceptance tests on a radiographic x-ray generator. However, several points require consideration for the procedure to be conducted in an efficient manner which will yield reliable results.

Having a strong working knowledge of the x-ray generator to be tested will be found most useful. This can be achieved by careful study of the vendors' manuals and closely working with service personnel during installation. An excellent understanding of x-ray generators may also be obtained from the series of papers by Kanomori (8, 9, 10).

Careful calibration of test equipment cannot be overstressed. Verify or have verified the calibration of all test equipment to be used. We have found that poorly calibrated oscilloscopes are the most frequent cause of discrepancies between vendor calibrations and our measurements.

Use extreme care in making and interpreting tube current measurements. Whenever a high voltage divider is inserted in a system such that the cathode jumper cable is in series with the transformer cathode cable, the tube current will decrease. The percent decrease will depend on the length of cables but for a typical installation with 50 feet high voltage cables, a five foot jumper may decrease the mA by 5 to 10% and a 15 foot jumper may result in a 20-25% decrease. With this decrease in mA, the kVp will increase. Typically this problem is overcome by the vendor's service representative by calibrating kVp and mA with the cathode jumper in and then removing the cathode jumper and readjusting mA to the proper values which will then yield the same kVp as with the cathode jumper in.

Try to maintain the x-ray tube at a reasonably constant head load as might be encountered under clinical use. The measured tube current will differ between a cold and hot tube. Also, it is possible that tube current will vary depending on the length of time the control has been on and whether cover panels are on or off due to thermal drift of electronic components.

When dealing with multiple tube installations in which the x-ray tubes have much different filament characteristics, it may not be possible to achieve the desired kVp and mA accuracy and tracking without the use of a special dissimilar tube adaptor kit available from the vendor. The reason for this is that the generator's filament and space change transformers are designed to cover a given range of tube filaments and require modification to cover other than normal tubes.

Summary

Instrumentation and procedures for conducting the initial acceptance tests on radiographic x-ray generators have been desired. These procedures when carefully followed will yield reliable data to determine compliance with purchase specifications and establish a baseline for future equipment performance.

References

1. Hebner, R.E., Electrical Measurement of High Voltage Pulses in Diagnostic X-Ray Units, NBSIR75-775, National Bureau of Standards, 1975.

2. Kieffer, L.J., Calibration and Related Measurement Services of the National Bureau of Standards, NBS Special Publication 250, 1980 Edition, National Bureau of Standards, 1981, pp. 45-50.

3. Direction 13288E, Bleeder, High Voltage Dual Type, T8005G and C1515A, Connections ... Applications, Medical Systems, General Electric Company 4-2977.

4. Operating and Maintenance Instructions for Dynalyzer II High Voltage Unit, (ST-2914-1), Macbleff Laboratories, Inc., 1976.

5. Lin, P.P., Kriz, R., A Four Conductor to Three Conductor Junction Box, Applied Radiology, Nov/Dec 1978, pp. 61-62.

6. Rauch, P.L., Block, R.W., Acceptance Testing Experience with Fourteen New Installations, Proceedings of the SPIE, Volume 96, pp. 12-18, 1976.

7. Brezovich, I.A., Barnes, G.T., An investigation and explanation of the fall-off of x-ray tube current during an exposure, Phys. Med. Biol. 25, 2, pp. 241-249, 1980.

8. Kanamori, H., Analysis of High-Voltage Waveforms in X-Ray Units: I Single Phase, Journal of the Franklin Institute, 279, 3, pp. 147-159, March 1965.

9. Kanamori, H., Analysis of High-Voltage Waveforms in X-Ray Units: II Abnormal Voltages, Journal of the Franklin Institute, 279, 4, pp. 246-253, April 1965.

10. Kanamori, H., Analysis of High Voltage Waveforms in X-Ray Units: III Three Phase, Journal of the Franklin Institute, 279, 5, pp. 347-359, May 1965.

PERFORMANCE CHARACTERISTICS OF DIAGNOSTIC X-RAY GENERATORS

Phillip L. Rauch, M.S.
Department of Diagnostic Radiology
Henry Ford Hospital
Detroit, Michigan 48202

Introduction

Diagnostic x-ray generators are designed to provide electrical energy to be converted into x radiation. The quality and quantity of x radiation is predetermined by the radiographer who sets the appropriate exposure factors at the control desk. There are many varieties of x-ray generators, and these can be classified according to their source of primary power (Table I), and the appearance of the secondary voltage waveform (Table II).

Table I. Sources of Primary Power for the X-ray Generator

Sources
Battery
Conventional A.C. line
Dedicated 1ϕ line
Dedicated 3ϕ line

Table II. X-ray Generator Secondary Voltage Waveforms

Waveforms
1ϕ (half-wave rectification)
1ϕ (full-wave rectification)
1ϕ Capacitor Smoothed
Condensor Discharge
3ϕ (6-pulse)
3ϕ (12-pulse)
500-1000 Hertz (inverter driven)
Constant Potential

The voltage applied to the x-ray tube can be observed during the exposure by inserting a calibrated voltage divider into the high tension secondary circuit, and displaying the waveform on a dual-trace storage oscilloscope.[1] The single-phase, half-wave rectified waveform (Fig. 1(a)) appears as individual voltage pulses varying from 0 volts to a peak potential, and returning to 0 volts. The voltage ripple is thus 100%. These pulses appear every 1/60th second (16.67 ms), which is one cycle of the mains supply voltage.

The single-phase, full-wave rectified waveform (Fig. 1(b)) is similar, with 100% ripple, except that 2 pulses appear per 1/60th second. These two voltage waveforms are inefficient for x-ray production, since both the quality and quantity of x radiation is greatly reduced as the voltage decreases below the peak value. In addition, the half-wave rectified waveform produces useful radiation for less than half of the exposure interval.

A single-phase power supply can be made to produce x radiation more efficiently by placing a capacitor in parallel with the x-ray tube. Once charged, the capacitor tends to maintain the voltage during the interval when the power supply drops to 0 volts (Fig. 2(a)). The voltage ripple is thus reduced , and

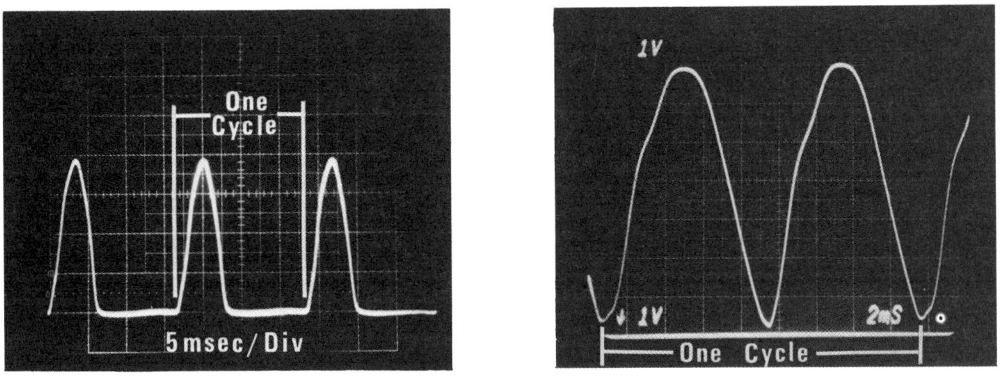

(a) (b)

Figure 1. Typical 1ϕ waveforms: (a) half wave rectified waveform displayed with horizontal sweep of 5 ms/div., (b) full wave rectified waveform displayed at 2 ms/div. Note that waveform (a) has only one pulse and (b) two pulses per cycle of line voltage (16.67 ms.).

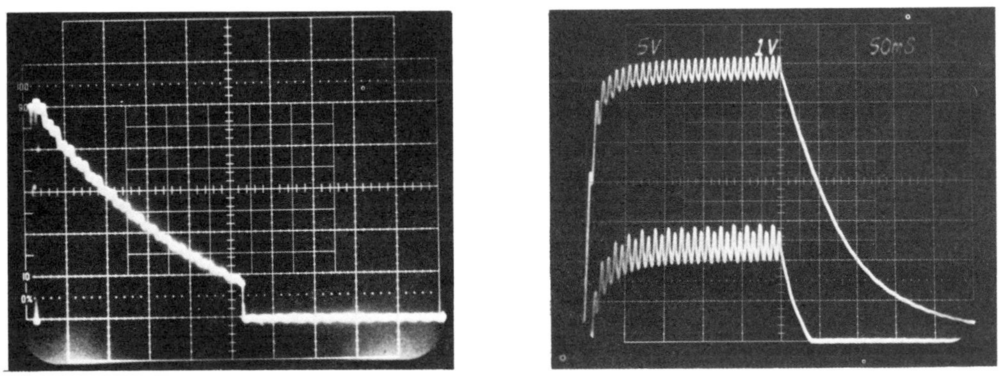

(a) (b)

Figure 2. A capacitor can be used to reduce 1ϕ ripple (a) or as the source of the potential applied to the x-ray tube (b).

Fig. 2(a) are waveforms from a 1ϕ-capacitor smoothed generator (Picker Mammorex^{TM}). Upper Trace: kV waveform at a vertical sensitivity of 5 kV/div., horizontal sensitivity of 50 msec/div. Lower Trace: Radiation waveform. Note that the condensor slowly discharges following removal of the primary voltage but that the radiation output is quickly blocked by a mechanical shutter.

Fig. 2(b) is a typical capacitor discharge waveform (Picker model DGC 1210). For a 1 μf capacitor, the voltage will drop 1 kV per mAs.

the quality and quantity of x radiation is greatly increased. Note, however, that the removal of primary voltage, at termination of the exposure, results in a slow discharge of the capacitor which may contribute unnecessary patient exposure. This unnecessary exposure can be reduced by using a lead shutter to block radiation produced after the exposure is terminated. The lower trace in Figure 2(a) is the radiation waveform, showing that radiation output is blocked by the lead shutter long before the capacitor is fully discharged.

A conventional A.C. line of relatively low power can be utilized to charge a capacitor to rather high potential (e.g. 100 kV). The voltage on the capacitor can then be applied to an x-ray tube to produce x radiation. For very short exposures of relatively low mAs, this method produces excellent results. However, for longer exposures both the quality and quantity of x radiation is severely reduced, as the capacitor is discharged (Fig. 2(b)). Typically, the voltage on the condensor drops at a rate of 1 kV for each mAs discharged through the x-ray tube. For this reason patient exposure will increase rapidly as the selected mAs is increased, whereas the exposure to the image receptor will increase very little.[2]

A dedicated three-phase line can be utilized to produce a relatively low ripple voltage, whose peak value remains constant throughout the exposure. Shown in Figure 3(a) is a 6-pulse waveform, with 6 voltage peaks per one cycle (16.67 ms). Voltage ripple can be further reduced by utilizing a transformer designed to produce 12 pulses per cycle (Fig. 3(b)). Even higher frequency pulses (1,000 Hertz - 2,000 Hertz) are available on some mobile radiographic equipment, which utilizes a battery as the primary power source, and a high frequency invertor to convert battery voltage to alternating potential.

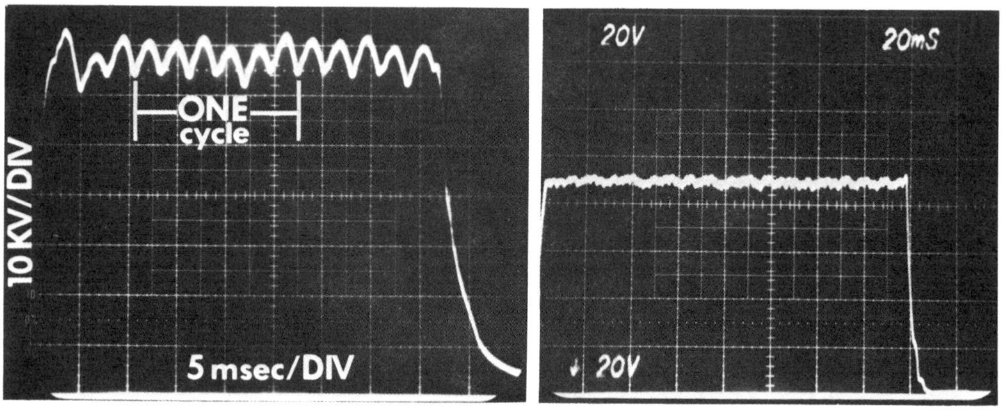

(a) (b)

Figure 3. kV waveforms from a dedicated 3ϕ power line, (a) 3ϕ 6-pulse, (b) 3ϕ 12-pulse.

Figure 3(a) is a 70 kVp waveform displayed at 10 kV/div. (Vert.) and 5 msec/div. (Horiz.).

Figure 3(b) is a 90 kVp waveform displayed at 20 kV/div. (Vert.) and 20 msec/div. (Horiz.).

The ideal voltage waveform is that having 0 ripple or "constant potential" (Fig. 4). This can be effectively achieved by means of triode or tetrode control valves placed in the high tension secondary. More will be said about this later. There is virtually no increase in radiation quality nor quantity with this waveform as compared with that produced by a 3ϕ 12-pulse generator; however, the advantage of this configuration lies in the ability of the triode to rapidly switch the high tension as necessary for high speed cineradiographic applications.

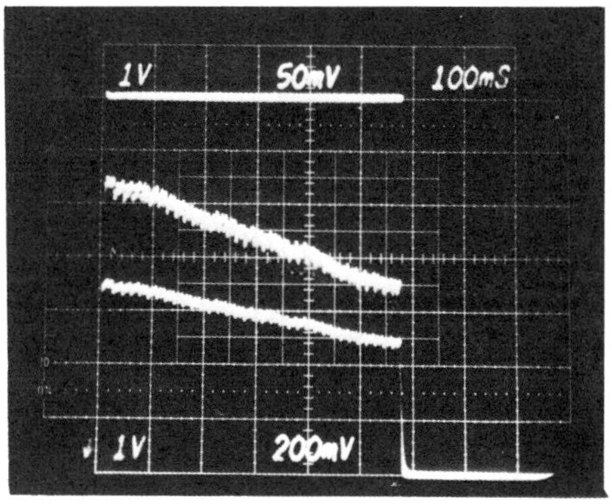

Figure 4. Constant potential kV waveform from a Siemens Pandoros Optimatic generator. Upper trace is the kV at 10 kV/div. Middle trace is the radiation waveform. Lower trace is the mA (continuously falling load) at 50 mA/div.

Besides the primary power source, and the secondary voltage waveform, generators are also classified according to the tube current waveform produced, and the type of exposure contacting utilized. The tube current may be applied as a constant load, a falling load (which may actually be a stepped constant load, or a continuously falling load), or a pulsed load. Exposure contacting may be mechanical or solid state.

Each of these types of generators, while significantly different in design, have two major factors in common: (1) Each supplies power to an x-ray tube via a transformation of the primary voltage and current, and (2) Each requires periodic calibration adjustment. The radiology equipment specialist will often be responsible for maintaining generator calibration, or at least overseeing vendor supplied calibration. The radiological physicist may be involved in the performance specification and acceptance testing of the x-ray generator as well as periodic calibration checks. It is therefore imperative that they develop an understanding of the requirements imposed on the generator by the x-ray tube, and be able to recognize characteristic deficiencies in the generator calibration. They must also realize that no generator is perfect, and some deviation from the nominal technique factors is to be expected.

The purpose of this presentation is to demonstrate a method of evaluating x-ray generator performance by means of waveform analysis. The method presented attempts to identify the underlying cause of the deviation of nominal tube potential and current from the expected behavior. Having identified the most likely source of the fault, the radiology equipment specialist can then determine which of the available service adjustments is appropriate to correct the deficiency. In this way adjustments are made specifically to optimize observed performance rather than merely to conform to a written protocol. I do not advocate that we totally dispense with the written calibration procedure; however, once it is completed, waveform analysis can be used to identify what further adjustments will be required in order to optimize generator performance. In order to understand this approach, it is necessary that one understand the requirements and limitations inherent in the x-ray tube and generator. The first limitation we will consider is associated with the problems encountered when large amounts of power are rapidly switched to the x-ray tube.

Exposure Switching

With modern x-ray generators tremendous amounts of power are switched on and off during an exposure. Typically, 30 kW to 100 kW must be rapidly applied to the x-ray tube within a very narrow time frame. Instantaneous switching of large voltages can produce large current surges in the primary circuit, and voltage overshoots in the high tension secondary.[3] These adverse effects can be reduced by loading one or more phases with a surge resistor at the initiation of the exposure, then shunting the resistor after a predetermined interval (Fig. 5).

This method of exposure contacting is essential for mechanical type contactors, which, due to mechanical inertia, cannot be made to close precisely at a zero crossing of the primary voltage. The result is a step start with the kV maintained at a lower potential until the surge resistor is switched out (Fig. 6). With some generators, the value of the surge resistor is constant, and therefore the voltage absorbed by the resistor is dependent upon the tube load. As a result, the voltage associated with the initial step will depend upon the tube load. In Figure 6(b) are shown three phototimed kV waveforms obtained with tube loads of 50 mA, 100 mA, and 380 mA, respectively. As expected, the higher the tube load, the greater the amount of voltage absorbed by the surge resistor, and the lower the voltage in the initial step on the kV waveform.

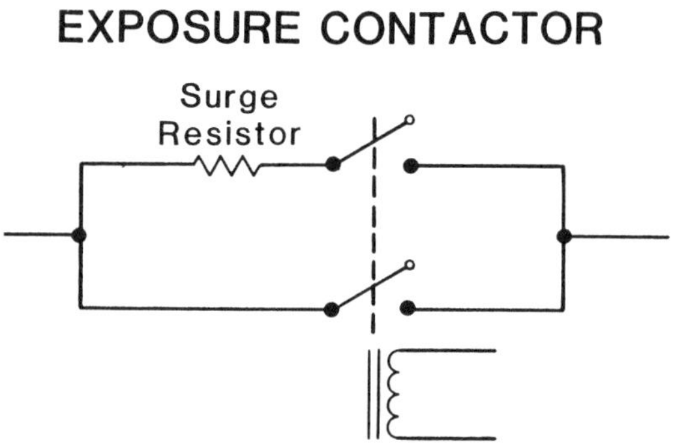

Figure 5. Surge suppression circuit (3). When the exposure is initiated the upper contactor makes first and the full power is not applied because of the surge resistor. The lower contactor then makes, shunting the resistor. This circuit reduces current surges and voltage overshoots at the beginning of the exposure.

Even with solid state exposure contactors, some surge suppression is required to avoid large overshoots in the kV at the start of the exposure (Fig. 7). This is especially important if the x-ray generator is to be used for high kV applications such as chest radiography. At Henry Ford Hospital, all chest radiographs are obtained at 150 kVp. The waveform in Figure 7, taken at acceptance testing of a unit which would be utilized for chest radiography, shows that the peak voltage is several kV higher than the maximum rated tube potential of 150 kVp. If allowed to continue, this condition could promote instability and arcing in the x-ray tube, leading to premature failure.

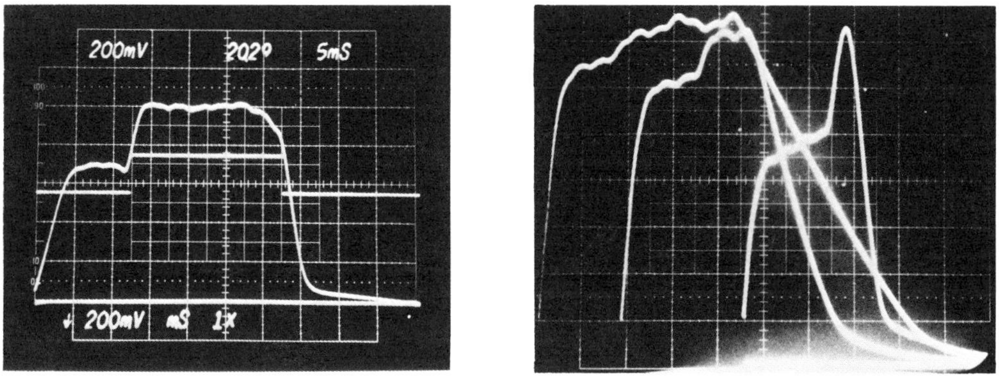

(a) (b)

Figure 6. "Step start" of kV waveform due to surge suppresion circuitry.

Figure 6(a): kV waveform for 100 kVp, 250 mA. The broken horizontal line is a gated signal which depicts the exposure time as defined at 75% of the kVp.

Figure 6(b): kV waveforms at 150 kVp with different mA settings. Left, 50 mA; Middle, 100 mA; Right, 380 mA. Vertical sensitivity 20 kV/div. Note that the kV during the initial step is highly dependent on the x-ray tube current.

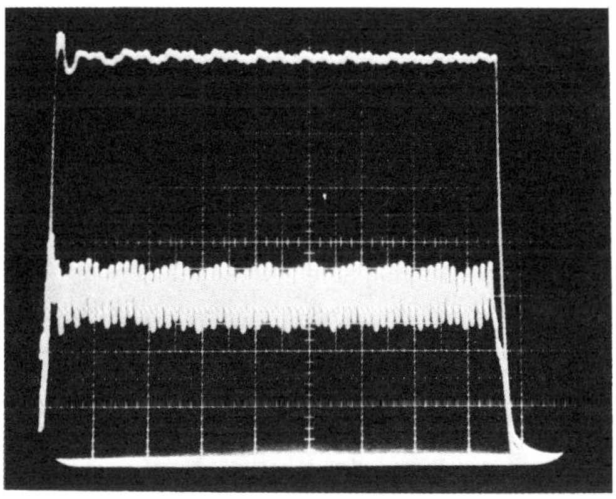

Figure 7. kV overshoot at the start of the exposure. Siemens Tridoros 712MP generator set at 150 kVp, 50 mAs. Upper waveform is the kV at 20 kV/div. The lower one is the tube current at 100 mA/div.

X-ray Circuitry

Figure 8 is a simplified diagram of an x-ray circuit which illustrates the fundamental requirements for the production of diagnostically useful x-rays. The primary purpose of the x-ray generator is to provide the radiographer with the means of consistently providing reproducible x-ray exposures of predetermined quality and quantity. Secondarily, the x-ray generator must provide constraints on the predetermined exposure factors, in order to prevent damage to the components of the x-ray tube. The x-ray tube is a thermonic diode with positive potential on the anode, and negative potential on the cathode. Voltage is supplied to the x-ray tube via transformation of primary potential. X-rays are produced when electrons, made available by thermonic action in the filament, are accelerated across the applied potential, acquiring kinetic energy which is converted to x-radiation in the tungsten target. Heating of the filament is accomplished by current flow which is solely dependent on the applied filament voltage. Quality and quantity of radiation are determined by the potential applied across the x-ray tube, and the filament current. The peak tube potential can be altered by changing the primary voltage applied to the high tension transformer. The filament current can be altered by changing the voltage applied to the primary of the filament transformer. Shown schematically are the components of a single phase generator. For three-phase generators, three autotransformers, three high tension primary windings, three high tension secondary windings, and 12 rectifiers are required.

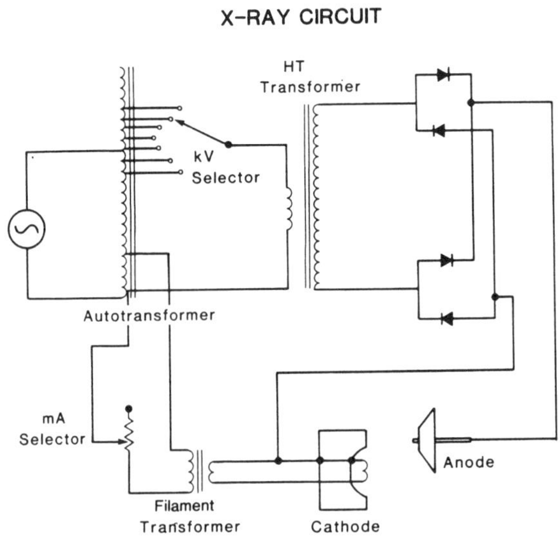

Figure 8.
Simplified schematic diagram of a single phase x-ray circuit. The primary power supply is directed, through the autotransformer, to the filament transformer and and to the high tension (HT) transformer. For three-phase generators, three autotransformers, three high tension primary windings, three high tension secondary windings, and 12 rectifiers are required.

For any generator, calibration of kV and mA refers to the adjustment of internal compensating networks which function to assure that filament temperature and primary voltage to the high tension transformer will result in an x-ray tube potential and current closely corresponding to the values preselected at the control desk. Calibration is important to assure that radiation quality and quantity can be accurately set prior to making the exposure. An important part of generator calibration involves the adjustment of tube overload protection circuitry. This circuitry is designed to block radiographic exposures for any combination of kVp, mA, and exposure time which exceeds the instantaneous heat loading of the x-ray target. Additionally, the generator must limit the maximum filament current, in accordance with manufacturer's specifications, since high currents result in very high filament temperatures and rapid deterioration of the filament.

The Primary X-ray Circuit

The basic components of the primary x-ray circuit are illustrated in Figure 9. Voltage to be applied to the primary of the high tension transformer is derived from the Mains supply via the autotransformer. This allows operator control of the kVp applied to the x-ray tube, by varying the taps on the autotransformer. Alternatively, the taps could be replaced by a moving contact forming a variac autotransformer. kV calibration consists of those adjustments which assure that the primary voltage will result in the selected kVp. Note however that if a tap is selected, and the Mains supply voltage changes, the primary voltage will also change, and the selected kVp will not be obtained. Therefore, it is necessary that means be available to monitor the line voltage and compensate for gradual changes in supply voltage throughout the day. Line-voltage compensation may be manual or automatic.

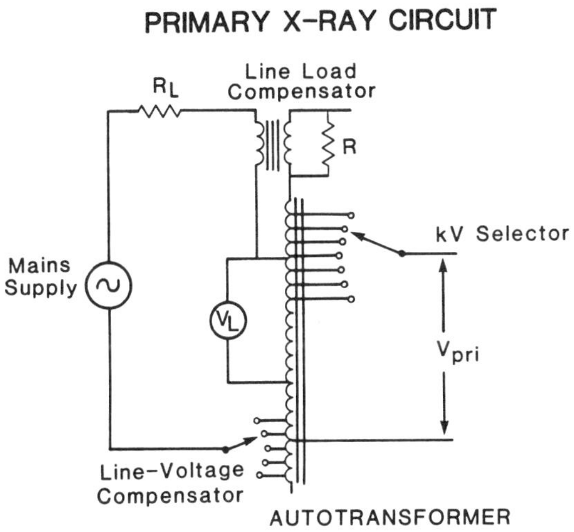

Figure 9.
The primary x-ray circuit must contain means to maintain the voltage applied to the autotransformer whenever the Mains Supply changes (Line Voltage Compensator), or to compensate for line losses during the actual exposure (Line Load Compensator).

Loss of voltage to the autotransformer will also occur during the exposure as current flows through the inherent line resistance. Typically, line resistance ranges from 0.2 ohms to 0.5 ohms.[3] A primary current of 100 amperes will result in a 20 to 50 volt loss in the power line. Not only would this loss affect the kilovoltage obtained, but would also change the voltage applied to the filament transformer. The effect on the filament current and, therefore, the mA, would be disastrous. Since the loss of voltage is porportional to the current in the power line, a transformer with primary windings in series with the power line can be used to provide a compensating voltage. The load on the output of the compensating transformer has a value which depends upon the line resistance. Because line resistance is not standardized, the manufacturer generally presets the line load compensator assuming a maximum line resistance of typically 0.5 ohms. During installation, therefore, the actual line resistance must be accurately measured and additional resistance added to bring the total equal to that used during factory calibration. For 3-phase generators it is necessary that this adjustment be accurately performed for each phase. Failure to balance the line resistance in three-phase generators, will result in distortion of the kV waveform. The greatest affect, however, will be in the filament supply which must be highly regulated.

The Filament Circuit

The current through the filament of an x-ray tube is determined by the applied voltage. A small change in filament current will result in a large change in the x-ray tube current (Fig. 10). For example, at 90 kVp a change in filament current from 4.58 A to 4.80 A (+ 4.8%) will result in a change in tube current from 300 mA to 500 mA (+ 67%). It is thus imperative that the filament current and, therefore, the filament voltage be highly regulated.

Even if the filament voltage was maintained at a constant value, the tube current would not remain constant if the tube potential is changed. As shown in Figure 11, changes in tube voltage for a fixed filament current will result in corresponding changes in the x-ray tube current. This is because as the x-ray tube potential is increased, additional electrons are drawn from the "space charge" surrounding the hot filament. Because of this phenomenon, compensation must be provided so that x-ray tube current will closely correspond to the selected value, regardless of the kVp applied to the x-ray tube. The x-ray filament circuit (Fig. 12) must therefore provide means of preselecting the desired tube current, and compensating for external factors that act to alter the current.

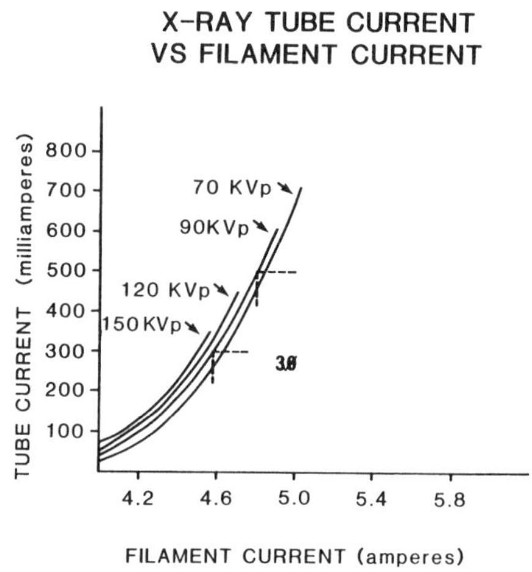

Figure 10. X-ray tube current as a function of the filament current for various tube voltages. Note that a relatively small change in filament current will produce a large change in tube current.

Maintaining the filament at a high temperature for prolonged periods of time can result in vaporization of tungsten, and rapid deterioration of the filament. Therefore, the filament current required for the exposure is applied only during "preparation", the interval during which the anode is made to rotate at its minimum required rotational velocity. During preparation, the filament current is determined by the voltage applied to the filament. This voltage is not applied directly from the autotransformer, but rather a voltage is obtained from the autotransformer and applied to a constant voltage transformer. The constant voltage transformer is designed to maintain the output voltage to very close tolerances, regardless of any fluctuations in the primary voltage. Part of the output from the constant voltage transformer is absorbed by a resistor (Fig. 12). The value of resistance selected thus determines the voltage applied to the filament. A variable resistor is connected in series to provide an adjustment for filament voltage. Adjustment is necessary to compensate for changes in resistance as the filament deteriorates with use. This is called a "pre-heat" adjustment because it determines the filament current and therefore the filament temperature prior to making the exposure.

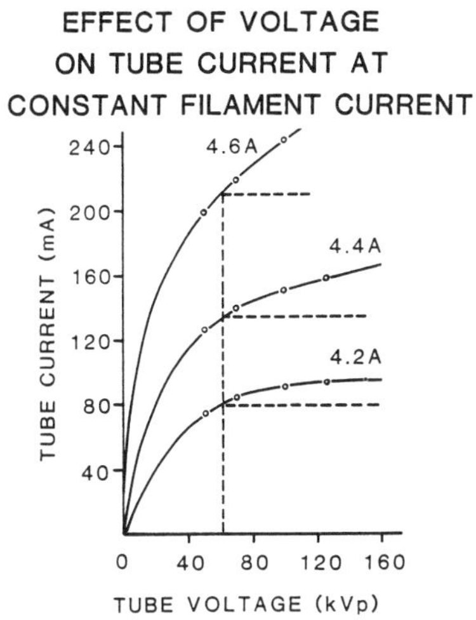

Figure 11. X-ray tube current at a function of tube voltage for various fixed filament current values.

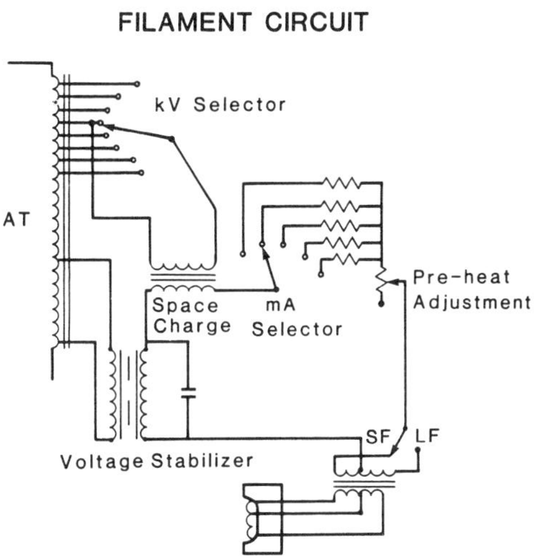

Figure 12. Schematic diagram of the x-ray filament circuit.

The filament voltage must be compensated to correct for changes in tube current which would result as the kVp across the x-ray tube is changed (i.e., space charge compensation). This can be accomplished by adding or subtracting filament voltage as the kV is changed. In Figure 12, note that for kV values below the value corresponding to zero compensation, a positive voltage is added to the filament circuit. This results in a larger filament current to compensate for the decrease in tube current which would occur as the kV is reduced. As the kV is increased, voltage is subtracted from the filament circuit. It should be noted that when making a space-charge compensation adjustment during calibration, the goal is to obtain the same mA value throughout the range of useful kVp's. Also, because space charge effects are greater at high filament temperatures (Fig. 11), space charge compensation should be adjusted at a relatively high mA, typically 500 mA. It is INCORRECT to adjust space charge compensation by merely setting a high kV and making the space charge adjustment to provide a mA value equal to that set at the generator. Proper space charge adjustment is made by measuring the mA at a low kV (60 kVp), and also at high kV (140 kVp), and adjusting the space charge compensation to provide the SAME mA, regardless of whether that mA corresponds to the value set. If, after the space-charge adjustment, the mA is not correct, the pre-heat and mA stabilizer should be adjusted.

The voltage applied to the x-ray tube is not dependent solely on the value of the voltage applied to the primary of the high tension transformer. Voltage losses occur across the internal resistance of the primary winding,and the amount of the loss varies as the primary current in changed. It is therefore necessary that the primary voltage, which is applied to the high tension transformer prior to the exposure, be higher than that which would be required if there were no internal losses. For example, referring to Fig. 13, at 200 mA a primary voltage of 200 volts would result in 80 kVp at the x-ray tube. If the mA were changed to 1,000 mA, with no change in the value of primary voltage, the resultant secondary voltage would only be 50 kVp. The primary voltage would have to be increased to 274 volts in order to maintain the x-ray tube potential at 80 kVp. The x-ray generator must therefore provide kV compensation for changes in tube load. It should be noted that the load lines in Figure 13 are not parallel, and larger changes in primary voltage, for a given change in tube current, are required at high kVp as compared to a lower kVp. A necessary part of x-ray generator calibration therefore is adjustment of the kV compensation for tube loading.

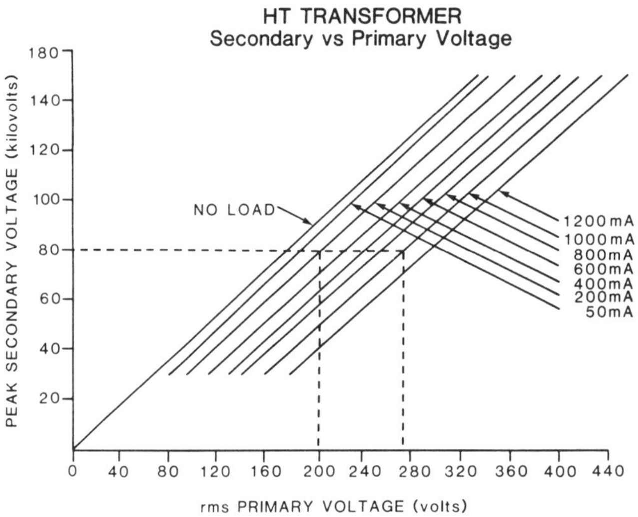

Figure 13. X-ray tube potential as a function of the primary voltage applied to the high tension transformer and the tube load (mA).

Waveform Analysis

Shown in Figure 14 are two kV waveforms obtained at a setting of 80 kVp. The left waveform was obtained at 100 mA and the right waveform at 700 mA. As shown, kV compensation for load was insufficient to maintain the peak kilovoltage as the tube current increased. This is the case corresponding to correct adjustment of the filament current but with inadequate compensation of the primary voltage. Conversely, the primary voltage can be correctly adjusted, according to the intended value of tube current, but, if the filament current was incorrectly set, the wrong tube potential would be obtained. That is, the wrong filament current would produce the wrong mA, and, from Figure 13, the value of tube voltage obtained would not correspond to the value selected. This condition is dramatically illustrated whenever the filament pre-heat is incorrectly adjusted. For the waveforms shown in Figure 15(a), the filament current, and therefore filament temperature, was too low at the beginning of the exposure, but then increased to the correct value by action of the mA stabilizer. The primary voltage on the other hand was correct, assuming the tube current corresponded to that selected. Because the tube current was low at the beginning of the exposure, the resultant x-ray tube potential was too high. Merely adjusting filament pre-heat therefore will correct both the low mA value and the high kV value at the beginning of the exposure (Fig. 15(b)).

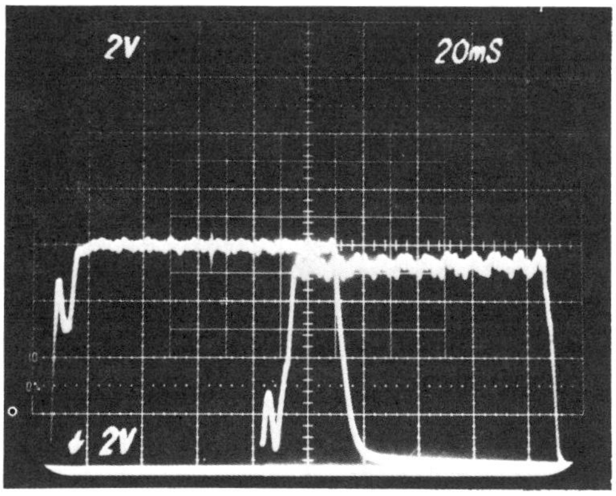

Figure 14. Waveforms obtained at 80 kVp. Left: Tube load of 100 mA, Right: Tube load of 700 mA. kV compensation for load was inadequate to maintain the tube potential as the tube current was increased.

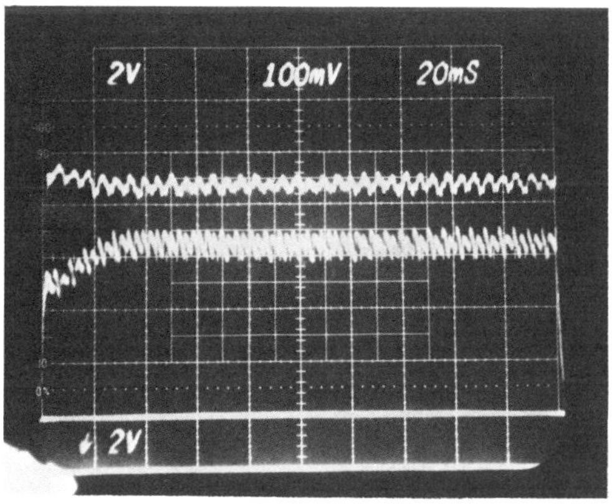

(a)

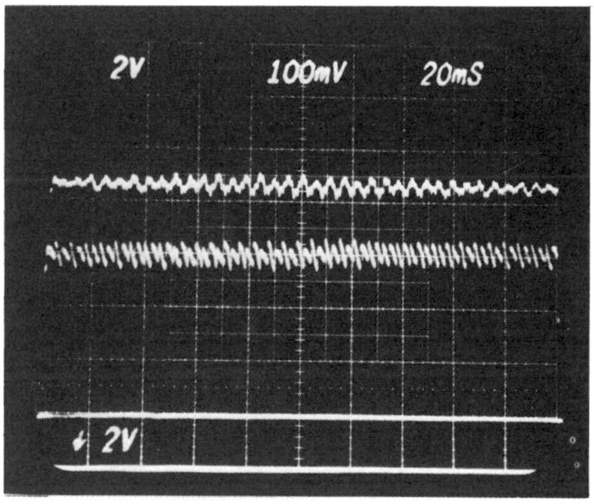

(b)

Figure 15. kV and mA waveforms obtained at 90 kVp, 400 mA.

15(a) Incorrent preheat adjustment resulting in the mA being too low and the kV too high at the beginning of the exposure.

15(b) Adjustment of filament preheat corrects both the mA and kV since the primary voltage applied to the high tension transformer was already correct.

The kV waveform is displayed at 20 kV/div. and its baseline is the second horiz. graticule.

The mA waveform is displayed at 100mA/div. and its baseline is the first horiz. graticule.

Another problem, very much resembling the results of incorrect pre-heat adjustment, is associated with the thermal inertia inherent in the filament. That is, filament temperature cannot instantenously change from one value to another, but changes only gradually. Thus, it is possible that both the filament current and the primary to the high tension transformer are correct, but the waveforms will indicate otherwise if the exposure is intitiated before the time required to achieve proper filament temperature. This condition is illustrated by holding the preparation phase of the exposure for various intervals (Fig. 16). The waveforms on the left were obtained by pushing the exposure switch all the way to the expose position so that the exposure is released as soon as the anode has reached its minimum rotational velocity. The middle waveforms were obtained by holding preparation until hearing the audible sound of the exposure release relay, and then completing the exposure. The waveforms on the right were obtained by holding preparation for 2 seconds before completing the exposure. Holding the preparation phase for longer intervals resulted in no further changes in the waveforms. Thus, the filament pre-heat and primary voltage were correctly adjusted, but the preparation phase of the exposure intitiation was not long enough to allow the filament temperature to change from the ambient value to that required for the exposure. This condition can be corrected by either increasing the ambient temperature (i.e. the standby filament current), or increasing the time delay before exposure release.

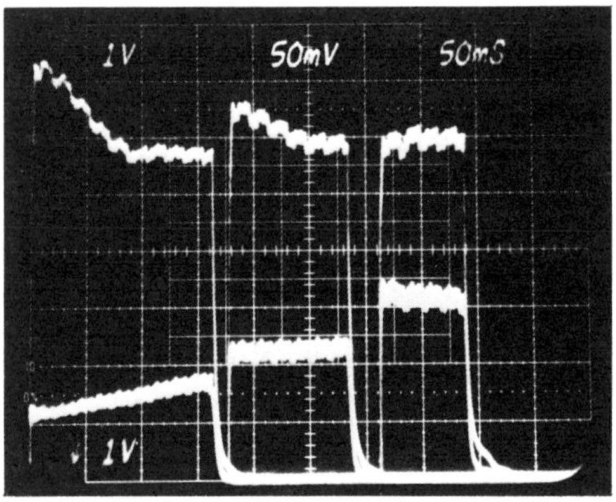

Figure 16. kV (upper) and mA (lower) waveforms from a Siemens Gigantos Optimatic generator set for 60 kVp, falling load.

Left: Exposure switch fully depressed so there was no delay in the exposure release.

Middle: Prep. held until just after the exposure release relay was energized.

Right: Prep. held for 2 sec. before completing the exposure.

These demonstrate that the filament current and primary voltage were correct, but that the filament was not reaching the required temperature before the exposure was released.

In general, the kV applied to the x-ray tube should consist of equal amplitude pulses whose peak value corresponds closely with that set at the control desk (Fig. 17). Large deviations from the nominal kV may adversely affect the quality of the resultant radiograph[4], and should be corrected, if possible. The origin of such deviations varies widely, but a few characteristic causes have been identified. For example, Figure 18(a) and (b) show periodic drops in the tube voltage which occur every 1/60 second, and which gradually decreases in magnitude. Such a pattern is usually associated with losses of magnetic flux in the high tension transformer. This can occur, for example, if the core becomes magnetically saturated. Premagnetization of the core, and phase sequencing are often employed to reduce this effect. Figure 18(c) and (d) on the other hand show kV waveforms which each have only one dip in the kV, occurring near the beginning of the exposure. For the waveforms in Figure 18(c), the loss of potential was found to be related to the synchronization of the exposure with the line voltage, and was corrected by adjustments in the exposure timing circuits. The voltage drop in Figure 18(d) on the other hand was due to contactor bounce. Contactor bounce can occur not only in the "make" contacts, but also in contactors associated with exposure inhibit circuitry. For example, this particular problem occurred in a contactor which was designed to inhibit the exposure until after the stator current had dropped to a level indicating that the anode was rotating at its nominal velocity.

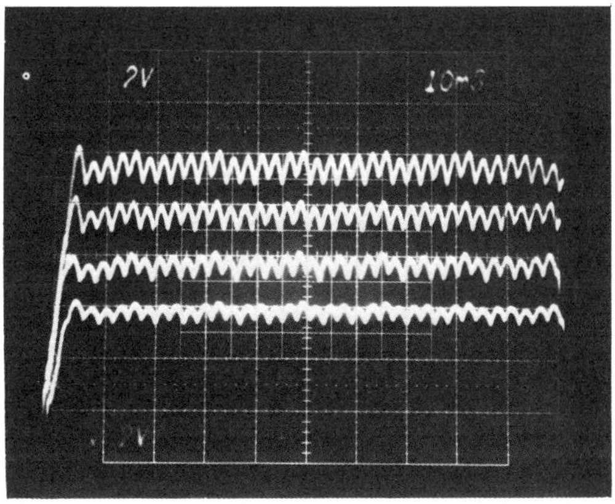

Figure 17. kV waveforms from a Philips SM-80 generator, displayed at 20 kV/div. (vertical) and 10 msec/div. (horizontal). Exposures were made at console settings of 60, 80, 100, 120 kVp. Note that the waveforms demonstrate little overshoot, and nearly equal amplitude pulses whose maximum value remains constant throughout the exposure.

Large deviations from the nominal kV will also occur with the loss of one phase in a three phase generator. This results in a nearly single phase appearance of the kV waveform (Fig. 19). In this instance, one phase was lost when one of the main line contactors broke off. Less severe variations of this form of voltage loss will occur if the voltage of one phase is considerably less than that of the other two phases. This situation can occur if the line resistance is not properly adjusted, if the line voltage is low on one phase, or if the tap on the autotransformer for one phase is improperly set.

The voltage applied to the x-ray tube should be evenly distributed between the cathode and anode, such as that illustrated in Figure 20(a). In this figure are the anode voltage (upper trace) and cathode voltage (lower trace) for a tube potential corresponding to 81 kVp. It should be noted that these waveforms are displayed at 20 kV per division, and that each reaches a peak of approximately

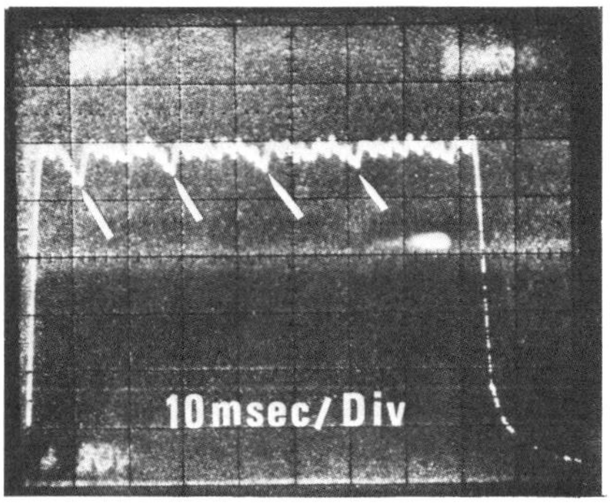

(a)

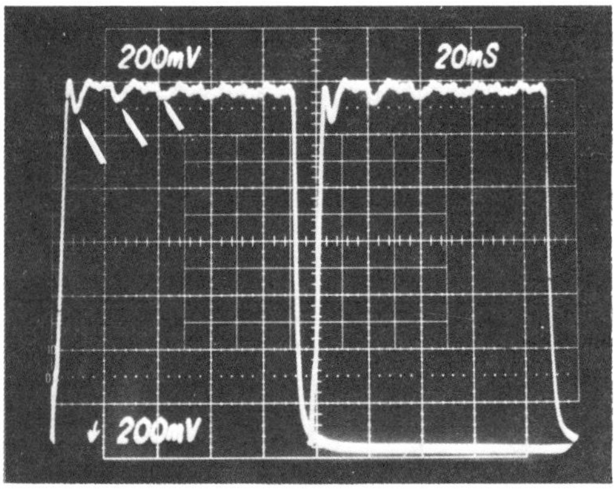

(b)

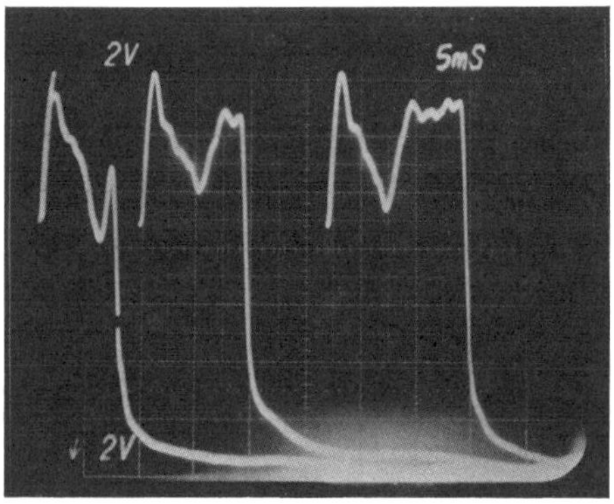

(c)

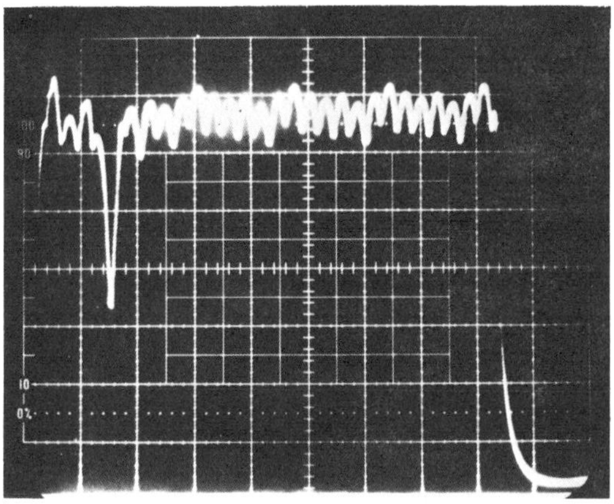

(d)

Figure 18. Waveforms which demonstrate some abnormal deviation from the nominal tube voltage.

(a) and (b) show the result of core saturation which causes a loss of magnetic flux in the transformer core. Note that the drop in potential occurs every 1/60 second (arrows) and gradually diminishes as the exposure continues.

The waveforms in (c) show a very large drop in the tube voltage due to poor synchronization of the exposure with the line voltage.

The loss of voltage in (d) is due to contactor bounce.

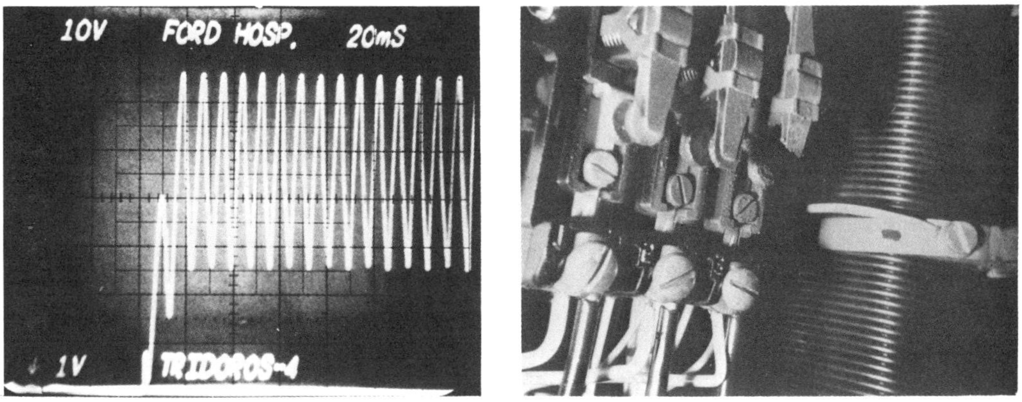

(a) (b)

Figure 19. A Siemens Tridoros 4 three-phase generator produced a waveform (a) which appears to be that from a single-phase generator. The problem was traced to a loss of one phase when a main line contactor broke off (right-hand contactor in (b)).

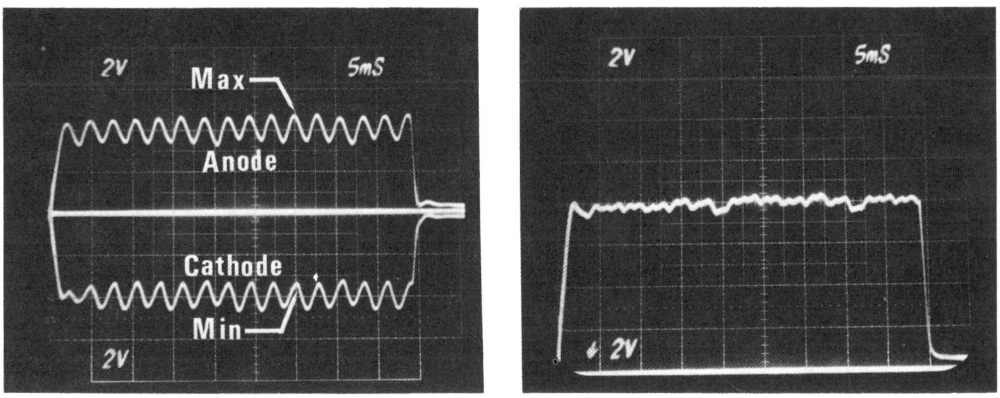

(a) (b)

Figure 20. Anode and cathode voltages must be added vectorially since they are 30 degrees out of phase.

Note that in 20(a) a maximum occurs in the upper trace (anode voltage) at the same instant that a minimum occurs in the lower trace (cathode voltage).

Thus, the composite voltage (20(b)) is less than the sum of the two peak values and also the net ripple is lower. Vertical sensitivity 20 kV/div.

46 kVp, which is considerably larger than half of the 81 kVp set at the control desk. This does not mean that the voltage was out of calibration, but illustrates the fact that anode and cathode voltages must be added vectorially, since they are 30 degrees out of phase. Note that a peak occurs in the anode voltage at the instant that a valley occurs in the cathode voltage. Addition of the two waveforms thus results in the correct kVp, with a simultaneous reduction in the total ripple (Fig. 20(b)). An exception to this occurs, for example, with the General Electric AMX-110 mobile radiographic generator. Shown in Figure 21 on the right is the cathode voltage (upper trace) and the anode voltage (lower trace). Note that the anode and cathode are <u>in phase</u>; thus, the composite voltage waveform (left trace) is equal to twice the voltage applied to either the anode or cathode alone. Note also that the voltage ripple is also twice that associated with either electrode alone.

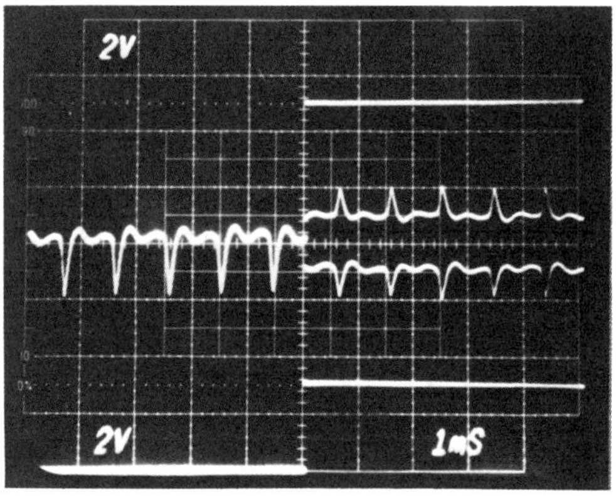

Figure 21. Left: Composite tube voltage produced by a General Electric AMX-110 mobile radiographic generator.
Right: Cathode voltage (upper trace) and anode voltage (lower trace). Note that the anode and cathode voltages are <u>in phase</u>. Thus, the net peak kV is the simple sum of the two voltages and the net ripple is greatly increased.

These effects must be taken into account when a voltage divider is used to monitor the kV during calibration procedures. That is, insertion of only the anode side of the voltage divider into the circuit will generally not produce a voltage waveform whose peak value is exactly half the nominal value set at the control desk.

As mentioned earlier, x-ray tube potential can be controlled by triodes in the high tension secondary so as to provide nearly constant potential to the x-ray tube (Fig. 22). With this type of generator, a very high potential (from 180 kV to 220 kV, depending upon the tube voltage and current selected) is applied across the triode. The grid of the triode is electronically controlled, and the triode becomes a variable resistor capable of absorbing large voltages almost instantaneously. Thus, x-ray tube potential can be maintained to nearly a constant value, regardless of the size of the tube load. In other words, kV compensation for load is virtually instantaneous and extremely large amounts of power can be switched very rapidly. This type of high voltage control is ideal for falling load generators (Fig. 4).

The Falling Load Technique

The principle of the falling load technique (Figure 23) is that the tube current starts at a rather high value, then continuously falls throughout the exposure.[5] If the falling mA curve were selected so as to maintain the anode

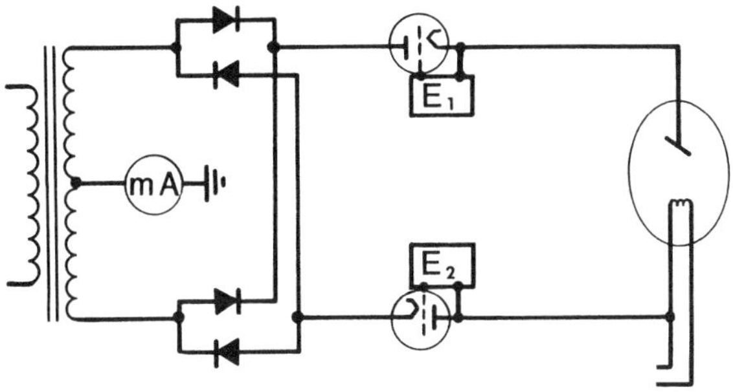

Figure 22. Simplified schematic of high tension secondary with triodes to control the voltage applied to the x-ray tube. The triodes act as variable resistors which absorb any variations in the transformer output, thus maintaining a constant potential on the x-ray tube.

temperature at the maximum allowable limit, then for any given mAs, the exposure time will theoretically be the shortest time possible. In Figure 23, from the conventional loading curve, the shortest possible exposure time for a 50 mAs exposure is 0.1 seconds. For 120 mAs, the shortest possible exposure time is 0.3 seconds, corresponding to 400 mA. Note, however, that if 400 mA were set but only 50 mAs were required, the exposure time would have to be 125 ms, which is 25% longer than necessary. To consistently obtain the shortest time, the radiographer would have to consult the tube loading chart for every exposure in order to determine the maximum allowable mA for the mAs value required.

To alleviate this requirement, the falling load curve is constructed such that, for any given exposure time, the mAs obtained is equal to the maximum allowed mAs. For example, at a tenth of a second the maximum allowed mAs is 50 mAs. Similarly, for

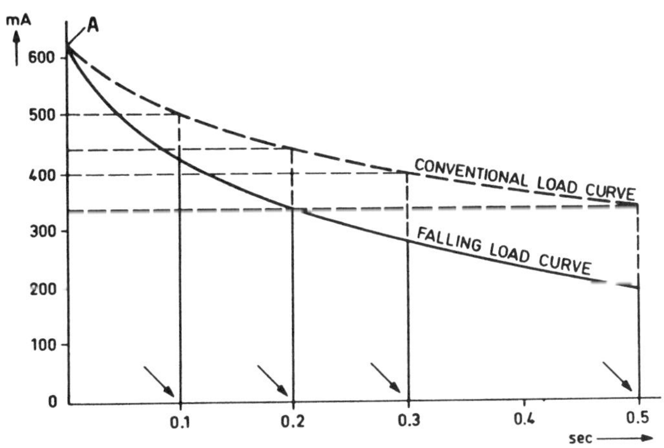

Figure 23. Comparison of the conventional anode loading curve with the falling load curve. For any given exposure time, the area under the falling load curve is equal to the product of the exposure time multiplied by the maximum mA allowable from the conventional loading curve.

the 0.3 second exposure the maximum mAs is 120, and for the falling load curve the area under the curve for a .3 second exposure is also 120 mAs. Thus, regardless of the exposure time, the maximum available mAs is utilized, and therefore, the exposure time is always minimized.

Since the falling load curve always produces maximum mAs regardless of the exposure time, it is identical to the maximum kW loading curve of the anode. In general, the kW of an x-ray tube is defined for each focal spot as the product of the maximum mA times the corresponding kVp at 0.1 seconds. One can also obtain the complete kW loading curve by plotting the kW loading (kV x maximum mA) as a function of the exposure time. It can be derived from the anode loading curves (Figure 24) by selecting one of the curves and calculating the kW as a function of exposure time. It should be noted that the kW for a given exposure time, is independent of the particular kV selected, with the exception of low kV values, where the maximum tube current is limited by a restriction on the maximum temperature of the filament, rather than on the temperature of the x-ray target. These limits are indicated by the truncated curves at low kVp. The resultant kW curve vs. exposure time (Figure 25) can then be used to evaluate whether the falling load circuitry is maintaining a consistent degreee of tube loading, without exceeding the thermal limits of the anode.

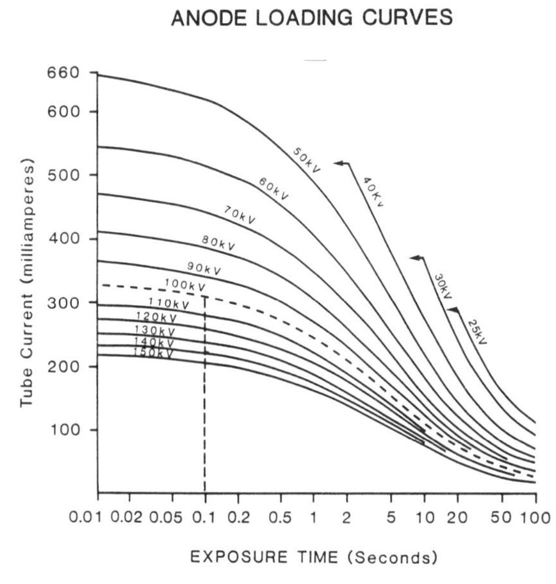

Figure 24.
Conventional anode loading curves. In general, the kW for a given focal spot is defined as the product of kVp x mA for a 0.1 sec. exposure.

It should be noted that it would not be possible to use the maximum anode thermal rating for every exposure, since these ratings are based upon a cold anode. If the maximum allowed exposure were always utilized, residual heat in the anode would soon raise the temperature of the target to a high level, and if the next exposure were also a maximum anode load, the temperature of the focal spot might exceed the melting temperature of tungsten. Therefore, falling load curves are not set to maximum but rather some percentage of maximum, typically 85%.

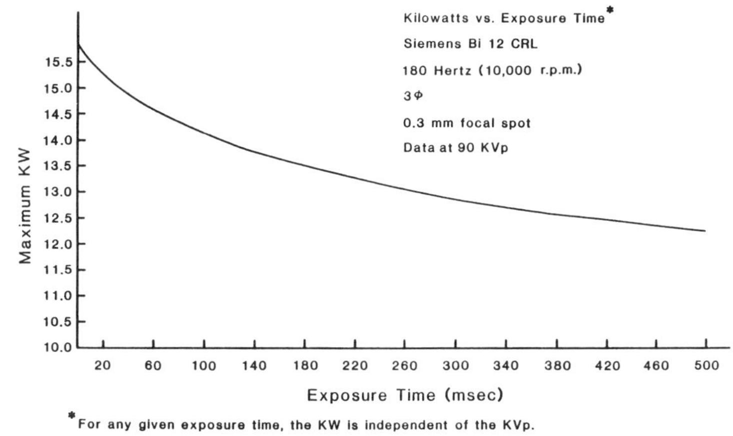

Figure 25. kW loading curve. The curve is derived from the product of kV x mA for a given exposure time, and is independent of the kVp, except where the mA is limited by the maximum filament current.

To evaluate whether a given falling load curve maintains a fixed percentage of tube load throughout the exposure, the following procedure can be utilized (Figure 26):

1. Select the focal spot and kVp and make a low mAs, short time exposure. Measure the exposure time, the mAs, and the maximum mA at the start of the exposure.

2. Verify that the maximum mA measured does not exceed the maximum mA allowed for the kVp which is selected.

3. From the kW loading curve find and record the maximum kW corresponding to the exposure time which was measured.

4. Calculate the maximum allowed mA which is equal to the maximum kW divided by the kVp.

5. Calculate the maximum allowed mAs which is equal to the maximum mA times the exposure time measured.

By comparing the calculated maximum mAs to the actual mAs measured, one can calculate the mAs measured as a percentage of the maximum allowed mAs. This procedure can be repeated, using progressively larger mAs values, to assure that the percentage of maximum mAs remains relatively constant. (See Figure 26).

When our constant potential generator was first installed, we obtained mA waveforms such as illustrated in Figure 27. Note that the mA overshoots to almost 1,700 mA at the beginning of the exposure, then rapidly drops to the correct value along the falling load curve. If the falling mA curve were extended to the left, we would obtain the value of mA which should have been

FALLING LOAD EVALUATION

(Continuously Falling mA)

DATE: 3/26/80

TUBE: B-Trans, AP Tube

81 kVp   168 Initial mA Measured

☐ Large Focus ______ mm

☒ Small Focus 0.3 mm

mAs (set)	mAs (Meas)	Time (Meas)	Maximum kW* at exposure time meas.	Max. mA†	Max. mAs	mAs (meas) as % of Max.
8	8.04	52.2	14.70	181	9.47	84.9
10	10.3	66.8	14.50	179	12.0	86.1
12	13.0	86.0	14.27	176	15.15	85.8
16	16.2	106.1	14.07	174	18.4	87.9
20	20.5	135	13.82	171	23.0	89.0
25	26.1	175	13.51	167	29.2	89.4
32	33.2	229	13.2	163	37.3	89.0
40	42	294	12.89	159	46.8	89.8
50	52.4	385.2	12.55	155	59.7	87.8
64	65.7	498	12.24	151	75.3	87.3
80	84.3	677.4	11.75	145	98.3	85.8

*Use the 90 kVp anode loading curve to obtain kW vs time. The kW at a specified time is independent of kVp.

†Max. mA = Max. kW ÷ kVp

Figure 26. Data form for evaluation of the continuously falling load technique.

produced at the start of the exposure. Adjustment of filament preheat failed to correct the large overshoot in mA. It was believed that the overshoot resulted from a large number of electrons being rapidly pulled from the space charge surrounding the filament at the start of the exposure. Since the kV was controlled by triode valves in the secondary of the high tension transformer, the proper voltage was maintained at the x-ray tube even though the tube load changed rapidly. As the electron emission from the filament came to equilibrium, the tube current rapidly dropped until it reached the value from which it was controlled by the falling load circuitry. It was thought that to eliminate this effect, the space charge around the filament would have to be reduced. However, merely reducing the filament current would not achieve the desired result since, after equilibrium, the tube loading would be far below the 85% value specified for falling load applications.

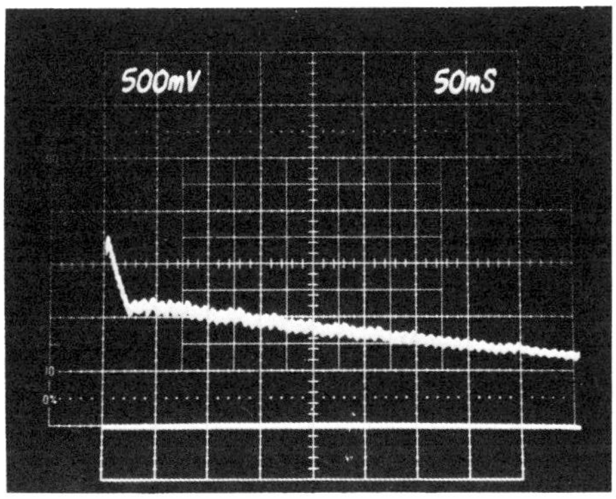

Figure 27. Tube current waveform from a Siemens Pandoros Optimatic generator. Vertical sensitivity 500 mA/div. Note that the initial mA overshoots to 1700 mA, then rapidly drops to 1100 mA. This overshoot was thought to be due to the large number of electrons which would be pulled from the space charge surrounding the filament at the beginning of the exposure.

The solution† was to place a limiting resistor in the filament circuit, along with a relay to shunt the resistor at the beginning of the exposure. The resistor thus limits the boost current and, therefore, the amount of space charge produced during preparation. The relay is closed at exposure release and the full filament current is available for heating the filament. By adjusting the timing of the relay with exposure release, the overshoot in mA can be eliminated. Shown in Figure 28 is the case where the relay is too slow and the mA is actually lower than that which should have been obtained. Figure 29 illustrates the case where the relay is properly adjusted and the mA (middle trace) follows a smooth curve from the beginning of the exposure.

A continuously falling mA generator may produce an mA waveform which remains relatively constant. In general, this may indicate a problem with the filament control circuitry. However, for a given kVp, the maximum mA which can be obtained may be limited by the maximum allowable filament current rather than by the kilowatt rating of the anode. In this case, the filament current must

†The circuit was conceived and built by Mr. Andy Zarrow of the Detroit Siemens office. It was completed prior to the publication by Brezovich and Barnes[6] which describes a similar circuit designed to reduce the fall-off of mA due to electron cooling of the filament.

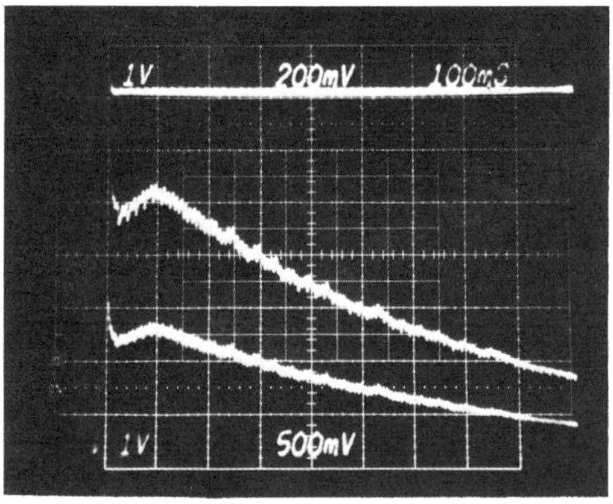

Figure 28. Waveforms obtained after installing the limiting resistor and relay in the filament circuit.
Upper: kV at 10 kV/div.
Middle: mA at 200 mA/div.
Lower: Radiation Waveform.
The relay did not close soon enough so that the mA at the beginning of the exposure was too low.

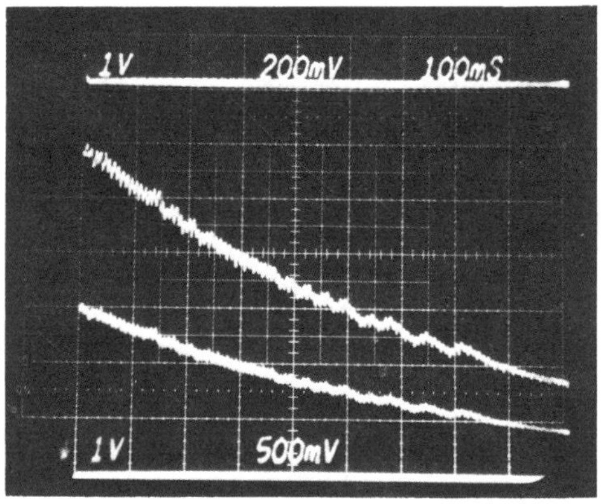

Figure 29. Waveforms similar to those in Figure 28, except the relay is correctly adjusted to provide the proper mA at the beginning of the exposure.

remain at the maximum value (and the mA remain constant) until it reaches the point where it would exceed the value associated with the falling load curve. After that point, the filament temperature will be reduced, and the mA will follow the falling load curve (Fig. 30).

Another form of "falling load" actually consists of a series of stepped constant loads[7] (Fig. 31). In the case shown here, the kV compensation for load is not properly adjusted and the kV does not remain constant as the tube current is stepped to a lower value. It should be noted that the primary voltage is instantaneously reduced to that required for the lower tube current in each step, whereas the thermal inertia of the filament does not allow the filament to cool instantaneously to the temperature corresponding to the lower mA. As a result, the mA slowly drops between steps. Thus, at the beginning of the

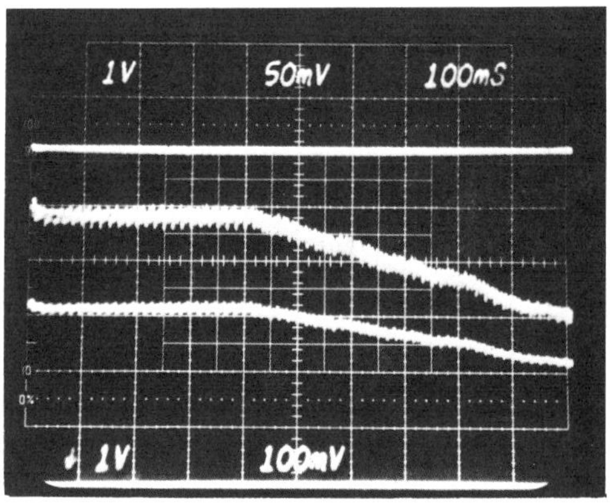

Figure 30. Waveforms demonstrating the result when the tube current (middle trace) is limited by the maximum filament current, rather than by the kilowatt rating of the anode. Note that after 400 ms, the anode loading becomes the limiting factor, and the mA continuously drops.

second step the tube load is too high for the new primary voltage, and the kV rapidly drops. As the mA changes to the lower value, the kV rises to the nominal value obtained in the second step. Figure 32 is a kV waveform with properly adjusted kV compensation for load, thus providing the same nominal kV value throughout the exposure. Also, the exposure is briefly terminated between steps to allow the filament to cool to the temperature corresponding to the lower mA in the second step. In this way, the effects of thermal inertia are reduced. Figure 33 illustrates our method of evaluating the step-wise falling load. From the tube loading chart, the maximum mA at 0.1 seconds is obtained for the kV value used in the exposure. The kV is evaluated at the beginning and ending of each step and expressed as a percentage of the set kVp. The mA at the beginning and ending of each step is also recorded and expressed as a percentage of the maximum allowable mA.

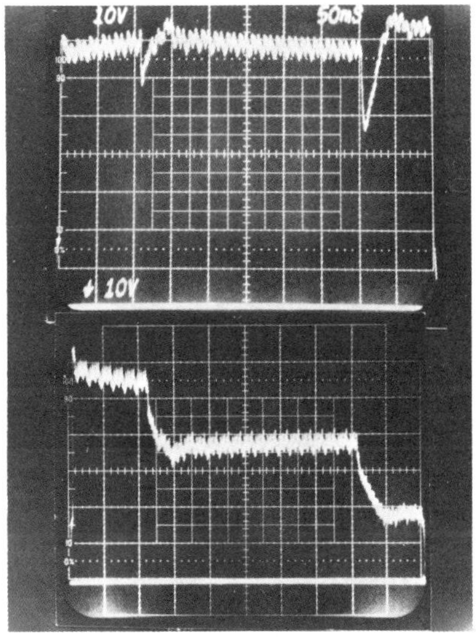

Figure 31. Example of a "falling load" which is actually a stepped, constant mA. Upper: kV waveform (10 kV/div.). Lower: mA waveform (200 mA/div.). Note that the mA is reduced in three consecutive steps each of which provides a relatively constant mA. The kV does not remain constant, especially at the transition between steps, because of the thermal inertia of the filament, and inadequate kV compensation for load.

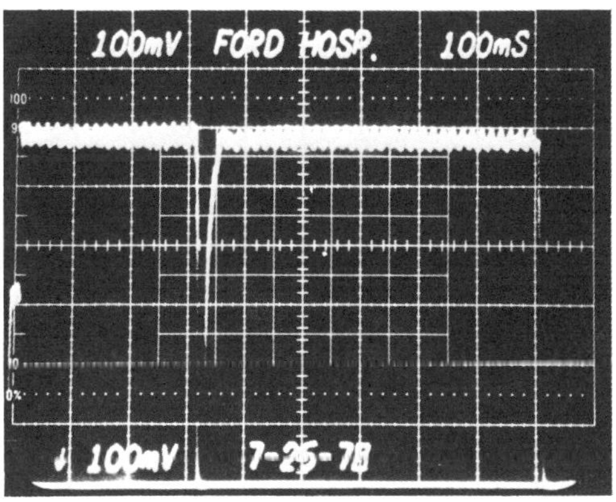

Figure 32. kV waveform for falling load (stepped constant mA) with good kV compensation for load. The exposure is automatically terminated between steps to allow for the filament to cool to the temperature corresponding to the lower mA. Thus, the effect of thermal inertia is reduced.

EVALUATION OF FALLING LOAD AT 60 kVp

Room W345 (X) Large Focus

Date 7/25/78 () Small Focus

Exposure Time: P.T. milliseconds

MAXIMUM mA* at 0.1 sec is 1400 mA

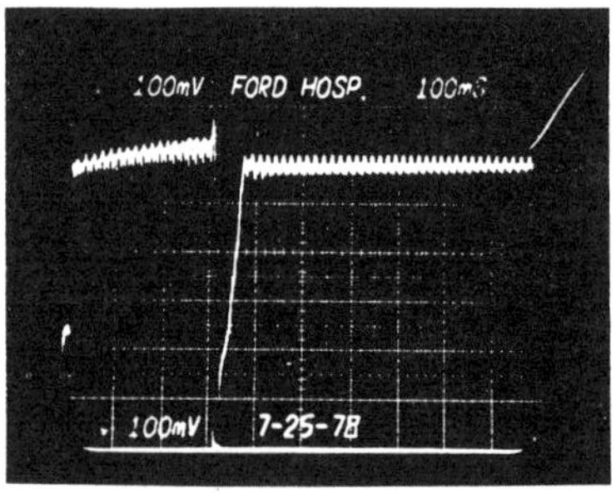

Kilovoltage Waveform

0.1 Volts/div.

(10 kV/div.)

0.1 sec/div.

$\Delta\% = \frac{kV_{(meas)} - kV_{(set)}}{kV_{(set)}} \times 100$

	DATA ANALYSIS			
	Kilovoltage Meas.		Time	
	Begin.	End.	Spec.	Meas.
1st Step	57	63	0.3	0.3
Δ %	−5%	+5%	±Ø	
2nd Step	59	59	N/A	
Δ %	−1.7%	−1.7%		
3rd Step	N/A	N/A	N/A	
Δ %				

COMMENTS: Ramping of KV due to drop in mA as resistors in filament circuit heat up.

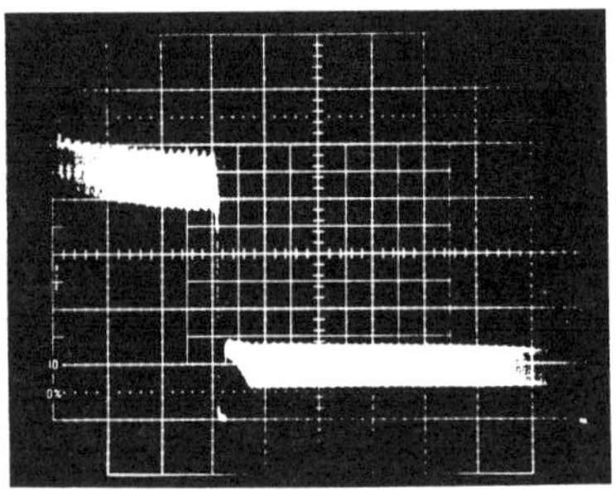

Tube Current Waveform

2 Volts/div.

(200 mA/div.)

0.1 sec/div

$\% \text{ of Max.} = \frac{mA_{(meas)}}{Max.\ mA} \times 100$

	DATA ANALYSIS	
	mA Meas.	
	Begin.	End.
1st Step	1140	1070
% of Max.	81.0	76.4
2nd Step	397	397
% of Max.	28.4	28.4
3rd Step	N/A	N/A
% of Max.		

COMMENTS:

*From the tube loading charts, for the kVp and focus set.

Figure 33. Form for evaluation of falling load waveforms with stepped constant loads.

The Tube Current Waveform

As previously illustrated, oscilloscope display of the tube current waveform is very useful in evaluating generator performance. We often use the Machlett Dynalyzer™ high voltage measuring instrument since it provides simultaneous kV and mA waveforms. In addition, the filament current may be monitored and displayed using this device. However, there are instances when we use other voltage dividers which do not provide this capability. In such cases, the mA waveform is observed by inserting a precision 10 ohm resistor in the mA metering circuit, and displaying the voltage drop across the resistor on a storage oscilloscope. For example, Figure 34(a) is the mA waveform obtained on a G.E. AMX-110 mobile radiographic unit. Unfortunately, with this type of display, it is difficult to determine the average tube current. We, therefore, developed a method of filtering the mA waveform to accurately obtain the average tube current.[8] This method is in excellent agreement with more conventional means of measuring tube current, such as a digital mA meter. The advantage, however, is that the average current can be obtained using short exposure times. The details of the method are described by Rossi in another paper in these proceedings, and will, therefore, not be discussed further.

Figure 34(b) is a typical mA waveform from a three-phase generator. The large spike at the beginning of the exposure is not abnormal; it is the initial surge of current which charges the inherent capacitance of the high tension cables. It can be easily distinguished from an overshoot in mA which is due to incorrect pre-heat adjustment, or from the overshoot described in Figure 27, since its duration is only 1 to 3 milliseconds, whereas the other overshoots will last several tens of milliseconds.

The inherent capacitance of the high tension cables will also prevent instantaneous removal of the voltage across the x-ray tube, with the result that the mA (and kV) waveforms will gradually, rather than instantaneously, drop to zero following termination of the exposure. This effect, as well as the effect of a surge suppression step in the kV and mA waveforms, presents a problem with the measurement of mAs. Generally, mAs is measured by inserting an integrating coulomb meter (mAs meter) in series with the mA metering circuit. However, the meter will integrate all signal present, including the area under the initial step, and the capacitive tail. Thus, the measured mAs will be larger than that calculated from the product of mA x exposure time. This discrepancy will be especially severe for short exposures since the initial step and the discharge tail will be a larger percentage of the useful mAs value. For this reason, we do not recommend using an mAs meter during generator calibration, except for generators which do not provide independent selection of mA and time, but only the product of these two. Observation of the mA waveform is preferred since it will provide direct measurement of mA, and may also help to identify the cause of abnormal generator performance.

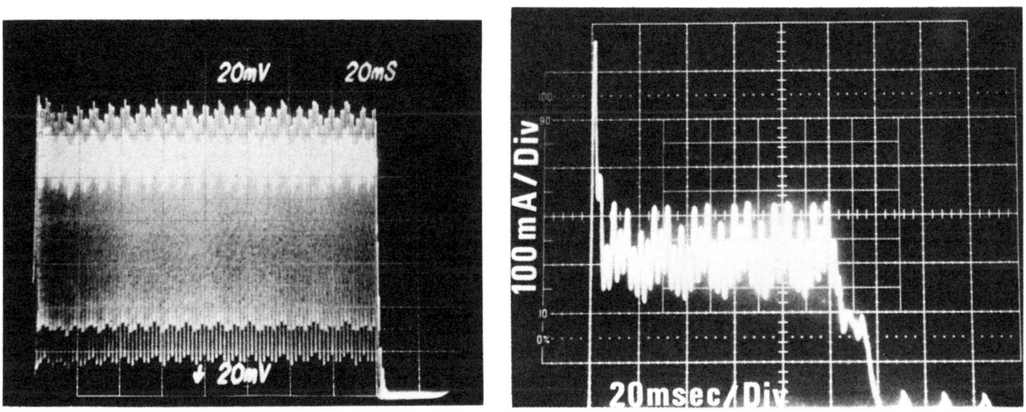

(a) (b)

Figure 34. Tube current waveforms obtained by inserting a precision 10Ω resistor into the mA metering circuit.

(a) Waveform from a G.E. AMX-110 Mobile Generator. Note that the mA ripple makes it difficult to determine the average tube current.

(b) Waveform from a three-phase generator. Note the initial spike due to the charging of cable capacitance, and the discharge tailing at the end of the exposure. Both of these are normal effects.

Multiple Tube Installations

The discussion thus far has assumed that the generator is used to control only one x-ray tube. If this is the case, the generator will probably have sufficient calibration adjustments available to correct any of the problems which have been presented. However, it is not uncommon for a single generator to control two or even three x-ray tubes in a given installation. In this situation, one would expect that the manufacturer would provide sufficient means to properly calibrate each tube and each focal spot on that tube. Lin[9] has shown, however, that this is often not the case, especially if the tubes have considerably different design characteristics. Typical design parameters which will affect calibration are anode-cathode spacing, filament size, filament diameter, and cathode design. In addition, the thermal characteristics of two different tubes may vary considerably. This may require that the tube overload protection circuitry be adjusted for the lower capacity tube, thus limiting the maximum tube load of the other tube to only a fraction of its capacity. The lack of independent calibration adjustments for each tube may force the service engineer to compromise between the optimal settings for each tube. This can result in poor calibration of both tubes. Two tubes which are controlled by the same generator should, therefore, be of identical electrical and thermal characteristics, unless separate calibration adjustments are provided.

Specification of Generator Calibration

The following are the specifications which we utilize for technique calibration of the x-ray generator. By utilizing the method of waveform analysis, we are able to target deviations from these specifications, and determine what service adjustments are required to correct these deviations.

Tube Potential (kVp)

Means shall be provided to adjust the primary voltage to compensate for changes in tube loading (mA). If this compensation is automatic, it shall be adjusted to provide kVp accuracy to within 5% of the indicated value throughout the range of useful tube currents, and for any kVp selected. With a given kVp selected, measurements made at any two mA settings shall not differ from each other by more than 5%.

NOTE: SINCE THE MAXIMUM DEVIATION OF THE MEASURED kVp IS 5%, IT IS EXPECTED THAT AT THE TECHNIQUES USED FOR CALIBRATION ADJUSTMENTS THE MEASURED VALUE WOULD HAVE TO BE SET MUCH CLOSER THAN 5%; TYPICALLY TO BETTER THAN 1 OR 2 kV.

Kilovoltage measurements shall be made using a voltage divider and a calibrated storage oscilloscope. The divider should be inserted in the high tension circuit at the x-ray tube. Alternatively, the divider can be inserted at the transformer, but this is less desirable. In either case the calibration must be corrected for the effect of the voltage divider in the circuit.

Whenever it is not feasible to use a voltage divider, the kVp calibration shall be measured with a kVp test cassette (modified Ardran-Crooks cassette) provided by Henry Ford Hospital. This cassette has been calibrated using a voltage divider.

kV overshoot at the start of the exposure shall not exceed 3 kV. The kVp shall not change by more than 5% during any exposure under 500 milliseconds.

Line resistance, phase rotation, phase voltage, pre-magnetization, exposure contractor timing and filament preheat shall be optimized. The resultant waveform shall contain equal amplitude pulses and shall be free of overshoots, dips, ramps, spikes, glitches or other unusual or abnormal changes from the nominal kilovoltage.

The high tension transformer must provide a balanced distribution of the secondary voltage to the anode and cathode of the x-ray tube. Imbalance, as measured by the difference between the peak anode voltage and the peak cathode voltage must be less than 2 kV.

Tube Current (mA)

Means shall be provided to adjust the filament current to maintain mA accuracy:

Filament Preheat shall be adjusted to provide mA accuracy to within 5% of the indicated value. Initial overshoot, excluding that due to cable charge, shall not exceed 10% of the indicated technique.

The mA stabilizer shall be adjusted such that, for any exposure less than 500 miliseconds duration, the maximum mA does not exceed the minimum mA by more than 10%.

Space charge compensation shall be adjusted such that the mA does not change by more than 5% as the tube potential is changed from 60 kVp to 140 kVp.

For falling load generators with stepped constant load, the mA during the first step shall be adjusted to not less than 85% of the maximum tube loading, but not more than 100% of the allowable load. The product of kVp X mA during the first step shall remain constant for all kVp settings, except where the mA would require more than the maximum allowable filament current.

For falling load generators with continuously falling load, the initial mA shall be adjusted to not less than 85% of the maximum mA allowed for a 10 ms exposure, but not more than 100% of the allowable load. The product of kVp times initial mA shall be constant for all kVp settings, except where the mA would require more than the maximum allowable filament current. The falling load curve shall be adjusted to provide the same percentage of maximum tube load for all mAs settings.

Tube current measurements shall be made using the Machlett Dynalyzer™, or by inserting a 10 ohm, 1% 50W resistor in series with the mA metering circuit and measuring the voltage drop with a calibrated storage oscilloscope. The use of an mA test point, which provides a voltage signal proportional to the tube current, is acceptable only after agreement with the Dynalyzer has been established. The use of an integrating coulomb meter (mAs meter) is not acceptable unless it can be shown that the results are identical to that obtained with the above methods.

Exposure Time

For single phase generators, exposure time is equal to the number of pulses times 8.33 ms. For three phase equipment, the exposure time is equal to the time from threshold crossing of the leading edge of the kV waveform to threshold crossing of the trailing edge of the kV waveform, where the threshold is specified by the manufacturer as a percentage of peak kV.

The exposure time shall be adjusted to within 10% of the indicated value for times less than 100 ms and to within 5% of the indicated value for times equal to or greater than 100 ms. Exposure times measured in three consecutive exposures shall not differ from each other by more than 5%.

The exposure time, except for mammography, shall be adjusted and measured at a technique of 80 kVp, and 200 to 300 mA so as to eliminate the effects of starting steps or cable capacitance on the kV waveform. The measured times less than 40 ms shall not change by more than 10% as the technique is increased to 140 kVp, 500 mA. For mammography the exposure time shall be measured at 30 kVp and maximum mA.

mAs Timer

For generators which do not provide independent control of tube current and exposure time, but only the product of these two (mAs), the exposure timer shall be calibrated to within 5% of the indicated mAs.

The method of measurement shall be the Machlett Dynalyzer^{TM}, or a digital electronic integrating coulomb meter (mAs meter) inserted in series with the mAs metering circuit.

Conclusion

In conclusion, one must realize that there is no such thing as the ideal generator. If it did exist, it would produce square wave kV and mA waveforms which would not deviate from the set technique (Fig. 35). The exposure time would be exactly defined, and the mAs would be the simple product of (exposure time) X (mA).

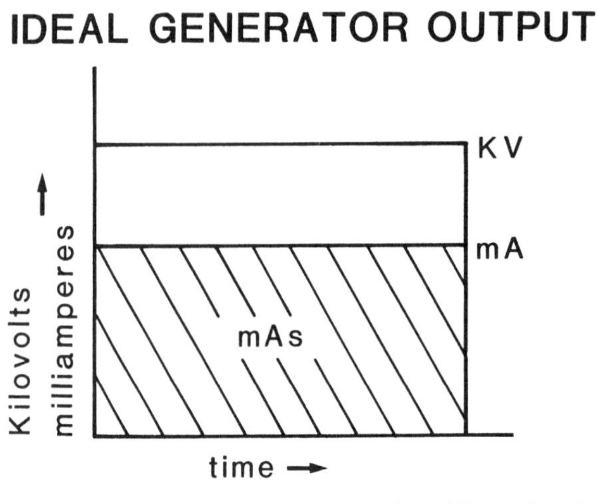

Figure 35.
The "ideal" x-ray generator would produce square wave kV and mA waveforms with no deviation from the set technique.

Realistically, however, the kV and mA waveforms are far from ideal (Figure 36). The kilovoltage waveform is characterized by a premagnetization pulse, a surge step, a possible overshoot at the beginning of the exposure, ripple, and a trailing edge which gradually drops to zero volts. The mA waveform is characterized by an initial overshoot due to charging of the high tension cables, a low mA value associated with the surge step, ripple, and a gradual drop to zero mA as the capacitance of the high tension cables is discharged. Exposure time is rather arbitrarily defined in terms of threshold crossing of the kV waveform. The mAs is not well defined since the simple product of exposure time times average mA excludes tube current present during the surge suppression step, and that during capacitor discharge of the high tension cables. This discrepancy can lead to a large deviation between the "calculated" mAs and that measured with an mAs meter. Specification of generator calibration should take into account these limitations of x-ray generator performance. However, recognition of these limitations, and application of waveform analysis can lead to optimization of x-ray generator calibration and increase the reliability of these devices for medical diagnostic applications.

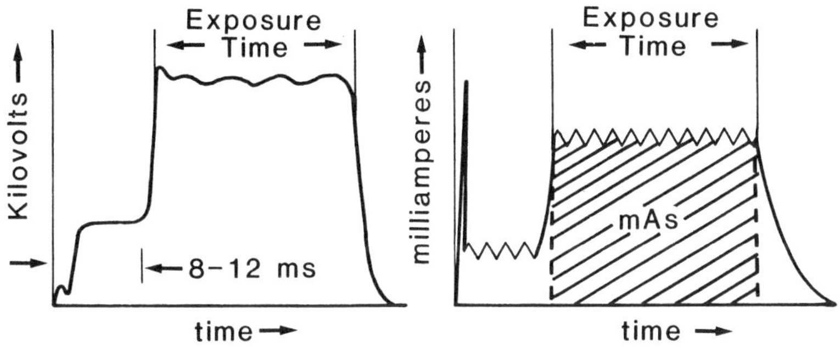

Figure 36. More realistically, the kV waveform (left) will be characterized by deviations from the nominal voltage, and the shape of the mA waveform (right) often makes it difficult to determine the "average" mA or to calculate the mAs.

References

1. "A Calibration System for X-ray Generators and Tube Factors" by T. Healey, et al, British Journal of Radiology, 52, 44-50 (1979).

2. "Selection of Technique Factors for Mobile Capacitor Energy Storage X-ray Equipment" by K.W. Weaver, et al, Radiology, Vol. 128, No. 1, 223-228 (1978).

3. "Principles of Diagnostic X-ray Apparatus" edited by D.R. Hill, 1975, Philips Technical Library, p. 162 & 340.

4. "Effects of Roentgen Tube Voltage and Current Waveforms in Radiography" by H. Kanamori, Acta Radilogica (Therapy) Vol. 4, 68-80 (1966).

5. "The Principle of Falling Load" by C.H. Levsen, Medicamundi, Vol. 14, No. 1 (1969).

6. "An Investigation and Explanation of the Fall-Off of X-ray Tube Current During an Exposure" by I.A. Brezovich annd G.T. Barnes, Phys. Med. Biol. Vol. 25, No. 2, 241-249, 1980.

7. "Performance Evaluation of the Falling Load Technique" by P.L. Rauch and R. Block, SPIE Vol. 127, 145-151 (1977).

8. "Acceptance Testing Experience with Fourteen New Installations" by P.L. Rauch and R. Block, SPIE Vol. 96, 12-18 (1976).

9. "Performance Specifications and Technical Pitfalls in the Equipment Purchase of Radiological Imaging Devices," P.P. Lin, Society for Radiological Engineering, Seminar 1981, Marina Del Rey, Calif.

X-RAY TUBE ACCEPTANCE TESTING IN THE FIELD

Keith J. Strauss, M. Sc.
Department of Medical Phyics
Michael Reese Hospital And Medical Center
Chicago, Illinois 60616
and
Robert J. Kriz, M. Sc.
Department of Radiology
University of Illinois
Chicago, Illinois 60612

Abstract

This paper describes an x-ray tube acceptance testing protocol used to test 20 x-ray tubes in 12 new room installations collectively supplied by 4 manufacturers. The characteristics of a typical focal spot are described. Effective focal spot size (Fef), equivalent focal spot size (Feq), focal spot growth (blooming), and leakage radiation are defined. Measurement methods and performance levels of each of these defined parameters acceptable to both the successful bidder and purchaser are listed and discussed. The paper concludes by presenting performance levels of 20 x-ray tubes and recommending changes in x-ray tube specifications which would be beneficial to both the radiologist and manufacturer.

Introduction

Acceptance testing of x-ray tubes in the field must include measurement of the focal spot size, focal spot growth as a function of tube current and voltage, and leakage radiation emitted from the diagnostic source assembly. Leakage radiation must be controlled to minimize needless patient exposure to radiation. Focal spot growth degrades image quality. Focal spot sizes which exceed tolerances due to improper construction degrade image quality. Focal spots which are smaller than specified can reduce tube life.

The design of an x-ray tube's focal spot size involves tradeoffs. Radiologists desire an x-ray tube with a high instantaneous heat load capacity (kW rating) and small focal spot. The high kW rating allows high mA and short exposure times which minimize unsharpness in the image due to patient motion. The instantaneous heat load capacity can only be increased by spreading the heat over a larger surface area of the anode provided the anode material, angle, rotational speed, and diameter remain fixed.

The small focal spot is required to minimize geometric unsharpness. Therefore, strict control of the focal spot size is required to avoid:

1. Loss of image quality due to a large focal spot
2. Anode surface damage from excessive temperatures due to a small focal spot.

Discussions (1,2) on acceptance testing suggest that the manufacturer and purchaser agree on the :

1. Parameters to be tested
2. Test method and measuring instruments to be utilized
3. Acceptable performance level

prior to the issuance of a purchase order.

Finding a measurement method acceptable to the manufacturer and purchaser can be difficult for three reasons. First, any measurement method has limitations. The theory involved in the method and cautions required in the interpretation of the test data must be understood by both parties. The unique training of each individual representing the two parties, ie. radiologists, physicists, technologists, engineers, and sales personnel, can complicate the communication required to develop these understandings.

Second, the different environments presented by the manufacturer's test lab and radiologist's field installation may dictate the choice of different measurement methods and tools. For example, the manufacturer has optical benches available in the test lab which insure correct alignment of relatively simple test jigs with respect to the x-ray tube. Since optical benches are not adaptable to field installations, the physicist or field engineer may be required to use a different jig which indicates correct alignment prior to data collection. Each jig design may introduce a unique set of alignment errors into the data requiring special interpretation or tolerances.

Third, a chosen measurement method may present information which is easily interpreted and which serves the primary needs of only one party. The cross sectional area of the focal spot imaged with a pinhole camera provides vital information to the manufacturer. If the focal spot size is smaller than designed, the kW rating of the tube will be smaller than specified. This could lead to premature tube failures and costly warranty repairs.

While the radiologist is concerned with equipment longevity, he must be primarily concerned with image quality. One important parameter which affects image quality is the resolving capacity of the focal spot (3,8,17). Since the resolving capacity of the focal spot is dependent on the composite of its size, shape, and intensity distribution, a pinhole measurment of size only, does not measure resolving capacity. Thus, a resolution pattern measurement of the resolving capacity of the focal spot provides more useful information for the radiologist than the manufacturer's desired size measurement with a pinhole.

In response to the radiologist's need of verifying the resolving capability of the focal spot and the manufacturer's requirement of measuring the cross sectional area of the focal spot, Michael Reese Hospital and Medical Center (MRHMC) adopted the following x-ray tube acceptance testing program.

1. Star resolution test pattern measurements of equivalent focal spot size,
2. Pinhole camera measurements of effective focal spot size,
3. Star resolution test pattern measurements of growth of the equivalent focal spot size as a function of tube current and voltage, and
4. Ionization chamber survey meter measurements of leakage radiation.

Focal Spot Characteristics

Since image quality is maximized by an infinitesimally small focal spot, the size, intensity distribution, and shape of an actual focal spot significantly degrade image quality. Standards allow (4) the physical dimensions of the focal spot to exceed the nominal size by 30% - 114%. The expected rectangular focal spot shape is sometimes twisted into a parallelogram. Figure 1 (5) contains illustrations of two three dimensional x-ray intensity distributions from typical edge band distribution focal spots. The intensity distribution of the focal spot is peaked at the edges, may or may not have intensity peaks which are of equal height, and has tapering edges resulting in poorly defined dimensions.

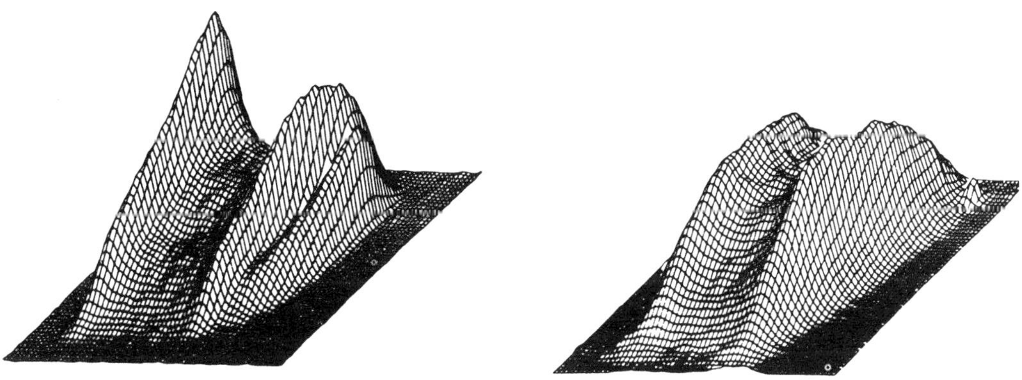

Figure 1 (5): Focal Spot X-ray Intensity Distributions (Courtesy of R.F. Wagner et.al, Medical Physics 1:14, Jn - Fb, 1974.)

The above comments suggest that actual focal spots are far from "ideal". The construction of x-ray tubes which perform within tightly specified tolerances is difficult. Positioning errors of the filament by as little as 0.001" can result in significant changes in focal spot size, shape, and intensity distribution (6,7). Obviously, the purchaser should protect his investment by measuring the performance of the size, shape, and intensity distribution of focal spots.

Definitions

For acceptance testing purposes the following definitions of effective focal spot size (Fef), equivalent focal spot size (Feq), growth of the equivalent focal spot size (blooming), and leakage radiation are adopted.

Fef is defined as (11):

> "the physical dimension of the electron beam impinging on the anode as viewed from the image plane in the central beam"

The definition of Fef dictates that the size of the focal spot be measured from a pinhole image obtained by placing the pinhole on the central ray of the x-ray beam.

The definition of Feq is adopted from Milne (3):

> "The equivalent size Feq of a focal spot F, of any shape and with any type of intensity distribution is equal to that size of a rectangular homogeneous focal spot which has the same resolving power..."

The Feq is measured by using a resolution test pattern and model which relates the resolving capacity of the focal spot in question to the size of a rectangular homogeneous focal spot with the same resolving capacity. Thus, when one measures Feq, one measures resolving capacity, not the size of the focal spot.

Focal spot blooming is defined as:

> the growth of Feq as a function of tube current and voltage.

Blooming is measured by comparing measurements of Feq obtained with different technique factors of tube potiential and current.

Leakage radiation is defined as (16):

> "radiation emanating from the diagnostic source assembly except for the useful beam..."

The permissible level (16) for leakage radiation "measured at a distance of 1 meter in any direction from the source shall not exceed 100 milliroentgens in 1 hour when the x-ray tube is operated at leakage technique factors." Leakage technique factors are defined (16) as " the maximum rated peak tube potential and the maximum rated continuous tube current for the maximum rated peak tube potential."

Discussion of Fef vs Feq

The debate over which measurement, Fef or Feq (pinhole camera or resolution pattern) is more appropriate for acceptance testing of x-ray tubes continues between manufacturer and purchaser. The unique size, intensity distribution, and shape of the focal spot can be measured with a resolution pattern or pinhole camera. The pinhole camera data can result in quantitative measurements of each of the three parameters, but requires computer assisted mathematical computations to obtain information on these parameter's composite degradation of image quality. [The Fourier Transform of the two dimensional point spread function (pinhole image) calculates the focal spot's Modulation Transfer Function (resolving capacity)]. The resolution pattern measurement provides qualitative information about each of the three parameters and provides a direct quantitative measurement of the resolving capacity of the focal spot. Since the radiologist is primarily concerned with the resolving capacity of the focal spot regardless of what the size, shape, or intensity distribution may independently be (3,8,17), MRHMC prefers to use a resolution pattern to measure Feq.

However, using Feq as a criteria for accepting an x-ray tube is rejected by the manufacturers in favor of Fef. The NEMA standard (4) defines acceptance of an x-ray tube on the basis of Fef for focal spot sizes greater than Ø.3 mm. The NEMA standard defines the Fef measurement protocol and tolerances. While a critical review of the NEMA protocol reveals a measurement method which is difficult to complete correctly as defined (33), all except one manufacturer insisted on using the NEMA protocol.

If a simple mathematical correction factor could be found to estimate Fef from Feq, one could provide acceptable information to both the manufacturer and radiologist by measuring only Feq. Bernstein, et al.(1Ø), presented data for converting from Feq to Fef which was developed from geometric ray tracing methods. However, this method requires a detailed knowledge of each focal spot's intensity distribution.

Since a single measurement method of focal spot size could not be found which provided information on the focal spot's effect on instantaneous anode heat load capacity and its effect on resolving capacity, MRHMC chose to measure both Feq and Fef for each x-ray tube. The comparison of collected Feq and Fef data leads to recommendations for change in existing standards. These changes should clearly define the independent issues of the focal spot size's effect on instantaneous heat load capacity and of the focal spot resolving capacity's effect on image quality.

Specification of Performance Parameters

Effective Focal Spot Size

The tolerances of Fef specified in the NEMA standard (4) and measurement protocol are adopted. These tolerances are listed in Table I. The measurement protocol is described in a later section.

Equivalent Focal Spot Size

Since the use of a star resolution pattern to measure Feq of nominal focal spot sizes greater than 0.3 mm is not defined by existing standards, reasonable Feq tolerances for focal spots are not defined. MRHMC attempted to establish reasonable tolerances for Feq for acceptance testing purposes by assuming that the shape and intensity distribution of the measured focal spots would fall within a defined range. Data collected during the actual acceptance testing revealed that this assumption was incorrect. However, this assumption led to the following acceptance testing specification for Feq which used other's comparisons of Feq and Fef measurements (9).

> Feq is measured with a star resolution pattern. Multiplicative correction factors of 1.25 for the length dimension (parallel to anode-cathode axis) and 0.85 for the width dimension (perpendicular to anode-cathode axis) are applied to Feq. The acceptance level of the corrected Feq values must be less than or equal to the NEMA tolerances listed for Fef.

The measurement protocol of Feq is described in a later section.

Focal Spot Blooming

While no standard exists which adequately deals with the growth (blooming) of Feq as a function of tube current and voltage, focal spot blooming is not a newly discovered phenomenon. The Federal Standard #83 (12) implied focal spot growth in 1956 by specifying the tube current at which focal spot size should be measured. Mattsson (13) noted growth of both dimensions of the focal spot as a function of the tube current in 1968. In 1974 Chaney and Hendee (14) investigated blooming of Feq of a number of x-ray tubes and developed a theoretical explanation of their results.

Figure 2 (14) illustrates the growth of Feq in the dimension parallel to the anode-cathode axis (length dimension) while Figure 3 (14) is similar data for the dimension perpendicular to the anode-cathode axis (width dimension). The theoretical derivation (14) of Feq blooming in the parallel dimension for space charge limited tube currents suggests that Feq is a function of the tube current, applied tube voltage, and the distance along the anode-cathode axis from the filament to the point where the electron beam begins to diverge. The perpendicular dimension is believed to exhibit less blooming than the parallel direction due to electrostatic beam focusing achieved by the geometric structure of the cathode assembly. Since the intensity distribution of a focal spot changes as a function of tube current (see Figure 4) blooming measured as Fef and Feq do not agree with one another(14,15).

MRHMC specified that:

> the variation of Feq of the focal spot, measured with a star resolution pattern over the rated kVp-mA range of the x-ray tube shall not vary by more than +/- 25% relative to the Feq measured at 75 kVp and 5Ø% of the maximum rated mA (or closest available mA station) at the highest anode rotation for a Ø.1 second exposure.

The technique factors chosen for the reference Feq are those specified by NEMA (4) for establishing Fef. The +/- 25% tolerance level should be achievable; it exceeds the range of blooming illustrated in Figure 2 by a factor of 3.

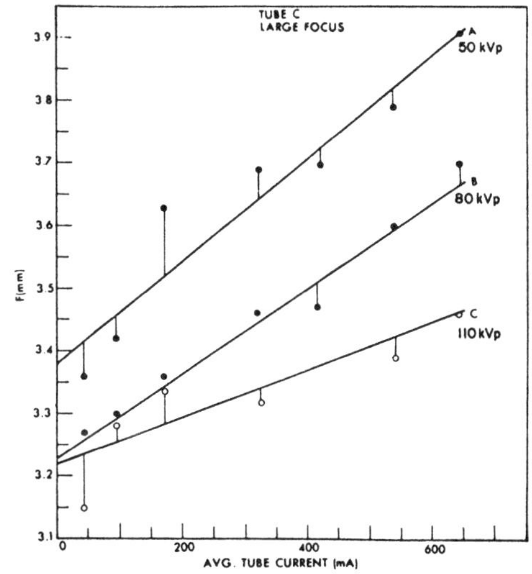

Figure 2 (14): Growth of Feq: Length Dimension (Courtesy of E.L. Chaney et.al, Medical Physics 1:142, My - Jn, 1974.)

The manufacturers were reluctant to commit their products to MRHMC's blooming specification despite the large tolerance level. One manufacturer admitted that his company had not thoroughly investigated blooming because existing standards did not address the question. Two of the four successful bidders refused to specify blooming performance. Again, the different concerns of the radiologist and manufacture are illustrated. Standards should be revised to include tolerances for focal spot blooming which are equitable to both the

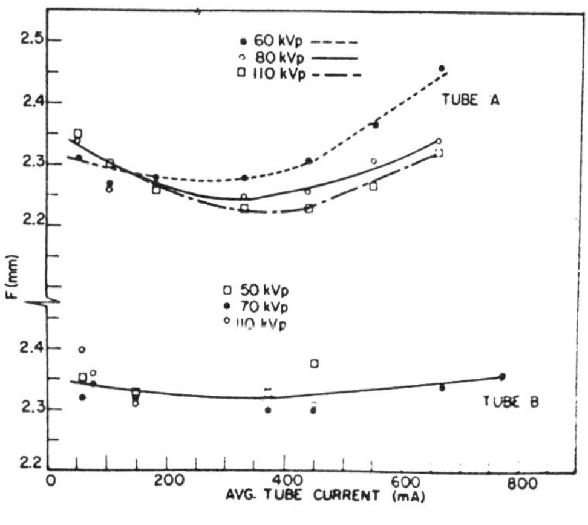

Figure 3 (14): Growth of Feq: Width Dimension (Courtesy of E.L. Chaney et.al, Medical Physics 1:143, My - Jn, 1974.)

manufacturer and radiologist. This is especially important for systems using falling load generators which are designed to provide the maximum rated tube mA during the entire exposure.

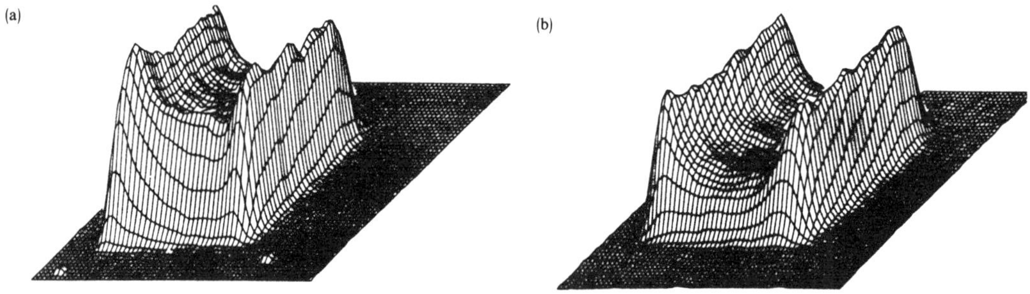

Figure 4 (5): Focal Spot X-ray Intensity Distributions: a) 200 mA b) 1200 mA (Courtesy of R.F. Wagner et.al, Medical Physics 1:20,21, Jn - Fb, 1974.)

Leakage Radiation

MRHMC requested that the manufactures provide data on the maximum leakage radiation from the diagnostic source assembly, its spatial distribution, and the technique factors utilized to obtain the data. MRHMC further specified that the permissible level of leakage radiation as previously defined be limited to 75 mR/hr at 1 meter. The manufacturers voiced no concern about this specification. However, all failed to provide information concerning the spatial distribution of their product's leakage radiation.

Table I lists the negotiated performance levels between MRHMC and the four successful bidders.

Table I
X-ray Tube Acceptance Testing Performance Levels

Dimension		Nominal Focal Spot Size < 0.8 mm	0.8 to 1.5 mm	> 1.5
Fef:	Width	+50 %	+40 %	+30 %
	Length	+114 %	+100 %	+86 %
Feq:	Width	Feq x 0.85 < NEMA std. for Fef		
	Length	Feq x 1.25 < NEMA std. for Fef		

	MRHMC	Successful Bidders A	B	C	D
Blooming	+/- 25%	+/- 50%		+/- 5%	
Leakage	75mR/h	75mR/h	50mR/h	100mR/h	
Technique	BRH	BRH	150 kVp 4 mA	150 kVp 4 mA	

In cases where the successful bidder failed to specify a performance level and objection to the measurement method proposed, the performance level originally specified by MRHMC became the acceptance level.

Measurement Methods

A number of different methods have been described which measure the focal spot's resolving capacity, Feq (17,18). MRHMC believes the star resolution pattern is the most straight forward device for measuring Feq in the field outside the research lab. As previously described, a pinhole camera is dictated for measuring Fef.

Effective Focal Spot Size

The use of a pinhole camera to measure Fef is defined by NEMA (4). This standard describes the:

1. Construction and physical dimensions of the pinhole as a function of nomimal focal spot size
2. Geometry of the pinhole with respect to the anode and film.
3. Appropriate film
4. Exposure techniques
5. Film development
6. Measurement of the x-ray image
7. Acceptable tolerances for Fef

A pinhole image of the focal spot is produced by following the NEMA standard. The NEMA required pinholes (0.03, 0.075, 0.1 mm diameter), stand, and 5-10x magnifying glass with graticule are commercially available for approximately $700. Individually wrapped pieces of dental film meet NEMA criteria and are the appropriate size. The stand, Figure 5, must be carefully positioned to insure placement of the pinhole on the central ray of the x-ray beam. The flouroscopic screen at the base of the stand allows this positioning to be verified with x-rays. The stand's adjustable legs allow the use of NEMA specified enlargement factors. The alignment holes in the lead plate spaced 1/2" apart allow measurement

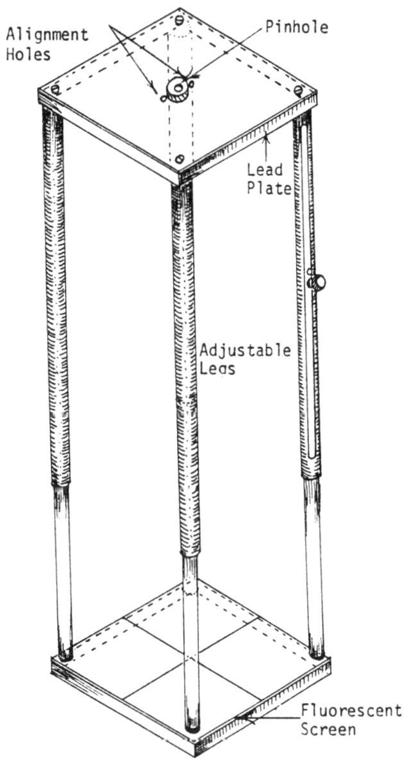

Figure 5: Pinhole Camera Stand

of the image magnification. Approximately 3000 - 6000 mAs is required at 75 kVp to produce the NEMA specified density on the dental film. The actual required mAs is primarily dependent on the focal spot size with respect to the NEMA specified pinhole size. Since the dental film is not designed for automatic processing, the exposed film should be hand processed. The magnifying glass with graticule allows measurement of Feq to the nearest 0.1 mm.

The calculation of Fef from the pinhole image corrects for the magnification, penumbra, and intensity distribution in the length dimension of the pinhole image. Equation (1) is derived from geometry considerations illustrated in Figure 6; it corrects for penumbra and magnification.

$$Fef = \frac{s}{m} - d\left[1 + \frac{1}{m}\right] \qquad \text{---(1)}$$

where s = the measured image size
m = enlargement factor = b/a
d = pinhole diameter

Using the NEMA standard results in penumbra corrections of approximately 6% of the nominal focal spot size.

The ICRU recommends applying a multiplicative factor of 0.7 (20) to the length dimension of the pinhole image. This correction supposedly converts the length of a centrally peaked intensity focal spot to the length of a uniform intensity focal spot which exhibits the same resolving capacity. This correction method ignores the existence of a unique intensity distribution for each focal spot. While more accurate corrections are not specified by NEMA, Prasad et.al. (19) developed more precise correction factors for the length intensity distribution. These author's correction factors are derived by assuming the centrally peaked intensity distribution parallel to the anode-cathode axis is modeled by either a cosine or Gaussian curve.

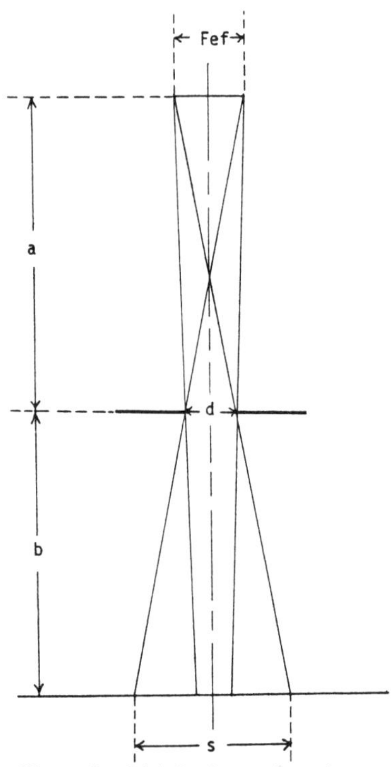

Figure 6: Pinhole Camera Geometry

Rigorous adherance to the NEMA

standard which is required by the manufacturers when acceptance testing the focal spot's Fef in the field, presents many practical difficulties. The commercially available pinhole stand is difficult to use. The lead plate, Figure 5, makes the small based stand top heavy and prone to vibrations when fully extended to 24". No means is provided on the base to allow leveling. The base design forces the use of film with dimensions smaller than the 4" base. This excludes the use of fine grain ready pack film designed for automatic processing. The pinhole stand is difficult to position and use on undertable tubes. Finally, the alignment holes in the mounting plate are too large resulting in excessive optical density on the film beneath the alignment holes. This excessive optical density hinders accurate measurement of the magnification factor. This magnification factor must also be corrected for the systematic error introduced if the alignment holes and pinhole are not in the same plane. If the pinhole lies in a plane between the alignment holes and focal spot the magnification factor is underestimated by triangulation resulting in an overestimated Fef. The magnitude of this error can range from 2 to 20% depending on distances a and b in Figure 6 (34) and the distance between the alignment hole and pinhole planes.

Excessive x-ray tube workloads are required to produce a pinhole image. The small NEMA specified pinholes (4), 0.03, 0.075, or 0.1 mm are required to minimize penumbra in the image (equation 1). These pinhole sizes require 3000-6000 mAs at 75 kVp to obtain an optical density of 0.8-1.2 on dental film. Since multiple exposures are required with cooling time in between, only two pinhole images can usually be obtained in one hour.

Controlling the optical density level of the image is difficult. Since the density is specified "at the most dense portion of the image" (4), the peak of the focal spot's intensity distribution must be measured. Thus, a microdensitometer is required to measure the density of the pinhole image. Unless an institution supports a research laboratory, a microdensitometer is probably not available. Incorrect optical densities due to incorrect exposure technique, require repeat testing and additional large tube workloads.

Dental film requires manual development which most institutions are not equipped to handle. As a substitute to manual development, the automatic processor can be used. The dental film must be securely taped to a leader film to insure adequate transport through the processor. The processor's cycle is interrupted prior to film drying. The dental film is placed back into the rollers within the fixer tank for 5 to 10 minutes prior to completion of the processor's wash and dry cycle. While this development process results in useable images, the radiographs will not be completely cleared of undeveloped silver. Alternatively, one can cut up small squares of wrapped x-omat direct exposure film in the darkroom and cover the edges with black tape.

Analysis of the pinhole image in the field is subjective. If the intensity distribution is similar to that of Figure 1b, the focal spot's edge is not well defined. Kuntke has shown that a change in density of the pinhole image can result in a significant change in the measured Fef (21). Also, the Fef measurement with a pinhole provides no DIRECT information about the resolving capacity of the focal spot which the radiologist requires.

Figure 7: 2° Resolution Star Pattern

While the measurement process is tedious, pinhole images are useful for measurements in the field. The pinhole provides an image of the focal spot and quantitative information about its physical size and shape. Since the measurement of focal spot size with a pinhole is defined, it can be used to reject focal spots which do not meet NEMA tolerances. Gray and Trefler (22) have developed modifications for the pinhole camera which make it easier to use. These modifications reduce the required x-ray tube work load per image and ease the positioning of the stand. However, these author's modifications violate NEMA protocol and probably would not be acceptable to the manufacturers.

Equivalent Focal Spot Size

The required resolution star patterns, Figure 7, are commercially available for approximately $200 each. Spoke angles of 1 ,1.5 ,and 2 degrees are the most common designs. Fine grain direct exposure film individually wrapped makes a suitable image receptor. A stand is required which allows positioning of the resolution pattern parallel to the image receptor, perpendicular to the central ray, and centered to the cen-

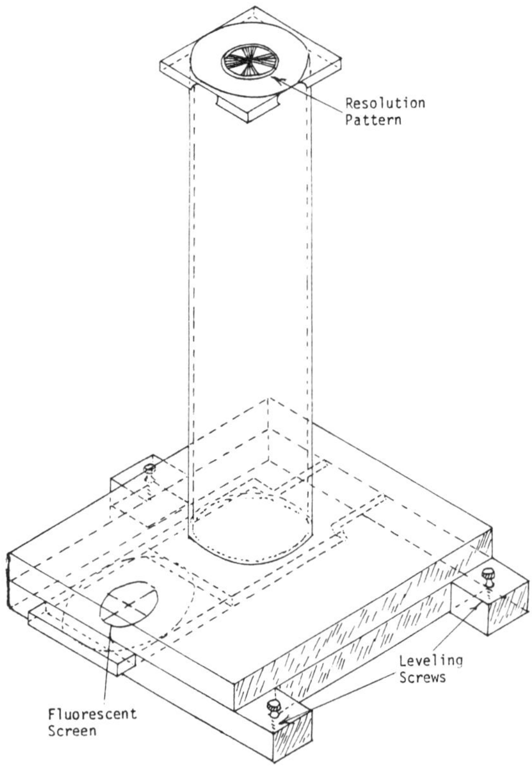

Figure 8: Resolution Star Pattern Test Stand

tral ray. While a stand is not available commercially, suitable stands have been designed (23) and can be constructed in a reasonably equipped machine shop, Figure 8.

Exposing the star resolution pattern and developing the image are straight forward. The stand (Figure 8) and x-ray tube are leveled. The magnification factor and spoke angle are chosen to result in the outermost blur diameter in the image with 1/2 to 3/4 the diameter of the spoke pattern. This insures that the outermost blur diameter is not cut off by the physical size of the test pattern. Table II lists appropriate choices for the spoke angle and magnification (24). The resolution pattern is

Table II

Nominal Focal Spot Size	Magnification	Spoke Angle
1.Ø < f < 1.5	1.8	2
Ø.6 < f < 1.Ø	2	1.5
Ø.4 < f < Ø.6	2	1.Ø
Ø.3 < f < Ø.4	4	1.Ø
f < Ø.3	8	1.Ø

placed in the stand. The fluorescent screen in the stand is used to insure the center of the resolution pattern is positioned on the central ray. An exposure of 5Ø mAs at 6Ø kVp is required when using direct exposure film and a magnification of 2 provided the distance from film to resolution pattern is approximately 16 inches. Automatic processing is used to develop the exposed film.

The Feq is calculated from the outermost blur diameter on the resolution pattern image. Figure 9 (25) illustrates the x-ray intensity expected in various planes beneath a repetitive bar pattern imaged by a finite size focal spot of dimension f and uniform intensity dis-

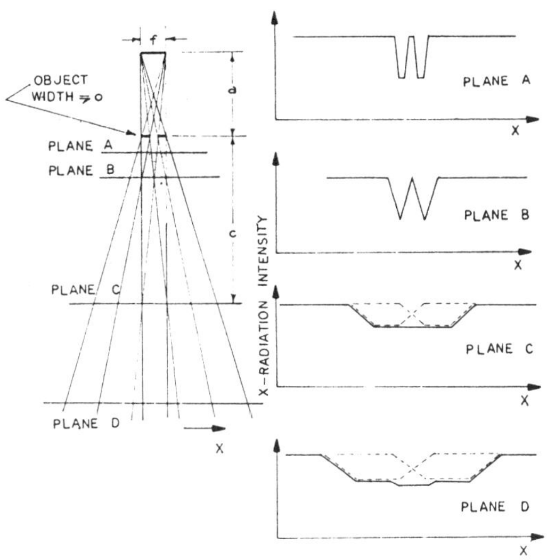

Figure 9 (25): Radiation Intensity Patterns Beneath Line Pair Test Pattern (Courtesy P. Spiegler et. al, Radiology 102:680, Mr, 1972.)

tribution. The object gap disappears at plane c due to the interaction of the penumbra of the two absorbers in the bar pattern. This results in the loss of resolution of the test pattern's spokes. Spiegler and Breckinridge (25) further explain the geometric considerations which lead to Equation (2):

$$\text{Feq} = \frac{\theta \pi D}{180 (m-1)} \qquad \text{---(2)}$$

where θ = angle of the star pattern's spokes
D = blur diameter on the star pattern image
m = image magnification of the star pattern

Using the outermost blur diameter, Equation (2) determines Feq: the size of a focal spot with a homogeneous intensity distribution which would have the same resolving capacity as the focal spot in question. The blur diameter measured parallel to the anode cathode axis results in the Feq width dimension. The length dimension is obtained from the measured blur diameter perpendicular to the anode cathode axis. The center of the region of blur (loss of contrast or phase shift) should be used when measuring the image blur diameter.

The star pattern image contains additional useful information. The resolving capacity of the focal spot in line pair/mm can be calculated from knowledge of the focal spot's Feq. From geometry considerations of Figure 7 the spatial frequency represented by the star spoke pattern at a particular radius is:

$$R(\text{lp/mm}) = \frac{90}{\theta \pi r} \qquad \text{---(3)}$$

where θ = spoke angle in degrees
r = radius in mm

Equation (3) can be combined with Equation (2) to obtain:

$$R(\text{lp/mm}) = \frac{1}{(m-1) \times \text{Feq}} \qquad \text{---(4)}$$

Thus, Equation (4) calculates the resolving capacity of the focal spot in lp/mm for any magnification of the object. This is useful to the radiologist when choosing appropriate screen film combinations or maximum magnification factors for a given focal spot.

Calculating a composite Modulation Transfer Function (MTF) of the focal spot, patient motion, and image receptor at the image plane is an alternate approach to setting up a clinical radiographic imaging system. This MTF calculation is straight forward if one knows the Feq of the focal spot (26):

$$MTF = \frac{\sin\ [\pi f_o a(m-1)/m]}{\pi f_o a(m-1)/m} \times \frac{\sin\ [\pi f_o vt]}{\pi f_o vt} \times \frac{1}{1 + (k^2 f_o^2/m^2)} \qquad \text{---(5)}$$

where f_o = spatial frequency in the object
a = Feq
m = magnification
v = velocity of patient motion
t = time of exposure
k = constant characteristic of screens

The three terms in Equation (5) respectively calculate the MTF for the focal spot, patient motion, and image receptor. The image quality from the composite imaging system is maximized by choosing a focal spot and image receptor which maximizes the composite MTF.

While the resolution pattern image is not an image of the focal spot, the resolution pattern im-

Figure 10 (27): Circles of Confusion of Rectangular Focal Spot (Courtesy of A.E. Burgess, SPIE 96:70, Sept, 1976.)

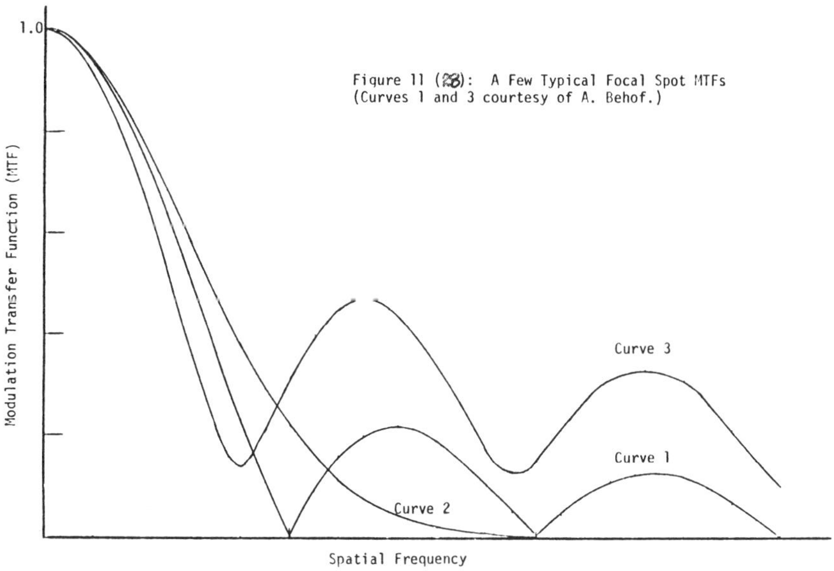

Figure 11 (28): A Few Typical Focal Spot MTFs (Curves 1 and 3 courtesy of A. Behof.)

age provides qualitative information concerning the shape of the focal spot. A rectangular shaped focal spot presents symmetric blur patterns with respect to the anode-cathode axis as outlined in Figure 1Ø (27). If the shape of the focal spot is twisted into a parallelogram, the lines of blur in Figure 1Ø are rotated about the center of the star image.

The resolution star pattern provides qualitative information about the intensity distribution of the focal spot. The circles of confusion or regions of loss of contrast (blur) in Figure 1Ø occur when the focal spot's MTF becomes zero. Figure 11 (28) illustrates a few of the different types of focal spot MTFs which can occur. If a typical focal spot's edge band intensity distribution is viewed along the direction parallel to the anode-cathode axis (along the length dimension), typical intensity distributions appear either uniform or Gaussian. If uniform,

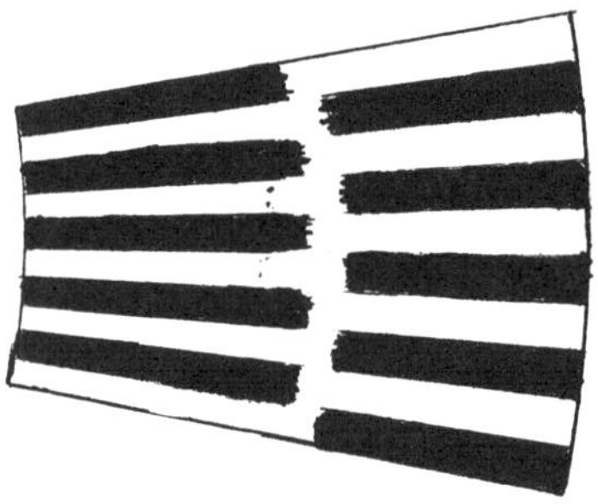

Figure 12: Star Resolution Pattern Image (Length Dimension) Uniform Intensity Distribution

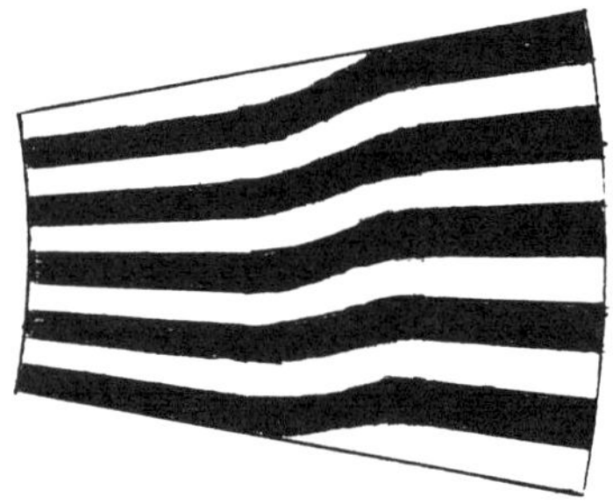

Figure 13: Star Resolution Pattern Image (Width Dimension) Unequal Peak Intensity Distribution

one expects a complete loss of contrast and complete phase shift on the star resolution image as shown in Figure 12. This corresponds to MTF curve 1 in Figure 11. If the intensity distributon is Gaussian, one expects no loss of contrast and no phase shift because the MTF, curve 2 in Figure 11, never reaches zero.

If the edge band intensity distribution is viewed along the direction perpendicular to the anode cathode axis (along the width dimension), typical intensity distributions appear as two equal or unequal intensity line sources connected by an area of weaker inten-

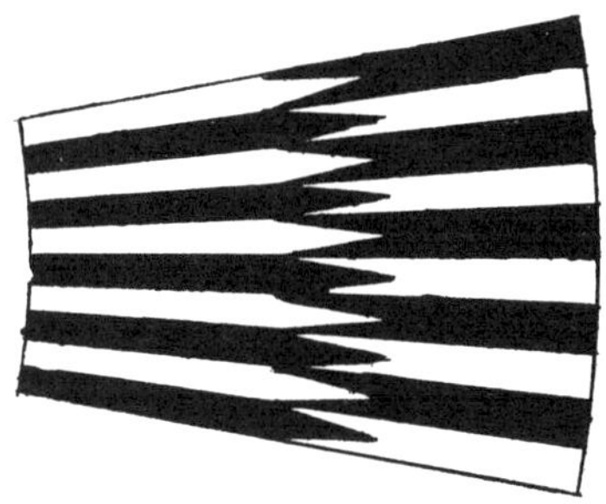

Figure 14: Star Resolution Pattern Image, Spatial Frequency Tripling

sity. The unequal intensity distribution results in the blur pattern shown in Figure 13, no loss of contrast, but a gradual phase shift. The corresponding MTF to the unequal case is illustrated by curve 3 in Figure 11. Although the MTF never reaches zero, the assymetry of the intensity distribution causes both the sine and cosine transform to exist. In addition to the well behaved cases illustrated in Figures 12 and 13, Figure 14 illustrates spatial frequency tripling, described by Gray et.al. (29), which often occurs in actual resolution pattern images.

The information provided by the star resolution pattern, resolving capacity, qualitative shape, and qualitative intensity distribution of the focal spot, is more useful to the radiologist and is easier to obtain than the Fef pinhole measurement. Positioning the resolution pattern for overtable or undertable tubes is easily done with the stand illustrated in Figure 8. 50 mAs or less is normally required per image. The reproducibility of the measurements by a properly trained individual is less than 5% (9). The image may be developed using automatic processing. The minimal effort required to measure Feq allows Feq mapping of the focal spot over its lifetime in a Quality Assurance Program.

The measurement of Feq can present difficulties. As previously discussed, the manufacturers are reluctant to accept this measurement technique as a basis for acceptance of an x-ray tube. The resolving capacity of focal spots with Gaussian MTF's can not be measured with a resolution pattern. However, the radiologist is more interested in the resolving capacity of the focal spot than its physical size. This suggests that existing standards should be updated to define acceptance testing on the basis of Feq.

Growth of Feq (Blooming)

The measurement of focal spot growth as a function of mA is measured using the same equipment and protocol described for Feq. Measurements are completed with the five techniques shown in Table III.

Table III
Blooming Techniques

Tube Potential (kVp)	Tube Current (mA)	Total charge mAs @ 32" SID
75	1/2 max rating	30
60	max	50
60	min	50
max-10	max	10
max-10	min	10

The Feq measured at 75 kVp is used as the baseline for calculating the percent growth of Feq. Since the kW rating of the generator and tube may not be well matched, tube rating charts must be followed when selecting the (maximum "kVp", maximum "mA") technique. Blooming measurements should not be completed until after verification of the generator calibration. Feq as opposed to Fef should be measured when measuring blooming for two reasons:

1. Feq measurements minimize the tube workload (40,000 mAs for Fef vs 320 mAs for Feq)
2. Feq measurements record changes in the focal spot's resolving capacity which monitors changes in the intensity distribution illustrated in Figure 4.

Leakage Radiation

Measurement of leakage radiation from the diagnostic source assembly is not trivial. Care is required to obtain meaningful results and prevent tube damage. The chosen ionization chamber should have a cross sectional area of 100 square centimeters which averages the reading over the recommened (16) area. The Federal Standard states the 100 mR/hr limit at 1 meter applies in all directions from the focal spot (16). An NCRP procedure attempts to measure all directions. It defines 14 primary measurement locations on a one meter radius sphere with the focal spot at the sphere's center (30). Accurately positioning a survey meter at these 14 different locations is difficult. Fluoroscopic tubes enclosed by the patient table provide additional positioning problems. Unless the tube in question is a fluoroscopic tube, the control will not allow operation of the x-ray tube at leakage technique factors (maximum kVp @ 3-5 mA).

Choosing an appropriate survey meter and correcting for the meter's characteristics is important. One must consider the chosen instrument's (31):

1. Energy response
2. Directional response
3. Intensity Response
4. Time constant when used in the rate mode
5. Chamber volume and cross sectional area
6. Precision
7. Accuracy and calibration

when making corrections. These problems are discussed further in the presentation entitled "X-ray Room Barrier Acceptance Testing" (32).

The following protocol is used to measure leakage radiation. A minimum of two leakage measurements are made on a radiograpic tube; a minimum of three are measured on a fluoroscopic tube. The two radiographic tube measurements are made 1 meter from the focal spot (measured with a tape measure) at the location of the input surface of the patient; one toward the patient's head and the other toward the patient's feet. The collimator blades are completely closed, but are not blocked by extra lead. (While this measurement may include some scatter radiation especially if the collimator does not completely close to zero field size, it is a better estimate of stray radiation to the patient than measurements with the collimator blocked by extra lead). The location of the additional fluoroscopic tube measurement is directed towards the location of the physician during an exam. If these measurements are significantly below 100 mR/hr, further measurements are not necessary. If measured radiation levels approach the 100 mR/hr level, additional measurements must be completed with the collimator blocked by extra lead.

Two people are required to collect the data. One positions and holds the instrument (wearing a lead apron) while the other person makes the exposure. All exposures are recorded in the integrate mode of the survey meter to minimize exposure times. A technique of maximum - 10 kVp, 100 mA, and 1 second is used with the large focal spot provided the instantaneous heat loading of the tube is not exceeded. The chamber of the survey meter is oriented to eliminate the need for directional response corrections. The instrument's precision is sufficient to avoid multiple exposures (excessive tube load). Equation (6) corrects the instrument reading to leakage radiation rate at 1 meter, at leakage technique factors, and at standard temperture and pressure.

$$LR(mR/hr) = R \times C \times ER \times DR \times IR \times AC \times 36I \times [(T + 273)/295] \times (760/P) \quad \text{---(6)}$$

where R = meter indication
C = calibration factor of meter = 1 +/- 5%
ER = energy response correction = 0.93 +/- 10%
DR = directional response correction = 1
IR = intensity correction = 1 +/- 5%
AC = chamber area correction = 1
I = leakage current at maximum kVp
T = room temperature (Centigrade)
P = barometric pressure (mm Hg)

The errors in the above parameters are small with respect to positioning errors of the survey meter. Therfore, one should assume the above calculation is accurate to +/- 20%.

Measured Performance

Focal Spot Blooming

Table IV lists the focal spot blooming performance of 40 focal spots measured for tubes supplied by 4 manufacturers collectively. The blooming values are calculated using Equations (7) and (8).

$$\text{Blooming} = 100(F_{max} - F_{std})/F_{std} \quad \text{---(7)}$$

Blooming = 100(Fmin - Fstd)/Fstd ---(8)

where Fmax = largest Feq
Fmin = smallest Feq
Fstd = Feq using 75 kVp, 1/2 rated mA

Fmax and Fmin are chosen from the 4 Feq measurements using the 4 techniques listed in Table III. All but 1 of the 40 focal spots measured complied with the +/- 25% blooming performance specified by MRHMC. (Three other focal spots exceeded the specification by 2%, but this is not considered significant due to measurement errors.) The data in Table IV shows no significant difference in blooming performance between the different manufacturers. Manufacturer C, however, failed to meet its unreasonable +/- 5% blooming specification listed in Table I.

Table IV
Blooming Data

	Manufacturer							
	A		B		C		D	
# Focal Spots Measured	12		8		8		12	
Maximum Blooming	+27	-18	+27	-25	+30	-18	+27	-16
Minimum Blooming	+12	-10	+14	-1	+9	-16	+7	-11
Average Blooming	+16	-14	+19	-16	+19	-18	+16	-12
1 Standard Deviation	5	6	7	7	7	5	6	3

Leakage Radiation

Table V lists the leakage radiation data measured from 20 x-ray tubes.

TABLE V
Leakage Radiation Performance
(mR/hr @ maximum continuous rated tube current)

	Manufacturer			
	A	B	C	D
# X-ray Tubes Measured	6	4	4	6
Maximum Leakage	6	10	21	5
Minimum Leakage	2	6	5	2
Average Leakage	3.8	7.3	14.3	3.7
Standard Deviation	1.8	1.9	7.6	1.0

All leakage radiation rates 1 meter from the focal spot in the direction of the patient from the x-ray tube housing were significantly below 100 mR/hr. No significant difference in the leakage radiation rate was measured between manufacturers A, B, and D. The leakage radiation rate from manufacturer C's radiographic tubes was significantly higher than the other manufacturer's tubes. This explains the large range between maximum and minimum leakage measurements for manufacturer C.

Fef and Feq

Figures 15 and 16 illustrate the correlation of the Fef and Feq measurements of 34 focal spots for the length and width dimensions respectively. Fef is plotted on the horizintal axis. Feq multiplied by the correction factor (Table I) used to estimate Fef from Feq is plotted on the vertical axis. The data could be fitted to a straight line to provide an "average" mathematical relationship between Fef and Feq. However, the scatter of the plotted data illustrates that one fitted line cannot accurately estimate the unknown parameter (Fef from Feq or vice versa) for an individual focal spot. Also, the data suggests that a characteristic relationship between Fef and Feq exists for each manufacturer.

Figure 17 illustrates the measured Fef values for 34 focal spots with respect to the maximum NEMA tolerance. The

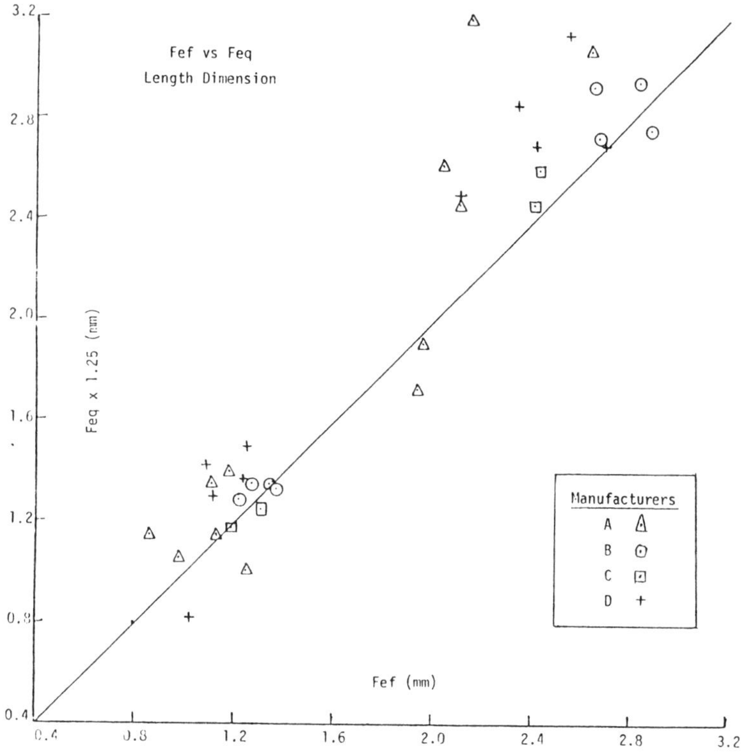

Figure 15: Effective (Fef) vs Equivalent (Feq) Focal Spot Size (Length Dimension)

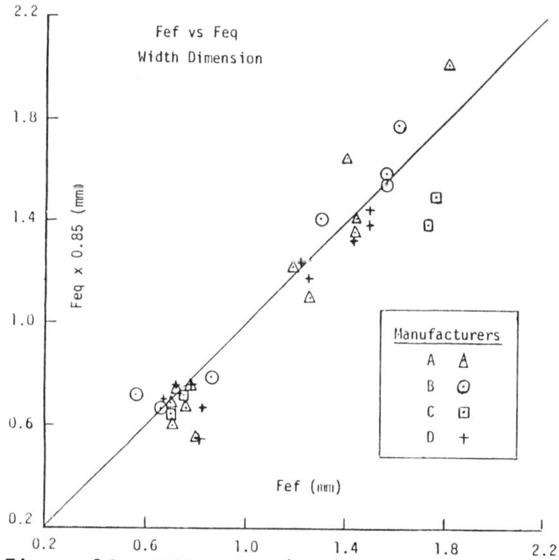

Figure 16: Effective (Fef) vs Equivalent (Feq) Focal Spot Size (Width Dimension)

length dimension is plotted above the width dimension for the same tube. All focal spots of the same nominal size are grouped together; the bottom scale labels the groupings. Three focal spots, two in the length and one in the width dimension were larger then the NEMA tolerance for Fef. (Five other focal spots exceeded the NEMA tolerances by less than 5%, but this was not considered significant due to measurement errors.) The two focal spots which exceeded NEMA tolerances in the length dimension were parallelograms instead of rectangles. The one focal spot which exceeded NEMA tolerances in the width dimension was powered by a falling load generator. Since this generator maintained the mA at levels exceeding 1/2 the maximum rated mA at 75 kVp (the NEMA specified technique), a portion of the measured focal spot size was attributable to blooming.

Conclusions

The tube performance data illustrates the different interests of the manufacturer and purchaser of x-ray tubes. As shown in Figure 17, the manufacturers design the focal spot size (Fef) to be within 80 to 100 % of the maximum NEMA tolerances in the length dimension and 70 to 90 % in the width dimension. This enables the manufacturer to provide the maximum instantaneous anode heat load capacity (kW rating) for the specified focal spot size. Thus, the manufacturer wishes to monitor and control Fef.

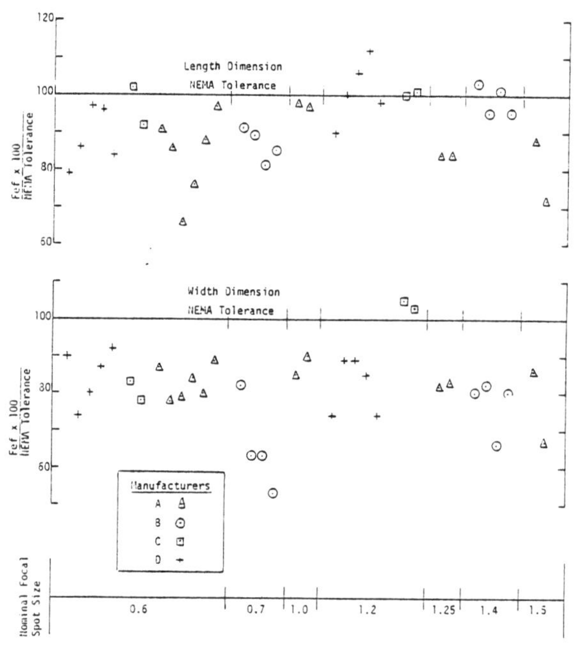

Figure 17: Effective (Fef) Focal Spot Size vs NEMA Tolerance

The purchaser is more interested in image quality and the resolving capacity of the focal spot. Since the test epuipment and computer required to measure resolving capacity from a pinhole image are not available at most field installations, a resolution test pattern image is required. However, existing standards do not define tolerances for the resolving capacity of x-ray tubes. The data in Figures 15 and 16 clearly verifies that the unique intensity distribution of each focal spot results in a unique correction factor between Fef and Feq. Thus, the NEMA tolerances for Fef cannot be scaled with an "average" correction factor to establish reasonable Feq performance levels.

Standards for focal spot acceptance testing must be changed. The current NEMA standard (4) speaks to the needs of the manufacturers, but neglects the concerns of the radiologist. The issues of:

1. Fef and its relationship to the kW rating of the tube
2. Feq and its relationship to the resolving capacity of the focal spot

are separate issues and should be specified separately. The specification for each tube should list the maximum actual Feq and the kW rating. The revised standard should also specify tolerances for Feq growth as a function of tube mA and kVp. These changes would allow the purchaser to eliminate Fef measurements which are time consuming and require excessively large mAs techiques. Current standards (16) for leakage radiation appear to be adequate.

The data presented in this study provides a performance specification of blooming and Feq. 98% of the focal spots tested passed a blooming specification of +/- 25% with respect to Feq measured at 75 kVp and 1/2 maximum rated mA. The Fef of 31 focal spots in this study met NEMA tolerances. The Feq of these 31 focal spots averaged (1.63 +/- 0.18) x nominal size in the length dimension and (1.32 +/- 0.13) x nominal size in the width dimension. Since one statistically expects 95% of these focal spot's Feq to fall within 2 standard deviations of the average value, an Feq performance specification of 2.0 x nominal size (length dimension) and 1.6 x nominal size (width dimension) is reasonable.

References

1. Hendee W.R., Rossi R.P: Performance Specifications for Diagnostic Radiological Equipment: A Delicate Interface Between Purchaser and Supplier. Proceedings of the Society of Photo-Optical Instrumentation Engineers 70:238-241, Sept 1975.

2. Joseph P.M: The Role of Hospital Physicists in Performance Specifications. Proceedings of the Society of Photo-Optical Instrumentation Engineers 70:244-247, Sept 1975.

3. Milne E.N.C., Roeck W.W: Determinants and Indices of Focal Spot Performance - The Concept of Three Dimensionality. Proceedings of the Society of Photo-Optical Instrumentation Engineers 56:124-133, Oct 1974.

4. National Electrical Manufacturers Association: Measurement

of Dimensions of Focal Spots of Diagnostic X-ray Tubes. No. XR5-1974.

5. Wagner R.F., Weaver K.E., Denny E.W., Bostrom R.G: Toward a Unified View of Radiological Imaging Systems. Medical Physics 1:11-24, Jan-Feb 1974.

6. Mahn G.R: X-ray Tubes. AAPM 1975 Summer School Proceedings: The Expanding Role of the Diagnostic Radiological Physicist, 1-25, Jl 1975.

7. Bernstein F: Specification of Focal Spots and Factors Affecting Their Size. Proceedings of the Society of Photo-Optical Instrumentation Engineers 56:159-163, Oct 1974.

8. Bergeron T.R: Manufacturers' Designation of Diagnostic X-ray Tube Focal Spot Size: A Time for Candor. RADIOLOGY 111:487-488, May 1974.

9. Robinson A., Grimshaw G.M: Measurement of the Focal Spot Size of Diagnostic X-ray Tubes - a Comparison of Pinhole and Resolution Methods. British Journal of Radiology 48:572-580, 1975.

10. Bernstein H., Bergeron T., Klein D.J: The Relationship of Radiant Intensity Distribution of Focal Spots to Resolution. RADIOLOGY 111:427-431, May 1974.

11. Rao G.U.V: Effective Focal Spot Size. Letter to the Editor. Investigative Radiology 7:352-353, Jl-Aug 1972.

12. American Standards Association: Federal Standard 83, 1956.

13. Mattsson O: Focal Spot Variations with Exposure Data - Important Factors in Daily Routine. Acta Radiologica (Diagnosis) 7:161-169, 1968.

14. Chaney E.L., Hendee W.R: Effects of X-ray Tube Current and Voltage on Effective Focal Spot Size. Medical Physics 1:141-147, May-June 1974.

15. Berstein H., Bergeron T.R., Klein D.J: Routine Evaluation of Focal Spots. Radiology 111:421-425, May 1974.

16. U.S. Department of Health, Education, and Welfare: Regulations for the Administration and Enforcement of the Radiation Control for Health and Safety Act of 1968. Bureau of Radiological Health, Sept 1978.

17. Rao G.U.V., Bates L.M: Effective Dimensions of Roentgen Tube Focal Spots Based on Measurement of the Modulation Transfer Function. Acta Radiologica (Therapy) 9:362-368,

Aug 1970.

18. Rao G.U.V: A New Method to Determine the Focal Spot Size of X-ray Tubes. American Journal of Roentgenology 3:628-633, March 1971.

19. Prasad S.C., Hendee W.R., Carson P.L.: Intensity Distribution, Modulation Transfer Function, and the Effective Dimension of a Line-focus X-ray Focal Spot. Medical Physics 3:217-223, Jl-Aug 1976.

20. International Commission on Radiation Units and Measurements: Report #10f - Methods of Evaluating Radiological Equipment and Materials, 1963.

21. Kuntke D.H: On the Determination of the Roentgen Tube Focal Spot Sizes by Pinhole Camera Roentgenography. Acta Radiologica 47:56-64, 1957.

22. Gray J.E., Trefler M: Pinhole Cameras: Artifacts, Modifications, and Recording of Pinhole Images on Screen-Film Systems. Radiologic Technology 52:277-282, Nov - Dec 1980.

23. Rossi R.P: Private Communication.

24. Swearingen F.L.V., Watts F.C: Importance of Magnification in Star Test Pattern Use. Medical Physics 8: 236-238, Mar-Apr 1981.

25. Spiegler P., Breckinridge W.C: Imaging of Focal Spots by Means of the Star Test Pattern. Radiology 102:679-684, March 1972.

26. Rao G.U.V: The Importance of Focal Spot Characteristics in Clinical Radiography. Proceedings of the Society of Photo-Optical Instrumentation Engineers 47:214-217, Aug 1974.

27. Burgess A.E: Focus Measurements Using the Star Test Pattern. Proceedings of the Society of Photo-Optical Instrumentation Engineers 96:68-75, Sept 1976.

28. Behof A: Private Communication.

29. Gray J.E., Capp M.P., Whitehead F.R: An Improved Technique for X-ray Image Evaluation: The Two Dimensional MTF. Investigative Radiology 9:252-261, Jl-Aug 1974.

30. National Council of Radiation Protection and Measurements: Report No. 33 - Medical X-ray and Gamma-ray Protection for Energies up to 10 Mev. Equipment Design and Use. Feb 1968.

31. National Council of Radiation Protection and Measurements:

Report No. 57 - Instrumentation and Monitoring Methods for Radiation Protection. May 1978.

32. Strauss K: "X-ray Room Barrier Acceptance Testing", in this Proceeding.

33. Gray J: Panel Presentation: X-ray Tube Focal Spots. Proceedings of the Society of Photo-Optical Instrumentation Engineers 56: 164-171, Oct 1974.

34. Nielsen B: A Source of Systematic Error When Measuring the Focal Spot Sizes of Roentgen Tubes with Pinhole Cameras. Health Physics 36: 726-727, June 1979.

AN ALIGNMENT TESTING DEVICE FOR COMPLIANCE TESTING OF PBL AUTOMATIC COLLIMATORS

Pei-Jan Paul Lin
Department of Radiology
Northwestern University Medical School, Chicago, Illinois 60611

and

Robert J. Kriz and Ruben Stortzum
Department of Radiology
University of Illinois Medical Center, Chicago, Illinois 60612

I. Introduction

In order to correctly align the radiation beam to the image receptor, a substantial amount of time and effort is required on the part of the service engineer to complete the installation of collimators in radiological imaging equipment. In an effort to reduce unnecessary radiation to patients, it is mandated that all new x-ray equipment be installed with a positive beam limiting device (PBL) so that the radiation field size is collimated to the size of the film employed for the radiological examination (ref. 1).

To maintain the PBL automatic collimator system in good working condition, various collimator testing tools have been made available in the commercial market (ref. 2, 3, and 4) as routine quality assurance instrumentation. The existing collimator testing devices, however, are not designed to efficiently evaluate the performance of the PBL collimator systems, and needless to say they offer little help in the initial set-up of the PBL collimators.

In view of this circumstance the authors have elected to design a test device that is specifically geared to evaluate the congruence and alignment of the radiation field with the image receptor(s) selected for the PBL automatic collimators. It should be pointed out, however, that this device is an improved version of a collimator testing device that has been described by the author previously (ref. 5).

II. Description Of The PBL Collimator Evaluation Device

The PBL collimator evaluation device consists of (1) the fluorescent screen fixture, (2) a 8"x10" cassette, (3) a copper ruler, and (4) two sets of extension legs. The details of each item appear in the following sections of this paper.

II-A The Fluorescent Screen Fixture

The fluorescent screen fixture is the main body of the PBL collimator evaluation device, and is depicted in Figures 1, 2, and 3. The fluorescent screen fixture is essentially constructed with two identically built templates as shown in Figure 1. Two such templates are joined together with two hinges as shown in Figure 2. The top template of the fluorescent screen fixture differs from the bottom template in two respects; (1) a fluorescent screen such as the du Pont E-2 fluorescent screen is attached to the front of the template, and (2) a plastic sheet is placed on top of the fluorescent screen which is scribed with lead markings and lead numbers. The lead markings and the lead numbers are arranged as shown in Figure 3. The plastic blocks are installed on both templates to serve as spacers so that a 8"x10" cassette can be placed securely inside the templates. When the cassette is to be placed in the bottom template, the hinges will let the top template swing out of the way while maintaining the alignment for proper evaluation of the PBL systems.

II-B. The 8"x10" Cassette

A 8"x10" cassette is required for proper recording of the radiation field size and documentation when necessary. This cassette is installed with only one detail (or slow) screen so that the exposure to x-rays will yield appropriate optical density for proper viewing as this film will be exposed to x-rays simultaneously with the film cassette placed in the spot film compartment or in the Bucky cassette tray. Since the film obtained from this cassette is used for the checking of radiation field size, an old used cassette may be employed for this purpose. However, if a phototiming cassette happens to be available, the lead foil backing should be removed from the cassette.

II-C. The Copper Ruler

The copper ruler in conjunction with the fluorescent screen fixture enables one to perform the measurement of source-to-tabletop distance of a standard type fluoroscopic x-ray equipment. The copper ruler is made of two pieces of plastic sheets with copper wires embodied between them as depicted in Figure 4. The copper wire employed in the ruler should be 1 mm or less in diameter.

II-D. The Extension Legs

There are two sets of extension legs required. The ½" legs are utilized when a standard type fluoroscopic equipment is evaluated. The 20" extension legs are required when a radiographic over-head type x-ray system is evaluated. The extension legs provide appropriate distance and physical support when the fluorescent screen fixture is positioned away from the examination tabletop.

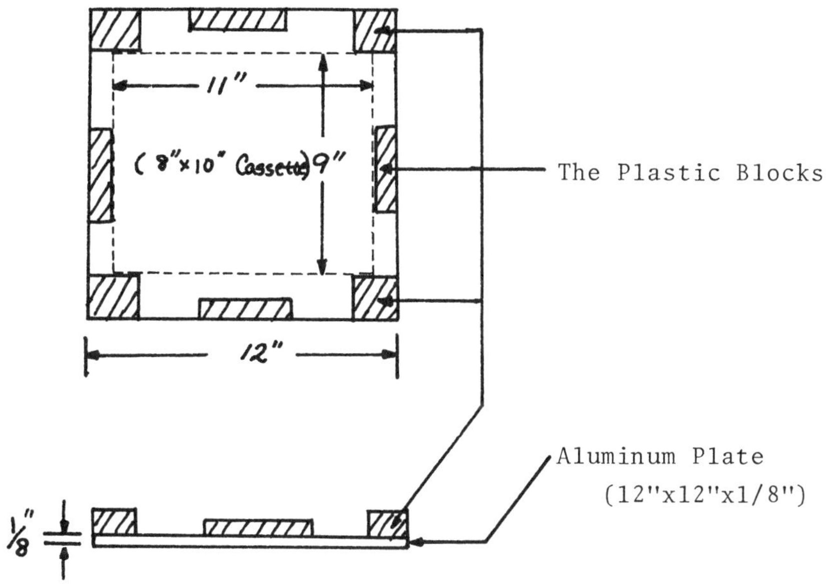

Figure 1. The Template For Fluorescent Screen Fixture

The aluminum plate serves as the physical support for the fluorescent screen fixture. Plastic blocks are attached to the aluminum plate as spacers so that a 8"x10" cassette can be placed inside of the template in two orientations. The aluminum plate serves also as attenuating material or as a phantom when the fluorescent screen fixture is placed in the radiation beam.

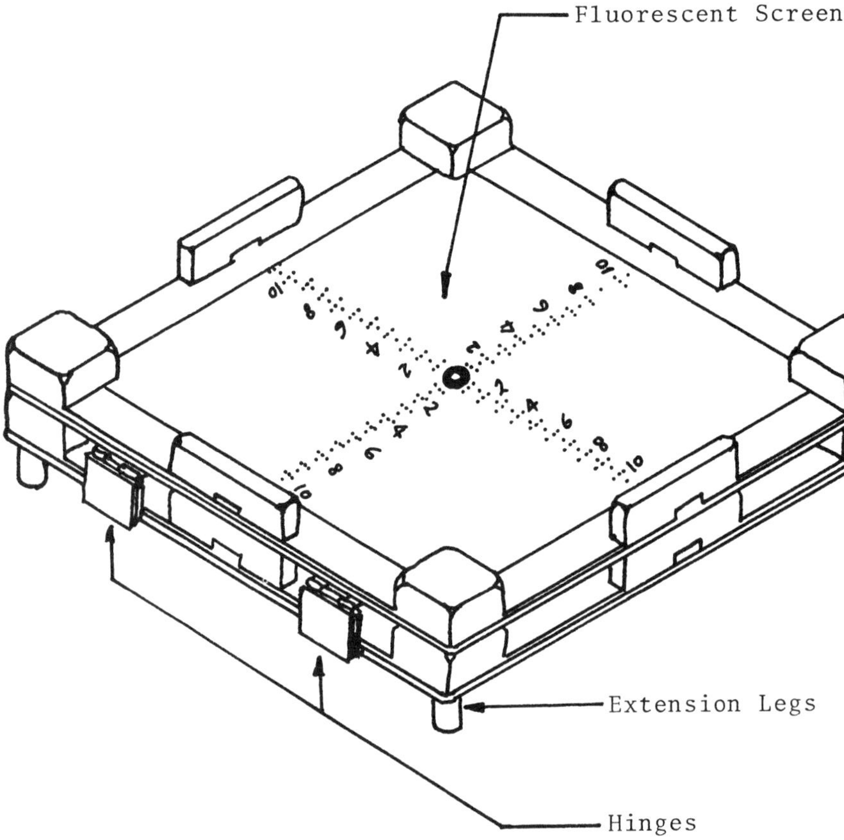

Figure 2. The Fluorescent Screen Fixture

The fluorescent screen fixture is shown here with the 1/2" extension legs. The top and bottom templates are essentially identical with two exceptions described in the main text. A 8"x10" cassette may be placed in the top template or in the bottom template when necessary. The hinges provide the access to the bottom template.

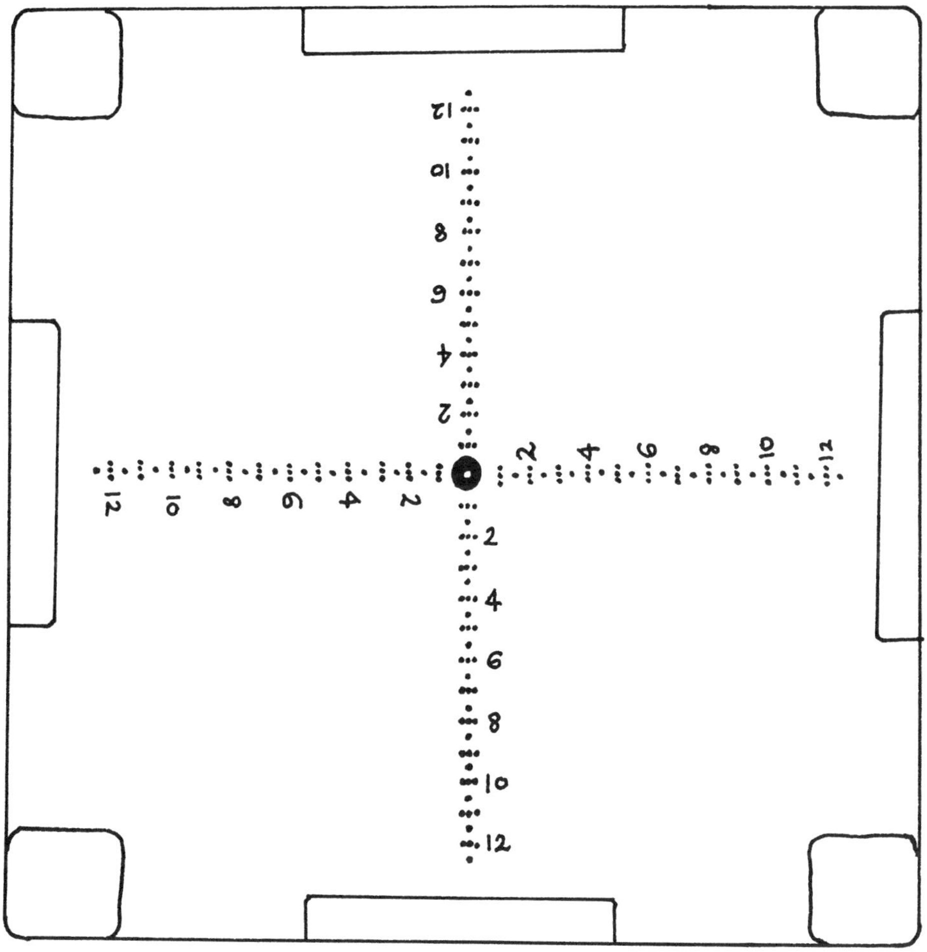

Figure 3. The Plastic Top Cover

The lead markings are equally spaced in 5 mm intervals. Note that the distance between two successive sets of three-lead dots is 10 mm. The lead numbers indicate the distance from the center to the lead dots. The Lead Ring in the center of the plastic top cover has a dimension of 5.0 mm I.D., and 10.0mm O.D.

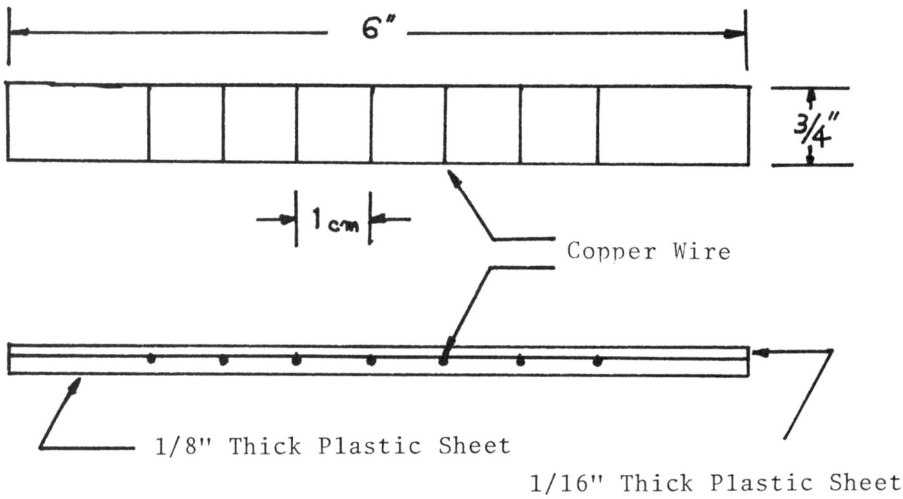

Figure 4. The Copper Ruler.

The copper wire should be 1 mm in diameter or less. The copper ruler has dimensions of 6"x3/4"x3/16". The copper wires are spaced equally with 1 cm intervals.

III. Operational Procedure

The use of the PBL collimator evaluation device is described for a standard type fluoroscopic unit in which the x-ray tube is located beneath the examination table and a radiographic x-ray system with the x-ray tube suspended from the ceiling. Appropriate procedural modifications are necessary to accommodate the differences in geometry and/or the image receptor locations for other systems.

III-A. Fluoroscopic Equipment

The fluorescent screen fixture is employed with the ½" legs so that the fixture can be placed on the examination table with the fluoroscopic tower pulled in for the normal operation of the equipment as depicted in Figure 5.

III-A-1. Alignment

The alignment of the fluoroscopic imaging chain can be evaluated with relative ease by using this fluorescent screen fixture. The mechanical center of the image intensifier is physically aligned with the center of the spot filming device by design. Therefore, one can assume that the center marking of the spot filming device at the bottom of the fluoroscopic tower is the center of the image intensifier. A small metalic washer is attached to this center with a piece of adhesive tape for the purpose of locating the center.

The fluorescent screen fixture is placed on the tabletop in the field of fluoroscopic view. With the fluoroscopy turned "on", position the fluorescent screen fixture in the center of the radiation field by visually looking at the fluorescent screen with the aid of the lead markings and the lead numbers.

For a fluoroscopic imaging chain with good alignment, the television monitor should show the magnified image of the lead ring (in the fluorescent screen fixture), and the image of the metalic washer (attached to the center of the spot filming device) in the center of the television monitor field.

The alignment should be kept well adjusted when the fluoroscopic tower was moved from the lowest height to the maximum height provided to the vertical travel of the fluoroscopic tower. This vertical travel is the TID (tabletop-to-image intensifier distance) shown in Figure 5.

The congruence of the radiation field with the image intensifier active input phosphor size can be verified by simply observing the radiation field size presented at the tabletop by the test device, and the image presented at the television monitor. The lead numbers and the lead markings are employed to identify the field sizes. The congruence of the radiation field with the image intensifier input phosphor size should be

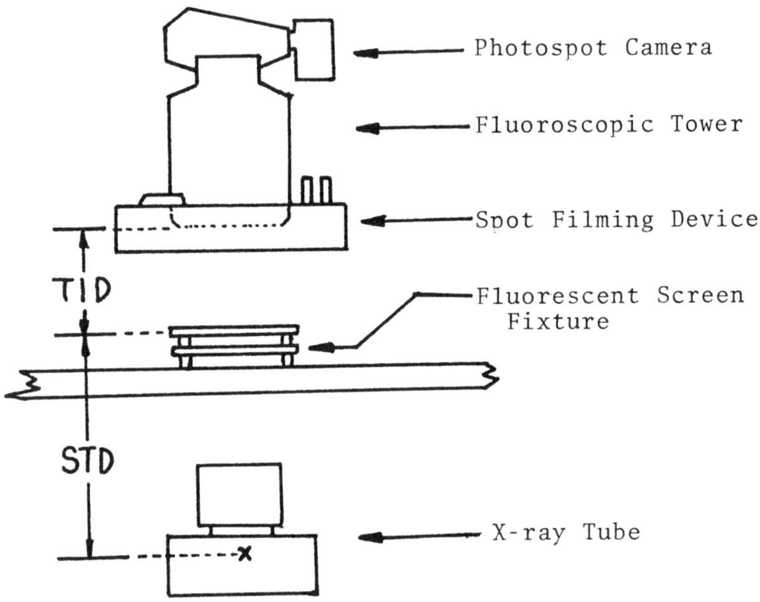

Figure 5. Schematic Arrangement For Fluoroscopic PBL Collimator Evaluation.

TID: Tabletop-to-image intensifier distance
STD: Source-to-tabletop distance

evaluated for all operational modes provided to the image intensifier; i.e. all imaging modes of multiple mode image intensifier. The TID should be varied to assure that the congruence is kept well within the tolerance specified by the manufacturer.

The congruence of the radiation field with the spot filming device can be evaluated in the same manner as for that of the image intensifier. However, since it is difficult to view the radiation field size during spot filming exposure, the the 8"x10" cassette is employed to verify the radiation field size at the tabletop. Prior to the spot filming exposure the cassette is placed inside the top template of the fluorescent screen fixture. This evaluation procedure should be applied to all types of cassettes if the spot filming device is designed to accommodate more than one type of cassette. The spot filming device is normally designed to record several images on one piece of film. Hence, this verification test should be conducted for the split spot filming modes as well. Two examples of this testing are depicted in Figures 6 and 7.

III-A-2. Source-To-Tabletop Distance

The STD (source-to-tabletop distance) can be determined by using the copper ruler shown in Figure 4, and the fluorescent screen fixture. The copper ruler is attached to the bottom of the spot filming device near the center. The fluorescent screen fixture is placed on the tabletop as shown in Figure 5. With the fluoroscopy turned "on", obtain an image as depicted in Figure 8 of the television monitor by maneuvering the fluoroscopic tower.

The intention of this maneuver is to obtain a 2X magnification of the fluorescent screen plate image at the plane of the copper ruler. When this condition of 2X magnification is achieved, the TID is equal to the STD by the rule of similar triangles. Hence, the measurement of the TID yields the STD. This is required to be at least 38 cm (15") on stationary fluoroscopic equipment.

III-B. Radiographic Equipment

The geometric arrangement in the evaluation of the PBL automatic collimation device for radiographic equipment is similar to that of the fluoroscopic spot filming situation. However, since the radiation source is now located above the fluorescent screen fixture and the maximum size of film is 14"x17", some modifications of the evaluation arrangement and procedure are necessary.

Depicted in Figure 9 is a typical arrangement on the use of the PBL collimator evaluation device. The 8"x10" cassette is placed in the bottom template of the fluorescent screen fixture. In order to provide an appropriate magnification of the template to be imaged on film placed in the table Bucky cassette holder,

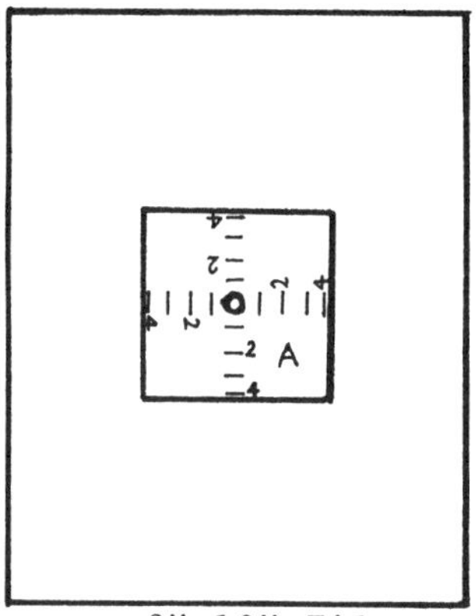

8"x10" Film
(For Field Size
Verification)

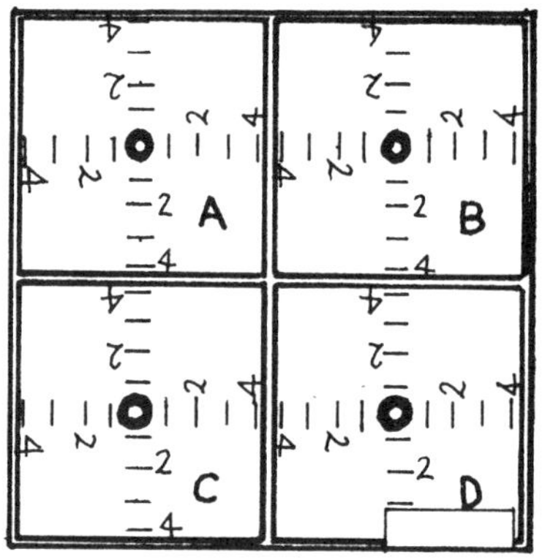

9.5"x9.5" Film
(The Spot Film from the
Spot Filming Device)

Figure 6. The Congruence Test Example.

The radiation field size at the tabletop is 4 cmx 4 cm, and the spot film shows the images correctly aligned. The lead letters, A,B,C, D are identifications of the four successive exposures for the 4 on 1 split spot filming.

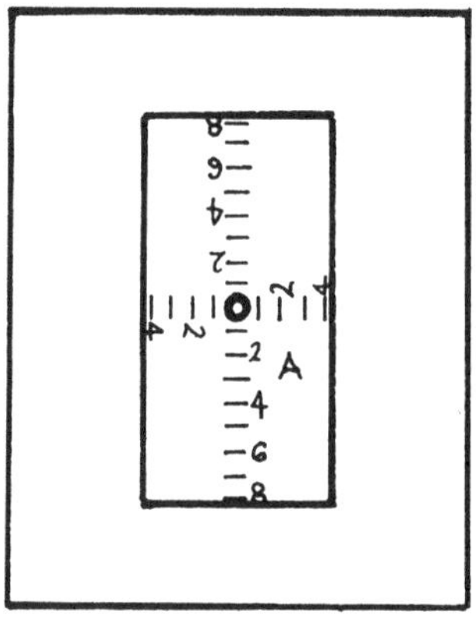

8"x10" Film
(For Field Size
Verification)

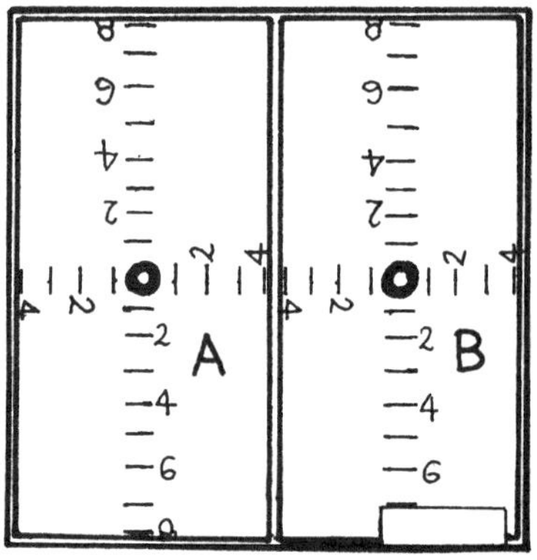

9.5"x9.5" Film
(The Spot Film from the
Spot Filming Device)

Figure 7. The Congruence Test Example.

The radiation field size at the tabletop is 4 cm x 8 cm, and the spot film shows the images are correctly aligned. The lead letters, A and B are identifications of the two successive exposures for the 2 on 1 split spot filming.

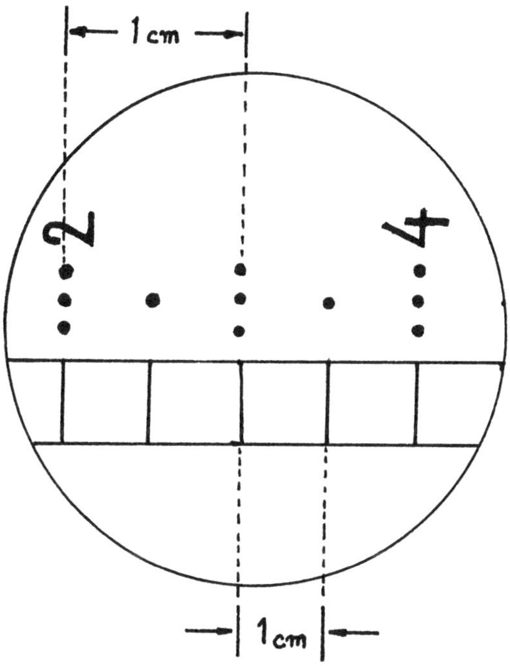

Figure 8. The Image On The Television Monitor For The Measurement of STD.

The copper wire interval is 10 mm (or 1 cm), and the interval between two successive three-lead-dots is 10 mm (or 1 cm) as well. Hence, the image shown here indicates that the fluorescent screen fixture image is magnified by a factor of 2.

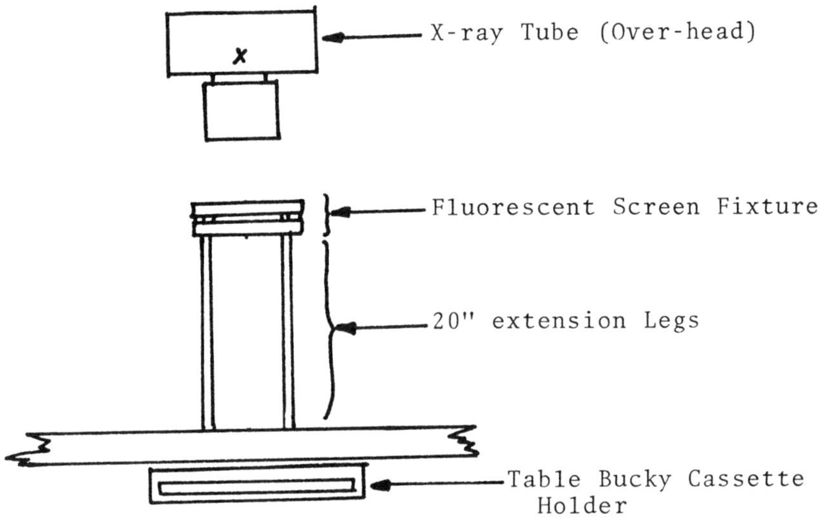

Figure 9. Schematic Arrangement In The Evaluation Of Radiographic PBL System.

the 20" extension legs are employed to provide the geometry required.

A few different types of cassettes should be placed in the table Bucky cassette holder to evaluate the congruence of the radiation beam with the cassette size under the evaluation. The 8"x10" cassette in the fluorescent screen fixture and the cassette in the table Bucky cassette holder are simultaneously exposed. Two films are obtained; the 8"x10" film records the size of the radiation field, and the film obtained from the table Bucky cassette holder demonstrates the portion of the radiation field captured at the image plane similar to those shown in Figures 6 and 7.

The congruence of the radiation field with the film selected and placed in the table Bucky cassette holder is then evaluated by comparing these two pieces of film as before. For a given cassette, there are two possible orientations in the manner the cassette is placed inside the Bucky cassette holder. Therefore, it is necessary to evaluate the congruence and alignment in both orientations.

When the image receptor is a wall cassette holder, the PBL collimator evaluation device must be positioned on its side with appropriate physical support. The geometry in this case can produce a 2X magnification or higher.

IV. Photospot Cameras

The alignment and the congruence of radiation beam with the image intensifier can be evaluated with the fluorescent screen fixture as described in the previous sections. Upon completion of such evaluations, one can proceed to check the alignment of the image intensifier output phosphor with the photospot cameras such as 100/105 mm camera and 35 mm cine fluorographic camera.

The arrangement of the geometry and the fluorescent screen fixture remain the same, but a photospot filming exposure is made. The fluoroscopic tower, and the x-ray tube support column of a remote control type fluoroscopic unit tend to be misaligned as the table is positioned in its up-right/vertical orientation. The alignment of the radiation field with the image intensifier should be checked both in the horizontal and vertical positions. The photospot camera alignment with the image intensifier output phosphor should subsequently be verified.

V. Summary

A device specifically designed to evaluate the PBL automatic collimators for fluoroscopic and radiographc equipment is described. A few examples are given with the operational procedure of the device. It is hoped that this device will become available through a commercial firm shortly, as the efficiency and ease of operation of this device, in the author's opinion, surpass most test tools available on the commercial market.

References

1. Regulations for the administration and enforcement of the radiation control for health and safety act of 1968. U.S. Department of Health, Education, and Welfare, Public Health Service, Food and Drug Administration.

2. Beam Alignment Test Tool, Model 162A, Radiation Measurements Inc. P.O. Box 44, Middleton, Wisconsin 53562.

3. Fluoroscopic Beam Alignment Device, Model 07-600 Nuclear Associates, Division of Victoreen, Inc., 100 Voice Road Carle Place, New York 11514.

4. X-ray Equipment Test Stand, The Machlett Laboratories, Inc. 1063 Hope Street, Stamford, Connecticut 06907.

5. P.P. Lin, "A device for checking alignment of fluoroscopic systems", Optical Engineering Vol. 16, No. 3, 302-304, 1977.

Acceptance Testing of Electromechanical Interface
For
Vascular Angiographic Imaging System

Pei-Jan Paul Lin and Harvey L. Neiman
Department of Radiology, Northwestern University Medical School
and
Department of Diagnostic Radiology, Northwestern Memorial Hospital
Chicago, Illinois 60611

Abstract

The electromechanical interfaces among various radiological imaging components and with their ancillary devices are important in carrying out a variety of tasks under a prescribed and organized manner. The orderly operation of all major components and ancillary devices ensures that the radiological examinations are properly performed to obtain required information for diagnosis. The acceptance testing of electromechanical interfaces is described.

Introduction

Suppose all major components in a given radiological examination room meet with all the requirements and performance specifications; yet the total operation of the examination room can be in a catastrophic status if the interfaces among these components are not correctly formulated. The acceptance testing of major components are designed to eliminate or minimize such difficulties.

In the case of special procedures' equipment, the electromechanical interfaces play a major roll in accomplishing the final goal of successfully completing the examination. More specifically, in the vascular angiographic imaging system the operation of the contrast agent injector, the rapid film changer, the generator/x-ray tube chain, and the motion of the examination table must be carried out in a organized and orderly fashion.

The complexity of the electromechanical interfaces becomes even more intricate when a bi-plane operation of the generators is utilized. The verification of such a bi-plane vascular angiographic equipment operation is described.

Acceptance Testing

To perform the acceptance testing, one must first understand how the entire system is designed to work. Although the acceptance testing described in this article is based on the vascular angiographic examination suite, parts of the description may be applicable to other radiological equipment with minor modifications. As long as the system operation is identified, the acceptance testing of all interfaces can be arranged and taylored to the specific radiological equipment to be tested.

The Bi-plane Vascular Angiographic Equipment

Depicted in Figure 1 is the schematic illustration of a typical bi-plane vascular angiographic system consisting of two generator/x-ray tube power trains, two automatic film changers, one power injector, two automatic film changer programmers and one tabletop motion programmer. The AP generator may be equipped with a stepping device for x-ray tube potential and up to four steps can be preprogrammed, resulting in a total of five different tube potentials that can be selected in a required manner. This feature is useful in the angiographic examination of lower extremities. Depending upon the angiographic studies to be performed, the bi-plane arrangement may not be necessary, or the tabletop motion may not be required. For illustraion purposes, however, all operational logics of the system will be employed to demonstrate the interfaces. Although, a bi-plane arrangement is not utilized for the angiographic studies of lower extremities, the equipment is set-up for bi-plane operation so that the interfaces can be evaluated simultaneously. Because of the selection in the use of 14" cut film, the tabletop must be shifted to move the patient during an angiographic run. Let's assume that the system is prepared and ready to start the examination, the power injector is filled with 92 cc of contrast agent and set at an injection rate of 8 cc/sec with a 3-second delay between the START initiation and the activation of the injection. Both automatic film changers are fully loaded and all other control factors such as x-ray tube potential, current, and exposure time, are already properly set. The automatic film changer programmers are pre-programmed with a punch-card to initiate x-ray exposures in alternating firing - in the following sequence:

1. 1 exposure/sec for 3 seconds for the AP plane.
2. 1 exposure/sec for 3 seconds for the LAT plane in alternating firing sequence.
3. delay 2 seconds and the tabletop motion is activated in the meantime.
4. repeat process 1, 2, and 3.
5. repeat process 4.
6. 1 exposure/sec for 1 second for the AP plane then the LAT plane.
7. delay 2 seconds (no tabletop motion).
8. repeat process 6.
9. end of operation.

Depicted in Figure 2 is a graphic illustration for the angiographic equipment operations of x-ray tubes, automatic film changers, the tabletop motion, and the power injector. The capital letters A through F denote:

A. The automatic film changer operation in the AP plane.
B. The anode rotation and the x-ray pulses in the AP plane.
C. The automatic film changer operation in the LAT plane.
D. The anode rotation and the x-ray pulses in the LAT plane.
E. The motor driven tabletop motion.
F. The power injector operation.

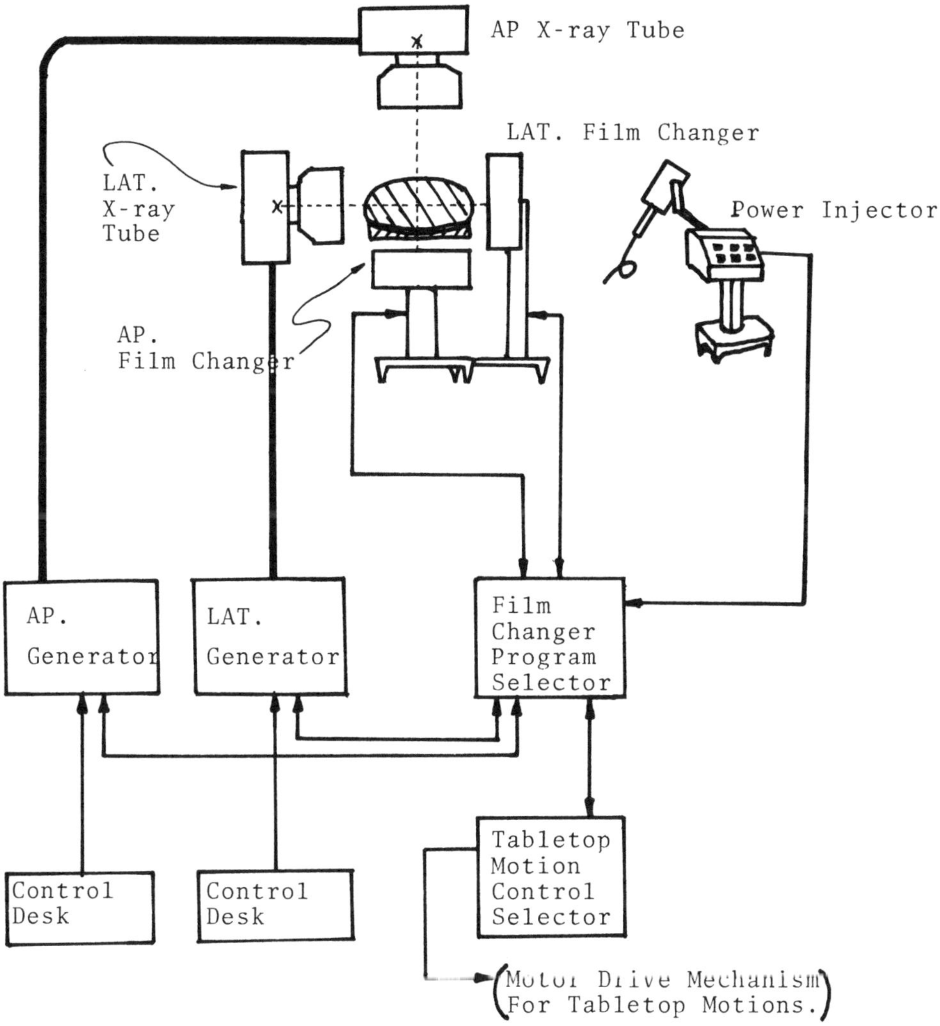

Figure 1. Schematic Illustration Of Bi-plane Vascular Angiographic Equipment And Operational Logics.

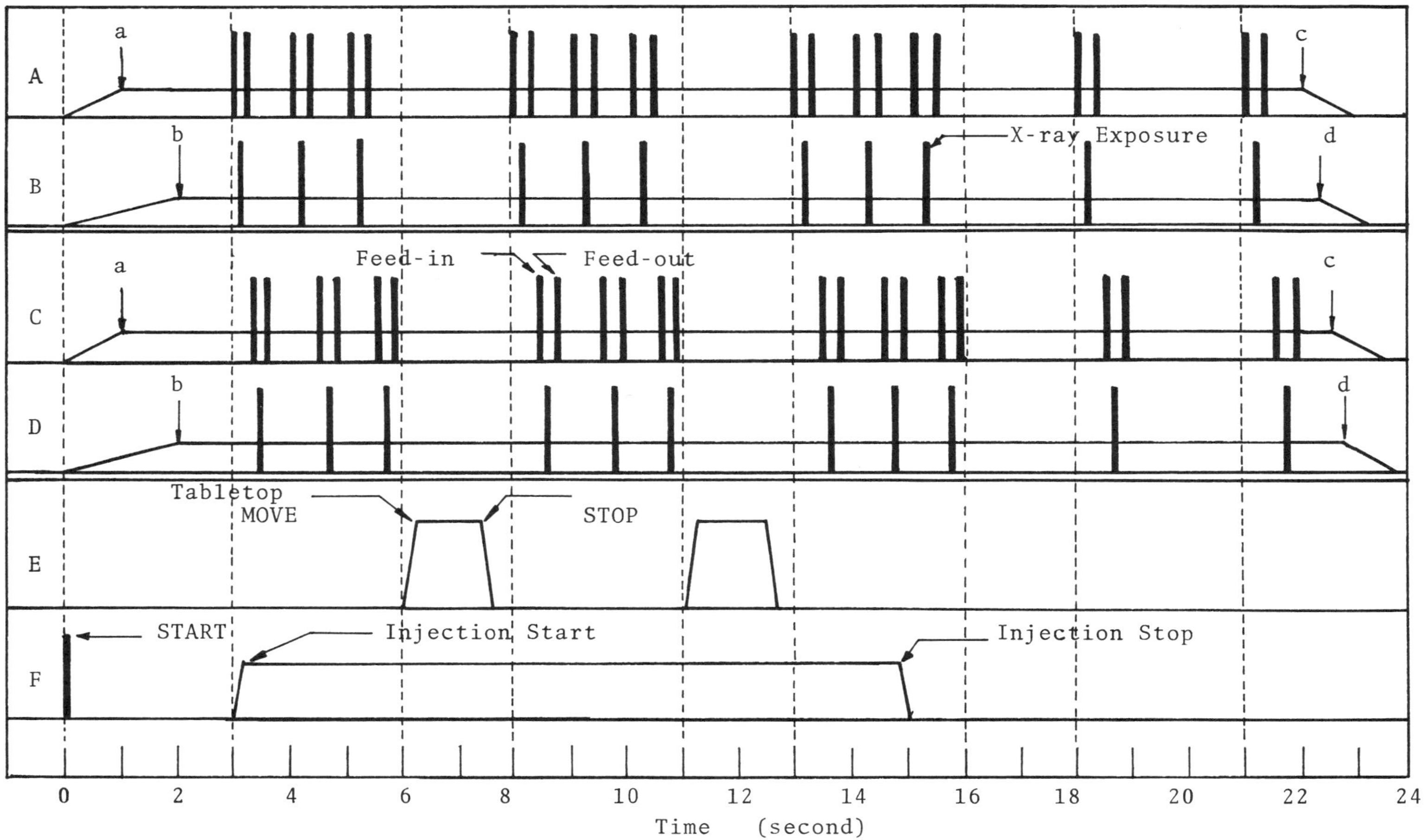

a: Automatic film changer motor at full speed, b: Rotating anode at full speed,
c: Automatic film changer motor power off, d: Rotating anode brake on.

Figure 2. Graphic Illustration Of Angiographic Equipment Operation.

In accordance with the program described, the sequence of the entire operation can now be demonstrated. At time T = 0 second the power injector start switch is depressed; a signal is sent to the automatic film changer program selector (see Figure 1). The power injector (F in Figure 2) is programmed to start injection after 3 seconds of delay and at 8 cc/sec injection rate. The automatic film changers (depending on the design) activate the motors, and approximately in one second the motors reach their operation speed and stand ready to transport the film. The anodes of x-ray tubes are activated and approximately in two seconds reach their maximum rotation speed and are ready to make x-ray exposures.

At Time T = 3 seconds, the AP automatic film changer activates the transport mechanism to bring the first film into the proper location between the intensifying screens. As the film is properly sandwiched between the screens (the film is now stationary), the relay in automatic film changer signals the AP generator to FIRE via the film changer programmer. As soon as the x-ray FIRE is completed the automatic film changer ejects the first film out to the receive magazine and the AP automatic film changer programmer switches the operation activation signal to the LAT automatic film changer. Similarily, the LAT automatic changer then repeats the process described above.

Therefore, at Time T= 6 seconds, 3 exposures have been taken for each plane at a rate of 1 exposure per second. The automatic film changer programmer signals the tabletop motion control and the patient is now shifted along with the tabletop by the motor drive mechanism according to the pre-selected amount of travel dialed on the tabletop motion control selector (see Figure 1).

The tabletop may be shifted for a total travel of 12" in one shift within 2 seconds including a time lag for the patient to stabilize from the vibration due to the transport of the tabletop. As the tabletop stops after the prescribed displacement, the tabletop motion control selector disengages an interlock mechanism that prevents the x-ray generators to FIRE through the automatic film changer programmers. This is done in order to prevent any x-ray exposures while the tabletop is in motion.

The operation of the entire system, thereafter, is self-explanatory. Notice that the power injector continues to inject for 11.5 seconds and ceases to operate automatically. The tabletop is shifted twice, only. At Time approximately T = 22 seconds, the x-ray tube anode rotation breaks are applied and the anodes stop to rotate within a second or two. It should be emphasized that the operation described is a typical example and other operational logics can be wired and/or programmed in a different manner. The power to the motors of the automatic film changers are also turned off at this time.

Verification of the Operation

To verify or perform the acceptance testing of the interface logic(s), substantial preparation is necessary, and depends, also, on how the verification is to be achieved. The x-ray exposures can be verified with a solid state detector attached to the collimator in the center of the primary radiation field. A high voltage divider may be employed to monitor the x-ray tube potential wave form in place of the solid state detector. The operation of the automatic film changer can be verified by monitoring the solenoid voltages for the feed-in, and

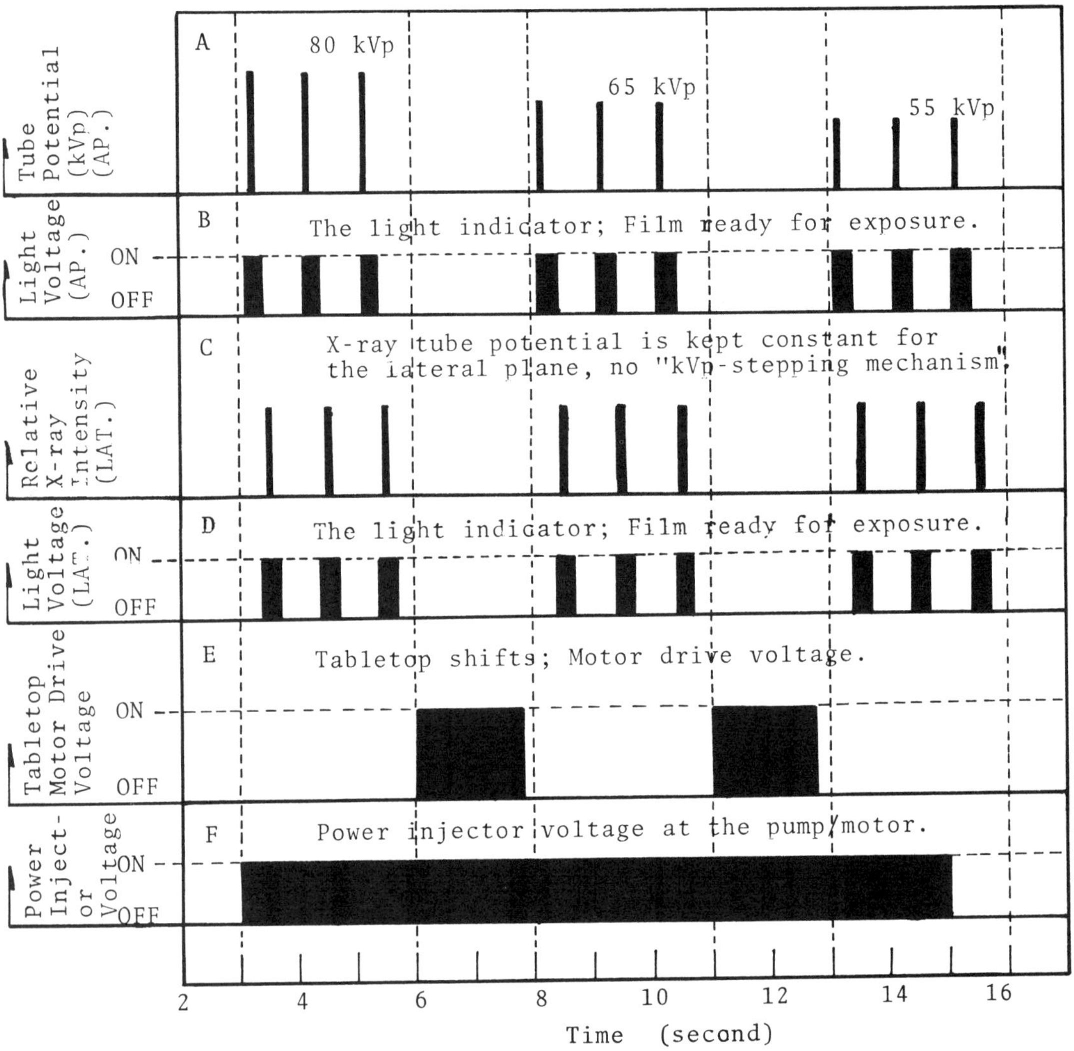

The wave forms for all physical factors under the monitoring can be displayed to observe the expanded wave patterns if so desired on the storage oscilloscope. The wave forms would look like the patterns illustrated above when the oscilloscope is made to display a relatively long time scale in one scan.

Figure 3. Examples Of Waveforms Observed In Acceptance Testing. (Redrawn and Synthesized from actual data and displays for the purpose of illustration.) For the time duration at T= 2 seconds to T= 17 seconds only.

the feed-out operation or the voltage supplied to the light indicator on the programmer.

Depicted in Figure 3 are the waveforms obtained during a typical acceptance procedure and the physical quantities being monitored are listed in Table I.

TABLE I PHYSICAL QUANTITIES MONITORED DURING ACCEPTANCE TESTING

Item	Description
A.	The x-ray tube potential waveform for AP plane.
B.	The voltage applied to the light indicator on the automatic film changer programmer for AP plane. When the light is ON the film is in place for exposure.
C.	The x-ray intensity waveform for LAT plane (a solid state detector is employed in place of the high voltage divider system).
D.	The voltage applied to the light indicator on the automatic film changer programmer for LAT plane.
E.	The voltage applied to the motor for the tabletop shift.
F.	The voltage applied to the power injector motor.

All pertinent operational factors are set the same as that for Figure 2.

Comparison of Figures 2 and 3 verifies that the operation is correctly performed. The x-ray tube potential waveform for the AP plane varies according to the x-ray tube potential stepping control. It should be pointed out that Figure 3 is synthesized from several sets of the same operation for illustration purposes.

Discussion and Conclusion

Monitoring of the operations for various subcomponents can be performed with numerous different approaches. The tasks here are to verify the system operation in which all subcomponents and accessories are working in a logical order and in concert with each other.

The motion of the tabletop should also be verified by placing radioopaque objects or markers to radiographically confirm the distance travelled by the tabletop. The travel of each tabletop shift should agree with the preselected travel within + 0" and -1". This means, for a 14"x14" field size, a travel of 12" should result in a overlap of at least two inches (1" at each end).

Ideally, all operations that are anticipated in actual clinical arrangement should be programmed and the correct operational sequences verified. The electronical monitoring should also be radiographically confirmed by the exposed x-ray films. The tabletop should also be loaded with the appropriate weight to simulate the patient.

There are time limitations and/or constraints due to the transport mechanism of the automatic film changer or the tabletop motion. The intervals between or among various operations should be verified to conform with the original specifications. Detailed discussion of this nature leads one into the acceptance testing of the automatic film changer or the power injector as a separate entity and will not be presented in this article.

Acknowledgement

The authors would like to express thier sincere thanks to the engineers from Philips Medical Systems, and Elema Schomander for their cooperation in carrying out the acceptance testing of the equipment and their helpful suggestions.

ACCEPTANCE TESTING OF AUTOMATIC EXPOSURE CONTROL SYSTEMS

Phillip L. Rauch, M.S.
Department of Diagnostic Radiology
Henry Ford Hospital
Detroit, Michigan 48202

Introduction

Prior to 1942, all radiographic examinations were performed on equipment which required the operator to make educated judgments with respect to the required value of tube potential, tube current, and time needed for a properly exposed radiograph. Even with the aid of technique charts these decisions were not standard for each projection, due to differences in patient density, variations in positioning, and the presence of anatomical or pathological abnormalities. As a result of these problems associated with the selection of proper exposure techniques, the need for a simple automated method of insuring consistent quality radiographs became apparent.

The initial step toward designing a "Radiographic Automatic Exposure Control" (R.A.E.C.), which would reduce much of the guesswork was introduced by Dr. Russell H. Morgan in a paper entitled, "A Photoelectric Timing Mechanism for Automatic Control of Roentgenographic Exposures."[1] His device incorporated a fluorescent screen mounted behind the plane of the image receptor. X-rays which passed through the patient and film would produce fluorescense in the screen. A phototube was optically coupled to the screen, producing a current signal which was proportional to the amount of fluorescense. The x-ray exposure time was controlled by a capacitive trigger circuit designed to terminate the exposure whenever a pre-set level of charge was accumulated. With the help of Dr. Paul Hodges and the facilities of the University of Chicago, the first automatic exposure control device, the Morgan-Hodges Phototimer, was constructed and introduced in the late 1940's.[2]

Today the "phototimer" has been widely accepted and has taken many forms and names depending on the type and application of the device (Table I).

Table I. Nomenclature For Automatic Exposure Control Devices

Radiographic Applications	Fluoroscopic Applications
Automatic Exposure Control	Automatic Exposure Rate Control
Automatic Exposure Termination	Automatic Brightness Control (ABC)
Automatic Density Control	Automatic Brightness Stabilizer (ABS)
Phototimer	
Ionization Timer	

Radiographic automatic exposure control (R.A.E.C.) will be used in this presentation to refer to any device used to automatically terminate the exposure to a plane image receptor such as a cassette or xerox plate.

The Purpose and Function of a R.A.E.C. Device

Purpose

We can define two key purposes of a R.A.E.C. System:

1. The first purpose is to provide to the Radiologist a radiograph which will enable him to visualize patient pathology. To accomplish this, the system must reproduce results with optimum density, for any given projection and any given patient, utilizing modern image receptors, including the rare earth intensifying screens.
2. The second purpose of a R.A.E.C. System is to simplify and speed the operation of making radiographs.

Unfortunately, not all R.A.E.C. Systems fulfill these two purposes. To understand why, we must distinguish between the purpose of phototimer, and its function.

Function

The function of a R.A.E.C. device is fourfold:

1. Sample radiation transmitted through the patient.
2. Integrate an electrical signal which is proportional to the accumulated radiation reaching the image receptor.
3. Set a reference signal which is dependent upon the exposure factors, field size, image receptor sensitivity, and required O.D.
4. Provide a terminate signal to the generator whenever the integrated electrical signal reaches the value of the reference signal.

It is important to understand that the R.A.E.C. device may perform each of the four functions and still not provide clinically acceptable radiographs.

In order to obtain acceptable results, the device must be properly calibrated, and the radiographer must have a clear understanding of its function so as to fully utilize it within design limitations. Calibration involves the adjustment of internal compensation networks by qualified service personnel, and setting the sensitivity to a level which provides acceptable radiographic density. A number of compensation networks have been identified, and may be grouped into INTERNAL and EXTERNAL control of sensitivity.

Internal compensation refers to changes in signal gain or reference voltage level within the R.A.E.C. electronics. These changes are either automatically produced in response to changes in technique factors, type of detector selected, cassette size, etc., or are adjustments to be set during calibration (Table II).

Table II. R.A.E.C. Internal Compensation Networks

1. Kilovoltage compensation	11. Auxiliary device compensation (several auxiliary devices with A.E.C. pickups may be connected to the same electronic control circuit)
2. Sensitivity of individual sensing areas in pickup device	12. Turn-off delay (Compensates for differences in delay between relay-controlled x-ray apparatus and thyristor-controlled)
3. Exposure rate compensation	13. Detector selection compensator (One system may utilize both ionization type, and phototube type of detectors, and compensation is required whenever the type of detector is switched)
4. Compression cone compensation	14. Thickness of ionization chamber (For systems which are capable for accepting two or more chamber sizes)
5. Overall density adjustments	
6. Cassette size dependent blackening correction	
7. Field size dependency compensation (e.g. Four-on-one spot film format)	
8. Leakage current compensation (Ionization type)	
9. Dark current compensation (Phototube type)	
10. Cable charge compensation (corrects for energy stored on capacitance of high tension cables)	

External compensation refers to changes in signal gain or reference voltage which are produced by operator controlled sensitivity adjustments (Table III). These provide the radiographer with means to change R.A.E.C. sensitivity for specific examinations, to meet individual radiologists' preferences, and to allow the use of two or more different image receptors. It may also be necessary to utilize external compensation to overcome design limitations which result in poor or inadequate internal compensation.

Table III. R.A.E.C. External Compensation Networks

1. Density control
2. Screen sensitivity correction
3. Patient thickness compensation

Basic Elements of an R.A.E.C. System

To determine the amount of radiation which reaches the image receptor, one could utilize either the ability of x-rays to cause fluorescence in selected phosphors, as did Dr. Morgan, or the ability of x-rays to ionize air. Regardless of the method used, the basic elements of an R.A.E.C. System are the same (Figure 1).[3] The difference is merely in the type of radiation sampling employed and the compensation networks employed.

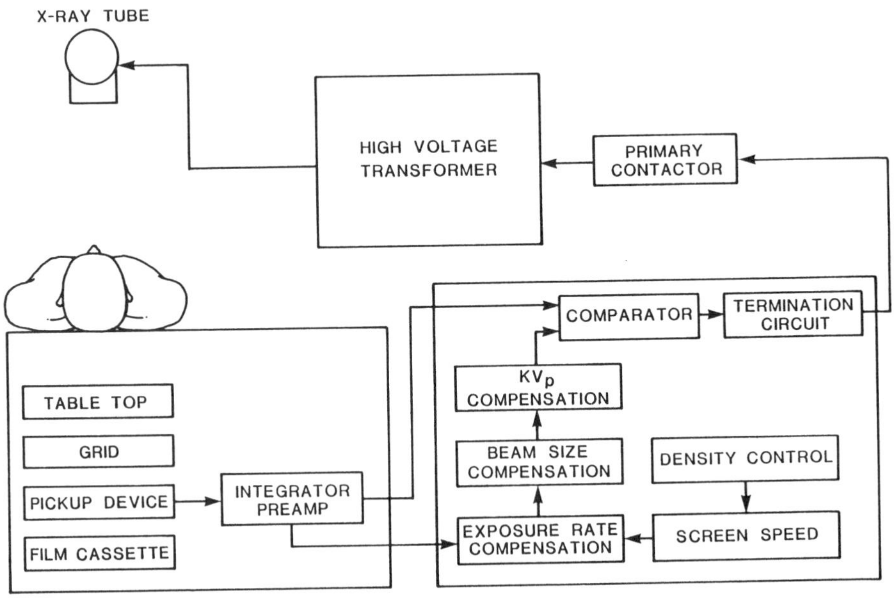

Figure 1. The basic components of an R.A.E.C. device. (Adapted from (3)).

A pickup, located near the plane of the image receptor, either anterior or posterior to the cassette, samples radiation by absorbing a finite number of x-ray quanta which has passed through the patient. An electrical signal is produced which is proportional to the intensity of the radiation, and an electronic circuit then integrates this signal. When the film has received just enough radiation to produce the proper density, that is, equal to what we consider a desired optical density, a signal is relayed to the generator and the x-ray switch is opened, terminating the x-ray emission. This is the standard textbook phototimer. In the real world, many physical problems must be overcome to allow the R.A.E.C. device to perform adequately, and the device must be properly calibrated.

Detectors

There are two types of sensor pickups used by the majority of equipment manufacturers to detect the transmitted radiation:

1. Scintillator Type
   Picker Phototimer
   CGR Luminix

2. Ionization Type
   Philips Amplimat
   Siemens Iontomat
   GE Quantamat

In addition, all manufacturers use photomultiplier tubes for brightness controls in cine radiogaphy and photo-fluoroscopic cameras, such as the 105mm or the 100mm cameras. Sensing for automatic exposure rate control in fluoroscopy is either by photomultiplier tube, or video sensing in the T.V. chain.

Scintillator Type

The pickup of the scintillator type device consists of a sheet of light conducting plastic, such as lucite, in contact with an x-ray sensitive phosphor, or scintillator. The plastic sheet transmits light, generated by the phosphor, to a photomultiplier tube which creates an electrical signal proportional to the intensity of the light. This represents two stages of energy conversion: 1. Radiation is changed into a light signal. 2. The light signal is changed into an electrical signal.

Since the photopickup consists of a laminate of luminescent screen, paper and plastic, it is much more rugged than the ionization chamber pickup. A single photomultiplier tube may be used along with mechanical shutters to control the light emitted from each of three scintillators in the pickup device. The three scintillators define three sensitive regions, "fields," or "dominants," in the plane of the pickup, and each field can usually be selected singly, or in combination. Alternatively, there may be a separate photomultiplier tube to sample the light from each dominant.

Ionization Type

The ionization pickup consists of thin metal electrodes separated by an air space. X-rays passing through this pickup produce free electrons, which are collected by the electrodes producing a small electrical current. This current, which is usually in the nano-amp range, is proportional to the x-ray intensity. Compared to the scintillator, this represents only one stage of energy conversion: Radiation is changed directly into an electrical signal.

Ion chambers, because of their low current generation must have fixed lengths of cable, not exceeding 35 meters, to carry the signal. Some require a first stage pre-amp just outside the chamber to increase signal strength and lower impedance.

Detector Location

There are three possible locations for a detector whether it be a scintillator or an ion chamber:[4,5] 1. Between the table top and the grid. 2. Between the grid and the image receptor. 3. Posterior to the image receptor. The first position is rarely used because it samples all radiation, including scatter, whereas the image receptor receives radiation which has been considerably altered in spectral characteristics and intensity by the grid. If the detector is in position (3), on the exit side of the cassette, it is called a "posterior detector." The x-rays which reach the posterior detector are of no further diagnostic value and can be absorbed by the detector with no added patient exposure. Because of the low amount of radiation which is available, after passing through the patient, the grid, the cassette, and the two intensifying screens, this method requires a highly sensitive detector.

Cassettes which are to be used with posterior detector R.A.E.C. systems must be of special construction and have little or no lead in the back portion of the cassette. A lead thickness of 0.0025" is maximum. If you have these "phototiming cassettes" in your department and use them for routine table top work, put a lead apron on the table to help absorb secondary radiation. Radiographic quality should greatly improve.

The radiation reaching the posterior pickup has a different spectral character because of the filtration effected by the phosphor in the screens. Radiation intensity is less because of the absorption in the screens. Posterior detectors have proven to be a disadvantage since the introduction of rare earth screens, because these screens absorb more quanta, than does the calcium tungstate, and the beam spectrum is altered considerably. However, one very successful application of posterior detectors has been for those imaging systems which do not use cassettes (i.e. automatic film loaders, such as the Picker Rapido™).

Since the R.A.E.C. device is used to control radiographic density, the detector should sample all of the radiation which contributes to that density. The logical place for the detector is, therefore, directly in front of the film, posterior to the grid. When the pickup is located between the grid and the cassette, it is called an "anterior detector."

With the pickup located posterior to the grid, there should be no need to compensate for secondary radiation from the patient. This is true for the most part; however, secondary radiation is generally different in quality from the primary radiation, and originates in subject details which are widely distributed in space. In general, the secondary radiation contribution varies according to the nature and the size of the object being radiographed, the field size, type of collimation and kVp used. Thus, some compensation for these effects may be required.

Detector Fields

Most manufacturers provide three fields or sensitive regions within the pickup device. These are generally operator selectable, singly or in combination. To effectively measure the amount of radiation being delivered to the image receptor, the body part of interest should be positioned over the sensitive area of the detector. This diagnostic area of interest is referred to as the "dominant." The radiation passing through the dominant and impinging on the sensitive field, causes the current which is used to terminate the exposure.

Both ionization type, and scintillator type detectors may have three separate fields. In anterior pickups the detectors may be somewhat radiopaque, casting a shadow on the radiograph at low kVp's. This effect is more a function of manufacturing design than whether the detector is an ion chamber or a screen phosphor. Both Picker (Scintillator) and Philips (Ion Chamber) have successfully designed pickups with detectors which are only slightly visible even at 35 kVp.

Whenever a phototimer is newly installed or if the pickup or electronics must be removed for service, it is important to verify that whenever a field is selected at the control desk, that corresponding field, and only that field is energized at the pickup device. This is relatively simple to check, and will relieve much frustration which may result if, for example, the upper left field is energized whenever the upper right field is selected, and vice-versa.

Desirable Features of an R.A.E.C. System

A well designed automatic exposure control system should have the following characteristics:

1. It should properly terminate the x-ray exposure, providing the desired radiographic optical density for a wide variety of conditions. This consistency of performance is called tracking. R.A.E.C. Systems should track under the following conditions:
    A. As tube potential is changed from 50 to 150 kVp.
    B. For variations in patient thickness.
    C. When cassette size is changed, or multiple format spot films are utilized.
2. The device should operate consistently, resulting in reproducible radiographic exposures.
3. The pickup should not degrade the radiographic image by producing additional scatter radiation, nor by appreciably attenuating the x-ray beam so that the exposure must be increased.
4. The pickup should not appreciable increase patient film distance.
5. The pickup should be rugged and not influenced by external forces such as humidity, temperature, and atmospheric pressure. It must be able to take some external abuse.
6. The system should offer a variety of dominant fields to select, when necessary.
7. The pickup must provide a uniform response regardless of the field or combination of fields selected.
8. The device should provide sufficient internal compensation adjustments to provide good tracking should the film-screen combination be changed.
9. It should provide sufficient operator controlled compensation adjustments.
10. It should not be used with a high kilowatt generator unless forced commutation, grid control, or secondary high tension switching is also provided.
11. It should include a backup termination system which will terminate the exposure in the event of failure of the R.A.E.C. System. This is necessary to preserve the x-ray tube and to avoid hazardous exposures to the patient.

R.A.E.C. Performance Evaluation

A large number of test methods have appeared in the literature.[6-10] These range from direct measurement of a system parameter (e.g. minimum exposure time) to indirect evaluation of performance (e.g. inverse square law verification). The exact testing protocol is a matter of personal preference but should consist of procedures which can be correlated with the clinical functioning of the system. In general, the following should be considered before evaluating the performance of an R.A.E.C. system:

1. Make radiographic exposures through a suitable attenuator and measure the resultant optical densities - a phototimer is intended to provide consistent film density, which is not necessarily synonymous with constant radiation exposure.[11]
2. Choose a suitable attenuator and be consistent. Water or acrylic is recommended. Masonite (pressed wood) does not produce a uniform radiographic density which makes interpretation of results difficult. Aluminum and copper are poor choices since their attenuation and scattering properties differ greatly from those of water (tissue)[12] and the thickness corresponding to a given amount of tissue is highly dependent upon the kVp.[13]
3. In a new installation, provide the vendor with your test methods and criteria.
4. Verify the reciprocation, centering, and alignment of the grid. Do not use an SID which exceeds the operating range of the grid.
5. If adjustment of the internal compensation circuitry is necessary, use the same cassette for all exposures and monitor the processing closely. This will assure that changes in radiographic density are only due to the internal adjustments.
6. If more than one film-screen combination is to be used, verify that the provided compensation adjustments are adequate to assure acceptable tracking for each image receptor.
7. Record photomultiplier high voltage, or comparator reference voltage, as a function of kV selected, and as a function of density or patient compensation controls. This will serve as a baseline if compensation adjustments are required or to confirm a suspected drift in sensitivity.
8. Record the change in O.D. as a function of photocell high voltage, or comparator reference voltage. This will help to establish the setting required for the specified optical density.

The testing methods which we have chosen are essentially the same as those proposed by Lin.[14] A specific testing protocol for R.A.E.C. systems, along with our acceptance criteria, is included in the appendix of this paper.

Fluoroscopy

The purpose of automatic exposure control in fluoroscopy is to maintain a constant luminance at the output phosphor of the image intensifier by controlling one or more technical factors. The image brightness can be monitored directly via a mirror or prism (Figure 2),[7] or indirectly via the photocathode current of the image intensifier, or the video signal of the television pickup tube. The technical factor(s) controlled depends upon the design or mode of operation. The commonly used modes are given in Table IV.

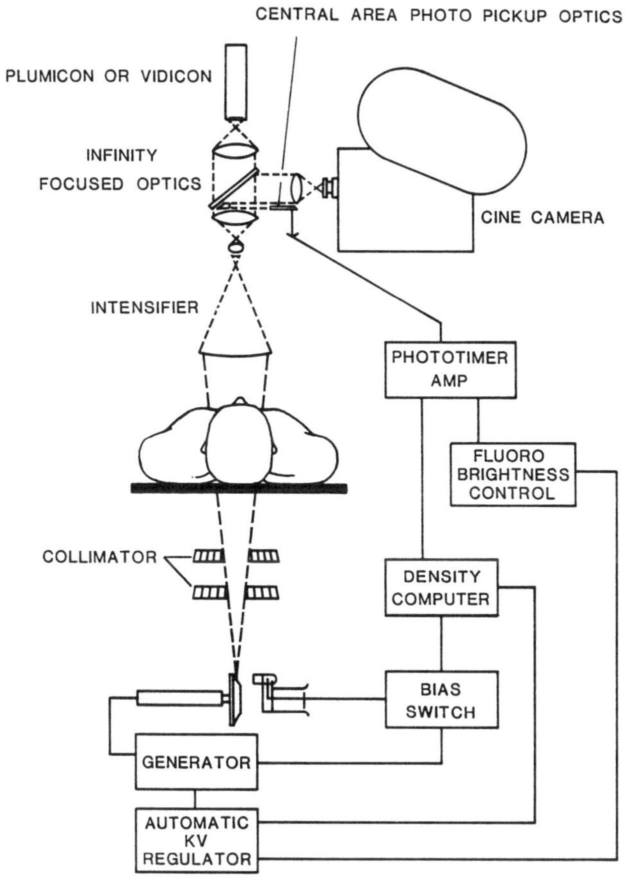

Figure 2. Automatic brightness control for fluoroscopic and fluorographic recording of the x-ray image. (Adapted from (7)).

Table IV. Variations in Automatic Brightness Control Mechanisms

1.	Operator controlled kVp, brightness controlled mA.
2.	Operator controlled mA, brightness controlled kVp.
3.	Brightness controlled mA with kVp override.
4.	Brightness controlled kVp and mA, with both increasing or decreasing together.
5.	Brightness controlled pulse width.
6.	Pulse width controlled with kVp override.

Exposure Rate Settings

Regardless of the method used, the fluoroscopic exposure rate at the image conversion plane (i.e., at the input phosphor of the image intensifier) should remain constant over a range of attenuator thicknesses which correspond to kV values between 70 kVp and 90 kVp.[15] Above or below these values, the input exposure rate will increase as the sensitivity of the input phosphor decreases. Typical values of input exposure rate range from 20 to 100 μR/sec (1.2 to 6.0 mR/min) for a 9-inch diameter image tube. The actual value depends upon the system design and acceptable noise limits, but typically is no greater than 60 μR/sec (3.6 mR/min). For magnified image mode, these values will increase by a factor which is approximately the square of the ratio of the image diameters. The integral dose, however, is generally less for magnified mode, especially for kV-variable brightness stabilizers. In general, an increase in radiation exposure rate is not an acceptable substitute for poorly designed or poorly adjusted imaging systems. The quanta of light available at the television pickup tube is a function not only of radiation exposure, but also screen efficiency, intensifier gain, intensifier focus, quality of optics, and aperture size. For a given exposure rate, the intensifier, aperture, pickup tube and video display must be adjusted for optimum signal to noise in the final image.

For serial fluorographic cameras (70mm and 105mm roll film or 100mm cut film) or for cinefluorographic cameras, the required exposure per frame is highly dependent upon the imaging system, film type, processing, aperture setting, and noise level which can be tolerated. In general, the required exposure can be minimized only by a concerted effort to optimize each of these parameters. The typical values of image intensifier input exposure are 0.05 to 0.20 mR/frame for serial fluorographic camera, and 0.01 to 0.04 mR/frame for cinefluorographic cameras.[16] Emphasis should be placed on setting the minimum exposure, consistent with the clinical needs of the system.

System Response Time

The ability of the automatic brightness control system to quickly respond to changes in x-ray attenuation is important. This is especially true for kV-variable systems which always start at a low kV value. If these systems do not respond quickly to raise the kV to the required value, the radiologists will soon be dissuaded from using intermittent fluoroscopy. It is, therefore, important to measure the time required for the kV to drive from the exposure off value (or the value corresponding to an attenuator thickness of 0.5" aluminum), to the value corresponding to an attenuator thickness of 2.5" to 3" of aluminum. The time required for the system to reach at least 90% of the stabilized kV value should be less than 0.5 seconds.

Maximum Patient Entrance Exposure Rate

The patient entrance exposure rate should be checked to verify that the maximum specified limits are not exceeded. These limits[17] are 5 R/min for units without automatic exposure rate control and 10 R/min for units with automatic exposure rate control. Exceptions to these limits are granted to those units with an optional high level control.

CONCLUSION

Radiographic Automatic Exposure Control has been used in roentgenography for over 30 years. However, successful application of this device requires a thorough understanding of not just the purpose, but also the function of a R.A.E.C. device. Both scintillator and ionization pickups are widely used, but regardless of the type of detector, adequate compensation networks must be incorporated if the device is to perform satisfactorily. The R.A.E.C. device must also be calibrated by qualified service personnel using the same film-screen combination and processing conditions as used clinically. Most pickups are located between the grid and the image receptor, and offer a choice of three fields or dominants. It is important to verify that the field selected at the control is in fact the one energized at the pickup.

The R.A.E.C. device should track well as the tube potential, patient thickness, and field size are changed. In addition, it must provide reproducible exposures. A specific testing protocol has been presented which is designed to give maximum information regarding the performance of the R.A.E.C. device. This information can be useful, not only to identify system malfunctions, but also to identify system limitations which should be taken into account by the operator of the system.

References

1. "A Photoelectric Timing Mechanism for the Automatic Control of Roentgenographic Exposure," R.H. Morgan, American Journal of Roentgenology, Vol. 48: 220-228 (1942).

2. "The Trail of the Invisible Light," E.R.N. Grigg, p.145, 415, Charles C. Thomas, Pub. (1965).

3. "X-ray Automatic Exposure Timing and Control Circuitry," E.S. Joseph and J.E. Schneble, SPIE, Vol. 70: 166-170 (1975).

4. "X-ray Automatic Exposure Control," G.P. Stanhope, SPIE, Vol. 70: 171-175 (1975).

5. "A Different Phototimer Placement in Automated Exposure Control," J.C. Hoeffel, et al, Radiologic Technology, Vol. 43: 347-349 (1972).

6. "Field Evaluation of Radiographic Phototimers, W.R. Hendee, et al, SPIE, Vol. 70: 183-188 (1975).

7. "Characteristics annd Evaluation of Automatic Exposure Controls," J.R. Craig, SPIE, Vol. 70: 189-195 (1975).

8. "Phototimers, User's Point of View," P. Lin, SPIE, Vol, 70: 196-200 (1975).

9. "Automatic Exposure Control in General Radiography," A.D. Smythe, Aust. Radiology, Vol. 10: 259-266 (1966).

10. "Test Methods for Diagnostic X-ray Machines for Use During Initial Installation," NEMA Standards Publication No. XR8-1979.

11. "Performance Evaluation of Automatic-Exposure-Control Systems," B.A. Horn, M.G. Randall, 20th Annual Meeting of the American Association of Physicists in Medicine, July, 1978 (Abstract).

12. "System-level Performance Testing of Image-Intensified X-ray Equipment," J.A. Mulvaney, et al, SPIE, Vol, 273: 153-159 (1981).

13. "Attenuation Phantoms for Patient Exposure Measurement During Radiographic Examinations Utilizing Phototiming Techniques," P. Lin, NBS SP456: 381-388 (1976).

14. "Acceptance Testing of Automatic Exposure Control Systems and Automatic Brightness Stabilized Fluoroscopic Equipment," P. Lin, Quality Assurance in Diagnostic Radiology, AAPM Monograph No. 4: 10-27 (1980).

15. "Variation of X-ray Tube Potential as a Function of Attenuator Thickness in kVp Variable Automatic Brightness Stabilization Fluoroscopy," P. Lin, R. Kritz, SPIE, Vol, 127: 152-157 (1977).

16. "Cameras for Image Intensifier Fluorography," ICRU Report 15 (1969).

17. "Regulations for the Administration and Enforcement of the Radiation Control for Health and Safety Act of 1968," HHS Publication (FDA) 80-8035 (1980).

APPENDIX

HENRY FORD HOSPITAL

ACCEPTANCE CRITERIA FOR RADIOGRAPHIC AUTOMATIC EXPOSURE CONTROL PERFORMANCE

The following procedures have been developed in order to ascertain the ability of a Radiographic Automatic Exposure Control (R.A.E.C.) system to maintain radiographic optical density in a consistent and reliable manner under a variety of conditions. The data obtained from these tests may be used to (1) Identify systems' limitations which may adversely affect radiographic quality, and (2) Determine what service adjustments are required to bring the system into compliance with these standards. Where limitations are found which cannot be corrected, the vendor will supply special instructions to the operator which will enable him/her to fully understand the capabilities and limitations of the system, and which will enable him/her to apply available compensation adjustments as required.

Preliminary Instructions

1. The attenuating medium, herein called the "attenuator" will consist of a plastic tank filled with varying amounts of tap water. Each side of the tank must be no less than 14" wide X 12" deep. For R.A.E.C. systems employed with vertical cassette buckys, a specially designed tank will be provided by Henry Ford Hospital.

2. All exposures will be made at 40 inches SID (Source to Image Distance), or at the SID for which the system is designed.

3. For all exposures use a 35cm X 43cm cassette, but adjust the collimator to provide a radiation field less than the dimensions of the attenuator (i.e. less than 14" X 14"). This is necessary to eliminate the effects of scatter on the performance of the system. The field size should be approximately 35cm X 35cm except when the field size test is performed, or whenever the R.A.E.C. device is being tested with serial format exposures (multiple views on a single radiograph).

4. The image receptor shall consist of a screen-type cassette loaded with fresh radiographic film. The film-screen combination shall correspond to that which will be utilized in clinical practice. The chief technologist shall be consulted to verify the proper selection of image receptor. The film used during the test shall all be from the same box to minimize sheet to sheet film variations.

   Important: The same cassette and the same orientation of the cassette must be used throughout the test, so that variations in screen output, screen contact, etc. will not influence the data.

5. Processing of all radiographs shall be done at the same facility which is to be used during routine clinical operations. Check with the supervisor, Radiological Engineering, prior to beginning the tests so that any required changes in processing conditions can be implemented before the commencement of the tests.

Henry Ford Hospital Acceptance Criteria:

Grid lines SHOULD not be visible on any of the radiographs, and MUST not be visible for exposure times equal to or greater than 6 ms. Non-uniformities in optical density due to grid misalignment must be corrected before continuing the tests.

2. SELECTION OF FIELD DOMINANT

Check for agreement between the sensing area or field dominant selected at the control console, and that located in the pickup device. To determine if the field dominant selected at the control console corresponds to the same region being energized in the pickup device, cover the other two sensors with lead, and make a low (50kVp) exposure with no additional attenuator in the field. If the correct sensor has been energized, the exposure will be very short. Select the other two sensors, one at a time, and verify that the first sensor is no longer energized. (i.e. the exposure should go to backup). Verify that the backup time meets manufacturer's specifications, and that the DHHS* backup limit is not exceeded. Verify the subsequent exposures cannot be made without first manually re-setting the limiting device.

Repeat the above for each field dominant.

Note: Having verified the DHHS required limits, and that manual re-setting is required, it is not necessary that subsequent exposures be allowed to continued to the full limit. Instead, one can terminate the exposure early, so long as it is clear that the exposure would not be terminated by the R.A.E.C. due to the presence of the lead attenuator.

Henry Ford Hospital Acceptance Criteria:

The selection of a field dominant at the console must energize the corresponding sensor at the pickup device, and only that sensor.

3. VISUAL DELINEATION OF FIELD DOMINANT

Using the means provided to delineate the location of the sensors in the pickup device, outline the location of the sensors using thin wire solder. With no attenuator in the field, and using 35-40 kVp, small focus, 50 mA, produce a radiograph with sufficient optical density to visualize the sensors. (Note: this test cannot be performed if the pickup device is located behind the plane of the image receptor). Observe the location of the sensors relative to the region outlined by the wire solder.

* DHHS: Department of Health and Human Services
The backup limit is 600 mAS for phototimed operation.

6. A Macbeth model TD 501 densitometer will be available for optical density measurements. (This instrument must be frequently zeroed).

7. The average optical density must be determined for five or more points near the center of the field. For successive radiographs, always measure the optical density at the same locations so as to elimate the effects of screen nonuniformity, or the density variations resulting from the shadow of the sensor in the R.A.E.C. pickup device.

8. All technique calibration adjustments must be completed prior to testing the R.A.E.C. device, and the Henry Ford Hospital performance criteria for kV and mA calibration must be met. Any deviation from the expected technique, especially kVp, may adversely affect the performance of the phototimer.

9. If Falling Load is provided, this technique shall be utilized for all R.A.E.C. exposures required in the following test procedures. Therefore, it is important to first verify that the falling load waveforms are consistent with design specifications, and that the peak kilovoltage does not vary significantly during the exposure.

Procedures

1. GRID RECIPROCATION

Verify that the grid reciprocates and that grid lines are absent from the radiograph for exposures of 3 ms or longer. To verify that the grid is reciprocating, place a "flag" on the grid and watch for reciprocation during the exposure.

Important: Listening for the reciprocation of the motor driving the grid is not a sufficient test of whether the grid itself is properly moving.

The absence of grid lines in the radiograph must be verified by exposing a 14" X 17" cassette with no attenuator in the field, using low kVp at 3 ms and 5-10 mAs (the optical density should be near 1.0).

Closely observe the radiograph for grid lines. Also look for nonuniformity in optical density which would indicate misalignment of the grid. If grid lines are present, repeat at progressively longer exposures until the grid lines are no longer visible. Submit the radiographs to the physics staff for approval.

Henry Ford Hospital Acceptance Criteria:

Displacement of the true location of the sensors shall not be more that 5 mm in any direction from the location that is delineated by the means provided on the equipment.

4. SAFETY (BACKUP) LIMIT

Verify the proper functioning of the exposure limiter (backup) device. Close the collimator or place a lead attenuator in the field and select a low (50) kVp. If the backup time is adjusted by setting the time on the generator, set the time to the maximum allowable. Set all other controls (i.e. density) to settings which will provide the maximum mAs. Make the exposure and record the backup mAs, and exposure time. Verify that the backup time meets manufacturer's specifications, and that the DHHS backup limit is not exceeded. Verify that there is visual indication that the limit was reached, and that subsequent exposures cannot be made without first manually re-setting the limiting device.

Repeat the above for each dominant.

CAUTION: Observe the thermal state of the anode and do not make any exposure which will exceed the heat storage capacity. Allow sufficient time for anode cooling after each exposure which is terminated at the limit.

Note: If the backup time can be changed to different values at the control, record the means available to do this and provide sufficient instructions to allow the operator to optimally set the desired backup. For the above tests, however, use the maximum available backup setting.

Henry Ford Hospital Acceptance Criteria:

The backup system must terminate the exposure at less than or equal to 600 mAs or 60 kilowatt-seconds and must give visual indication of the fact that the limit was reached.

Successive exposures must be blocked unless the limiter is manually reset.

Reference: Title 21, CFR § 1020.31 (a), (3).

5. MINIMUM EXPOSURE TIME

To determine the minimum exposure time, use a voltage divider and cathode ray oscilloscope to visualized the kV waveform, and use an mAs meter to measure the total mAs under the following conditions:

1. With the grid in the field, but with no additional attenuator, make an R.A.E.C. exposure at 50 kVp, large focus, 80-90% of maximum tube load. Observe the kV waveform. If the exposure time is less than 25 ms, add sufficient attenuator to provide an exposure slightly greater than 25 ms duration. Record the exposure time, as measured at 80% of the kVp. Record the mAs. Photograph the kV waveform and note any surges, overshoots, or other deviations from the nominal kVp. Record any abnormal findings and report the result to the physics staff.

2. Increase the kVp by 10 kV and make another R.A.E.C. exposure. Photograph the kV waveform and record the mAs. Observe the waveform for surges, overshoots or other deviations from the nominal kVp. Record any abnormal findings and report the result to the physics staff. Record the exposure time, as measured at 80% of the kVp.

3. Repeat step 2 up to the maximum kVp (several waveforms can be recorded on one photograph, so long as they can be distinguished for waveform analysis). Determine the minimum exposure time and record. Submit the oscillograph recordings to the physics staff for approval.

   Note: The exposure time must be sufficient to allow the kV waveform to reach the peak value set. The minimum exposure time must be measured from a waveform which reaches at least 90% of the set kVp.

4. Repeat steps 1 to 3 for each sensor in the pickup device.

Henry Ford Hospital Acceptance Criteria:

The minimum exposure time shall be equal to or less than 2 milliseconds. The kilovoltage overshoot at the start of the exposure shall not exceed 3 kV. The kVp shall not deviate from the selected value by more than $\pm 5\%$ throughout the exposure.

6. RADIOGRAPHIC OPTICAL DENSITY

The radiographic optical density shall be determined using 20 cm of water as the attenuator, with the center sensor and a technique of 90 kVp† and 80-90% of maximum tube load with the large filament selected. The R.A.E.C. sensitivity must be adjusted to provide an average optical density of 1.25.

Note: All external compensation controls must be set to the normal (N) or zero position, and screen compensation controls set for the screen and film being utilized.

†For a chest bucky, use 16 cm of water and 150 KVp.

Henry Ford Hospital Acceptance Criteria:

$1.2 \leq$ Optical Density ≤ 1.30

(Averaged over five or more points near the center of the field)

7. OPTICAL DENSITY FOR EACH FIELD DOMINANT

Check the agreement in optical density provided by each sensor or combination of sensors provided.

Using 20 cm of water as the attenuator, and using the large focus at 90 kVp,† 80-90% of maximum tube load, make an R.A.E.C. radiograph for each sensor, or combination of sensors, provided at the control console.

†For a chest bucky, use 16 cm of water and 150 KVp.

Henry Ford Hospital Acceptance Criteria:

The average optical density provided by each sensor or combination of sensors shall not deviate from that provided by the center sensor by more than 0.1 O.D. The optical density for any two radiographs in the series shall not deviate from each other more than 0.15 O.D.

Note: If system design dictates that such agreement is not intended, the vendor must furnish sufficient documentation (schematics, description of function, etc.) to demonstrate to the physics staff that the manufacturer did not intend for each sensor to provide the same optical density.

8. REPRODUCIBILITY OF R.A.E.C. EXPOSURES

Using the following method, determine the reproducibility of R.A.E.C. exposure.

a. Using 20 cm of water as the attenuator, and using the large focus at 90 kVp†, 80 - 90 % of maximum load, make five R.A.E.C. exposures for each sensor. For each exposure measure and record one or more of the following:

   1. Exposure time (milliseconds)

   2. The product of the tube current and the exposure time. (mAs)

   3. Average radiographic optical density.

   †For a chest bucky, use 16 cm of water and 150 KVp.

b. For each of the sets of five exposures, calculate the coefficient of variation of the parameter measured, defined as the ratio of the standard deviation, σ, divided by the mean $\bar{x}$.

$$\text{coefficient of variation} = \frac{\sigma}{\bar{x}},$$

where $\bar{x} = \frac{1}{5} \sum_{i=1}^{5} x_i$ is the algebraic mean and

$$\sigma = \left(\frac{\sum_{i=1}^{5} (\bar{x} - x_i)^2}{4} \right)^{\frac{1}{2}}$$ is the standard deviation

Henry Ford Hospital Acceptance Criteria:

For each sensor in the pickup device, the R.A.E.C. exposures shall provide a coefficient of variation which is less than or equal to 0.05.

9. kV COMPENSATION

Check the kV compensation to determine the ability of the R.A.E.C. system to maintain constant radiographic density as the kilovoltage is varied from 60 kVp to the maximum kVp. Provide a brief written explanation of the means used to provide kV compensation, and the limitations of the method (e.g. "No kV compensation provided from 40 to 90 kVp; between 90 and 96 kVp amplifier gain is discontinuously increased. Same occurs between 117 and 125 kVp. No adjustments are provided").

For attenuator thicknesses of 15 cm, 20 cm, and 25 cm of water, and using the large focus at 80-90% of maximum tube load, obtain R.A.E.C. radiographs for kilovoltages from 60 kVp[†] to maximum kVp in 10 kV steps. (e.g. 60, 70, 80, 90 . . .)

[†] For a chest bucky, use 11 cm, 16 cm and 21 cm of water and 150 KVp.

For each radiograph record the average optical density (five or more points) and plot the O.D. verses kVp on the linear graph paper.

> Note: If there is a separate kV compensation adjustment for each sensor in the pickup device, this check must be repeated for each sensor.

Henry Ford Hospital Acceptance Criteria:

1. The difference in optical density between <u>any</u> two radiographs produced with technique settings 10 kV apart (e.g., 80 and 90, or 90 and 100, or 120 and 130, etc.) must be less than 0.10 O.D.

2. The difference in optical density between <u>any</u> two radiographs in the series must be less than 0.30 O.D.

   Note: If system design dictates that these criteria cannot be met, the vendor must furnish sufficient documentation (schematics, description of function, etc.) to demonstrate to the physics staff that the system cannot be adjusted to meet Henry Ford Hospital requirements.

10. OPTICAL DENSITY VERSUS FIELD SIZE

Using 20 cm of water as the attenuator, large focus, center sensor and a technique of 90 kVp, 80 - 90% of maximum tube load, obtain R.A.E.C. radiographs for field sizes corresponding to each cassette size available.

If serial format (multiple exposures on a single radiograph) is available, repeat the above test using the serial mode and with the field size adjusted to correspond to that portion of the film being exposed.

Henry Ford Hospital Acceptance Criteria:

The difference in optical density between any two radiographs produced with different field sizes must be less than 0.20 O.D.

11. DENSITY CONTROL

Using 20 cm of water as the attenuator, and using the large focus, center sensor, and a technique of 90 kVp[†], and 80 - 90% of maximum tube load, measure and record the mAs, and the radiographic optical density for each position of the density control provided on the generator. Calculate the percentage change in mAs and in the optical density with respect to that obtained with the density control in the normal or zero position. The vendor must provide a short explanation of the method employed to change the density (e.g. change in photomultiplier high voltage, change in reference or comparator voltage, etc.).

[†]For a chest bucky, use 16 cm of water and 150 KVp.

Henry Ford Hospital Acceptance Criteria:

The density control available on the generator shall provide sufficient compensation to allow the operator to adjust the R.A.E.C. sensitivity for patient size and pathology. Typically, if the density control has ±4 steps, each step should be 20-25%; but if only ±2 steps are provided, each step should be 30-50%. The density control must be sufficient to make up for design limitations which prevent adequate kV compensation.

12. RECORD OF FINAL SETTINGS

After the above tests have verified that the R.A.E.C. system is functioning properly, and that no further adjustments are required, record the final value(s) of all internally adjustable compensation networks (e.g., link or jumper connections on printed circuit boards or density programming boards, power supply voltage, chamber collection voltage, photomultiplier high voltage, etc.).

Measure and record the photomultiplier high voltage (if this is used in the system) and the reference or comparator voltages under the following conditions:

1. As the kVp is adjusted in 5 kV increments from 35 kVp to maximum kVp. Do this for each of the cells in the pick-up device.

   Note: Even if the kV compensation is independent of the cell selected, it is important to verify that the measured voltage does not change significantly as the sensitive region or sensor of the pickup device is changed.

2. As the density control at the generator is varied from minimum to maximum in the increments provided on the control.

Record the change in radiographic optical density as a function of photomultiplier high voltage or as a function of the reference or comparator voltage. Use 20 cm of water as the attenuator, and use the large focus at 90 kVp†, 80 - 90% of maximum tube load.

†For a chest bucky, use 16 cm of water and 150 KVp.

13. R.A.E.C. PERFORMANCE WITH AUXILIARY BUCKY'S

Repeat procedures 1, 2, 3, 6, 7, and 12 for each bucky equipped with an R.A.E.C. pickup device.

Acceptance Testing of Cine Imaging Systems for Cardiac Catheterization Laboratories

Pei-Jan Paul Lin
Department of Radiology, Northwestern University Medical School
Chicago, Illinois 60611

Abstract

A procedure for the acceptance testing of the cine imaging system is described. The emphasis is placed on the operational logics of the automatic cine exposure control circuitries to verify the functions of the imaging chain and the interface of the cine photofluorographic camera and the generator.

Introduction

The advent of microprocessors and their application in radiological imaging equipment has changed the operational aspects of various generators and imaging devices. Incorporation of such sophisticated imaging components enabled the medical team to pay more attention to the patient and fulfill their clinical duties rather than spending their valuable time on the determination of technical factors for proper radiological imaging.

As these automated and/or preprogrammed devices free the medical team from setting-up the technique factors, it becomes even more important that all automatic features provided with the imaging system be checked and evaluated for their accuracy and correct functions in accordance to their design and operational logics. The acceptance testing procedure described in the paper focuses on the automatic features and the electromechanical interfaces for the generator, the control console, and the cine imaging chain.

The image performance characteristics such as the resolution, the low contrast visibility, etc. are not dealt with in this paper. Although the imaging characteristics of the image intensifier and the cine camera are important, acceptance testing of these devices may be found elsewhere.

The Automatic Cine Exposure Control System

The automatic exposure control (AEC) circuitry of a typical cine photofluorographic system (cine system) is designed to maintain the level of light output at the image intensifier output phosphor during cine filming. The basic operational purpose of the cine AEC circuitry is, therefore, similar to that of the photospot camera AEC and the radiographic AEC devices. The complexity and the sophistication of the AEC systems deserves a careful study prior to the acceptance testing.

Because of the generator designs and the methods in achieving the constant light output at the output phosphor of an image intensifier, the acceptance testing procedure may have to be taylored to each specific equipment. Attempts are made in this paper to encompass all possible cine AEC logics by describing the cine system parameters that must be taken into account in the acceptance testing. A description of the cine system operation is also provided so that the actual acceptance testing procedure can be easily explained later.

Cine System Parameters

The complexity of the cine AEC system is primarily due to the number of variables and/or parameters involved in the cine filming. A change in any of the parameters listed in Table I will alter the light output of the image intensifier and/or result in a difference in the cine film exposure.

TABLE I PARAMETERS OF CINE FILMING

Item	Physical Parameters	Typical Range in Clinical Case
(a)	X-ray Tube Potential	60 - 120 kVp
(b)	X-ray Tube Current	100 - 800 mA
(c)	Cine Pulse Width	1 - 5 mSec
(d)	Cine Filming Rate	24 - 90 frames/sec
(e)	Focal Spot Size (Power Rating)	0.3 mm - 1.2 mm (10 kW - 100 kW)
(f)	Cine Run Time *[1]	8, 10, 12, 15 seconds
(g)	Image Intensifier Mode	4", 5", 6", 9" (single or dual modes)
(h)	Cine Film Size *[2]	16 mm, 35 mm

*[1] For some special applications,a cine run time of 120 seconds may be used, but is not applicable for cardiac catheterization.

*[2] While 16 mm film may require less radiation than 35 mm film, practically speaking, all cardiac catheterization works are now recorded with 35 mm cine cameras today.

In cine filming, x-ray intensity is pulsed so that the production of x-ray is synchronized with the cine camera shutter movements. The description of such synchronization is well demonstrated in standard radiological physics textbooks and is not described here.

A typical cine system equipped with an AEC device may utilize one or two of the cine technique parameters to attain the constant brightness and such parameters are normally (a) the x-ray tube potential, (b) the x-ray tube current, or (c) the cine pulse width; other parameters include (d) cine filming rate, (e) focal spot size, and (f) cine run time may be considered as the secondary parameters and are often preset or pre-determined in the programing boards for cine filming.

Description of Cine AEC System

The exact AEC circuitry varies with the manufacturers and the particular models of generators. A schematic block diagram of a typical cine AEC system is depicted in Figure 1 and is based on a cine system that utilizes the x-ray tube potential as the principal variable of the AEC. It should be noted that in many cases one of the parameters in BLOCK A is employed as the principal variable of the AEC system.

Figure 1 is divided in to three columns: the hardware on the left, the electronical and the electro-mechanical controls in the center, and the user selectors on the right. BLOCK B contains the microprocessors and/or the programing boards which interface the selected factors in the right column with the respective electro-mechanical circuitries in the center column.

Let us follow the sequence of setting up the cine filming technique factors for an adult patient undergoing a cardiac catheterization examination.

1. CINE FILMING SELECTOR: set to 30 frames/sec.

   The CINE CAMERA CONTROL AND SYNCHRONIZATION stand ready for cine filming and the acutal cine filming rate along with the synchronization pulses are sent to BLOCK A for timing adjustments for x-ray pulses. The cine filming rate will significantly affect the total heat units to be stored at the anode, and hence the selected cine filming rate information is fed to BLOCK B for pre-exposure calculations.

2. The FOCAL SPOT SIZE may also be selected with the same push button as the cine filming rate selector. The power rating of the focal spot size determines the instantaneous heat loading allowable to the anode, therefore, this information is fed to the kVp-CONTROL UNIT, and the mA-CONTROL UNIT in BLOCK A as well as the POWER RATING CHECK in BLOCK B. Let us assume that the large focal spot size is selected and is rated at 1.2 mm nominal and 100 kW. The interface between BLOCKS A and B will become apparent when the cine exposure is initiated.

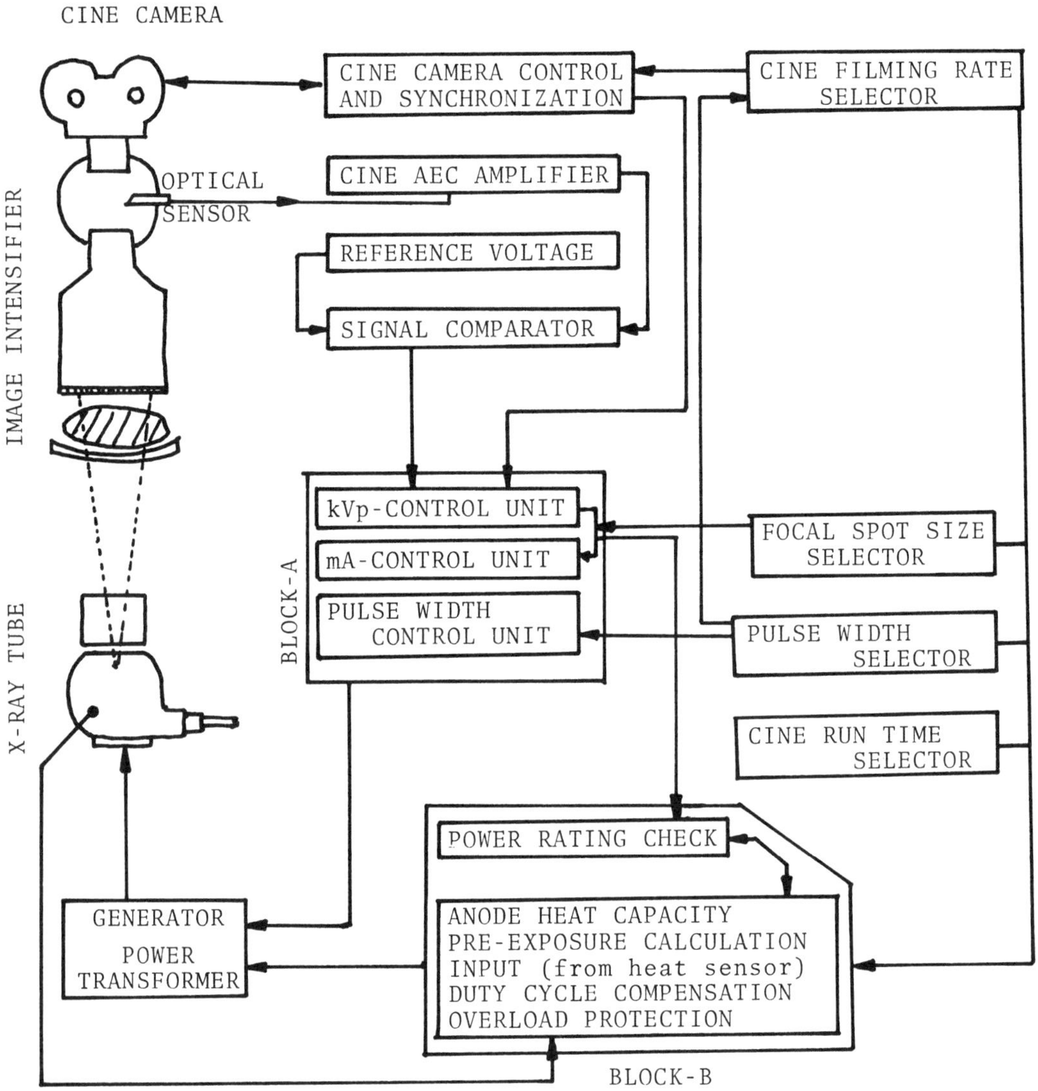

Figure 1. Schematic Block Diagram Of A Cine AEC System.

3. The PULSE WIDTH SELECTOR determines the cine pulse width to be employed for each cine frame exposure but may have some limitations due to the cine filming rate that is selected previously. The limitations are caused by the time required for proper film transport in the cine camera, and the number of frames to be exposed in a given time period. With the cine filming rate selected at 30 frames/sec, the cine pulse width can be as long as 5 mSec/pulse. The cine pulse width thus selected is then presented to the PULSE WIDTH CONTROL UNIT in BLOCK A and to BLOCK B as the pulse width will also affect the total heat units to be accumulated in the anode.
4. The CINE RUN TIME is normally preset by the installation engineer for the particular facility to meet the user's application requirements. Hence, usually this factor is not available for selection by users at the control panel. The cine run time is usually set at 10 to 12 seconds. At the end of the cine run time of 10 to 12 sec., a warning buzzer is activated and the cardiologist is then given another few seconds before the generator is automatically turned off for x-ray tube protection.

The BLOCK A is the heart of the cine AEC system. Prior to initiation of a cine run, lets assume that (a) the kVp-CONTROL UNIT drives the tube potential circuit for cine filming to the lowest potential setting of 60 kVp, and (b) the mA-CONTROL UNIT drives the tube current circuit for the cine filming to the highest current setting of 600 mA. The PULSE WIDTH CONTROL stands ready to provide 5 mSec pulses as selected. The optical sensor may be a small prism or a mirror guiding the light to a photomultiplier. The prism is so designed that it looks at the output phosphor with a field view in the range of 40 to 60% of the output phosphor area. The signal from the optical sensor is amplified at the CINE AEC AMPLIFIER and fed to the SIGNAL COMPARATOR. The SIGNAL COMPARATOR receives the signal from the REFERENCE VOLTAGE source and compares the signal voltage from the CINE AEC AMPLIFIER, and subsequently sends out a signal to activate BLOCK A in accordance with the brightness presented to the optical sensor.

Assume that the brightness is too low for proper cine film exposure and requires a higher x-ray tube potential of 68 kVp. The kVp-CONTROL UNIT activates the generator to increase the tube potential, and the AEC system may reach a stable state if the tube current rating allows 600 mA for cine filming. As the x-ray tube potential is increased the POWER RATING CHECK circuit maintains monitoring of the techniques and will activate the mA-CONTROL UNIT to decrease the tube current to 400 mA. Meanwhile the optical sensor sends another signal out to increase the tube potential again because the brightness at the output phosphor may have been acceptable at 68 kVP, 600 mA settings but is not sufficient at 68 kVp, 400 mA settings due to decreased tube current. The same process repeats itself and may reach the final steady state at 72 kVp tube potential with 350 mA tube current.

The description given above is that of an automatic iso-watt cine AEC system. Replacing the mA-CONTROL UNIT with a manual tube current selector, the cine AEC system described here becomes that of an "automatic kVp" cine AEC system with constant tube current during cine filming.

Replacement of the kVp-CONTROL UNIT with a manual tube potential selector yields an "automatic mA" cine AEC system - although this type of cine AEC system is not suitable for practical application.

Both tube potential and tube current may be manually selected and the exposure time (the pulse width) may be automatically varied. This type of cine AEC system is similar to the radiographic phototiming system, if one considers the image intensifier/optical sensor as the phototube, and each cine frame as the single radiographic exposure.

For each of these types of cine AEC systems, a minor change is required in Figure 1 in accordance with its operational logic. There are other features provided with the cine AEC systems depending on the level of the equipment sophistication.

Acceptance Testing Procedure

The acceptance testing procedure is essentially designed to (a) verify and confirm the accuracy of the physical parameters listed in Table I with the exception for items e, g, and h, and (b) evaluate and confirm the automatic/preprogrammed features of the generator/cine camera operation based on the functional and operational designs of the logic circuitries. A high voltage divider and a dual beam storage oscilloscope are the essential instrumentation to carry out such an acceptance testing. The basic instrumentation and the testing arrangement are depicted in Figure 2.

Upon confirmation of the accuracy for the primary control factors, such as the tube potential, tube current, and pulse width, one can then proceed to evaluate the operation and function of the automatic control logics. It should be noted that a few special procedure imaging systems are equipped with an array of self-monitoring sensors including a built-in high voltage divider so that there is a servo-mechanism in the regulation of the generator controls. When such a sophisticated self-monitoring, self-regulating loop is incorporated in the imaging command chain, the physical parameters that are preprogrammed can be continuously up-dated. Confirmation of such a loop mechanism must also be carried out.

In order to demonstrate the acceptance testing procedure of a highly sophisticated generator, lets utilize the Optimus M200 generator manufactured by Philips Medical Systems as an example. Without a sepcific example one cannot clearly identify some of the aspects described in previous sections. Other generators, such as the MPX generator manufactured by General Electric Company or the Pandoros 1200 A generator by Siemens Corporation, and others may otherwise be chosen as the examples as well.

Lets assume that a cine program has been selected on the control panel of the Optimus M200 generator and has the following factors preprogrammed:

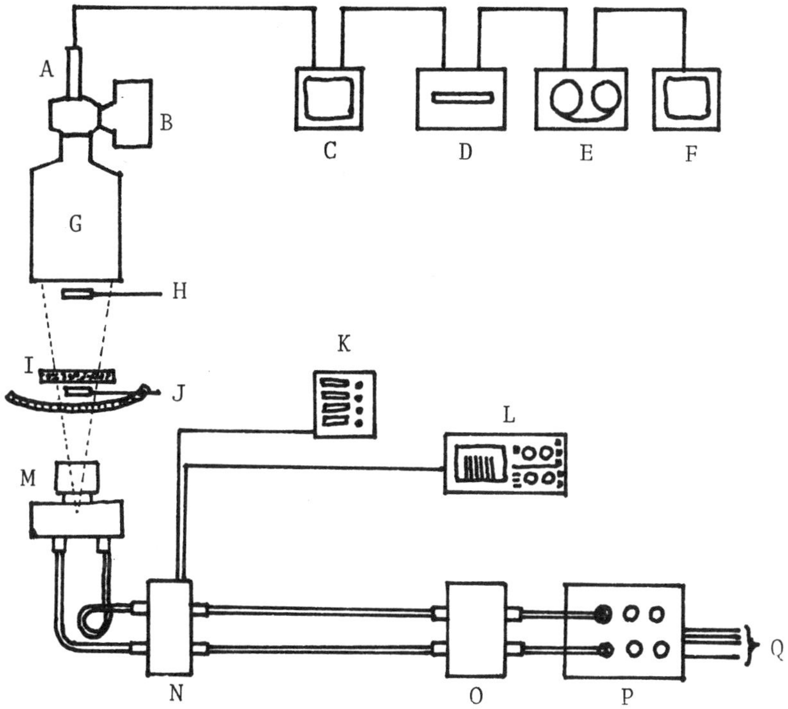

Figure 2. Schematic Illustration Of Acceptance Testing Arrangement.

A: Television Camera, B: Cine Camera, C: Television Monitor, D: Video Disc Recorder, E: Video Tape Recorder (Cassette Type), F: 2nd Television Monitor, G: Image Intensifier, H: Ionization Chamber (Inpiut Exposure Per Cine Frame), I: Copper Phantom, J: Ionization Chamber (Patient Entrance Exposure Rate), K: Digital Readout Module, L: Dual Beam Storage Oscilloscope, M: X-ray Tube, N. High Voltage Divider, O: Pulsation Transformer, P: High Voltage Transformer. Q: To 3-phase Power Supply.

1. Cine Filming Rate: 30 frames/second
2. Cine Run Time: 10 seconds
3. Cine Pulse Width: 3.2 milliseconds
4. Small Focal Spot is used and rated at 26 kW (cine power rating under these specific conditions)

Two of the advanced features that are unique with this particular generator are: the automatic optimum tube potential search logic prior to the cine exposures, and the iso-watt loading logic for cine AEC. The evaluation processes of these two features are illustrated in the following segments.

The Automatic Optimum Tube Potential Search Logic

This automatic search logic calculates the lowest possible tube potential with the first five pulses based on the preprogrammed factors listed in Table I, the cine power rating of the focal spot and the light output at the image intensifier output phosphor. The x-ray tube current during these five pulses is at a reduced current value determined by the focal spot power rating so that these five measurement pulses are obtained at minimal radiation exposure to the patient.

Under the acceptance testing arrangement depicted in Figure 2, with an appropriate amount of copper phantom placed in the x-ray beam, one can verify the operation of this automatic feature.

Depicted in Figure 3 is a series of illustrations redrawn from the photographs obtained from the storage oscilloscope. In Figure 3-A the first five measuring shots (pulses) and the first several frames of cine exposures can be observed. The wave forms shown are that of the x-ray tube potentials. The measuring shots are at (a) 63 kVp, (b) 74 kVp, (c) 96 kVp, (d) 90 kVp, and (e) 90 kVp. The cine filming is, then, to be exposed at 90 kVp. There is a 0.4 - 0.5 second duration between the end of the fifth shot (e) and the beginning of the cine run (f). This duration is needed to synchronize the generator with the cine camera. Since it takes approximately 0.5 - 0.6 seconds for the cine camera to attain the prescribed filming rate of camera speed (g) and stabilize (h) at a constant speed, there is a 0.6 second duration between the first cine exposure (f) and the first stabilized cine exposure (h). The peak (g) is due to the active automatic exposure circuitry seeking the optimum x-ray tube potential during the cine run while the cine camera filming rate is gradually gaining the speed. The x-ray tube potential required and the cine filming rate are coupled together to form a loop that seeks the lowest possible tube potential (hence, the highest possible tube current) at a given cine-filming rate. Figure 3-A reflects the functions of the kVp-search logic and the automatic control circuitry of the generator.

Figure 3-B depicts the x-ray tube potential, a few seconds after initiation of cine filming. The tube potential is at 90 kVp, consistent with the last measuring shot (e) of Figure 3-A. The filming rate is at 30 frames/sec. Figure 3-C depicts the first five measuring pulses, and Figure 3-D a few pulses corresponding to that of Figure 3-B. This series of illustrations shown in

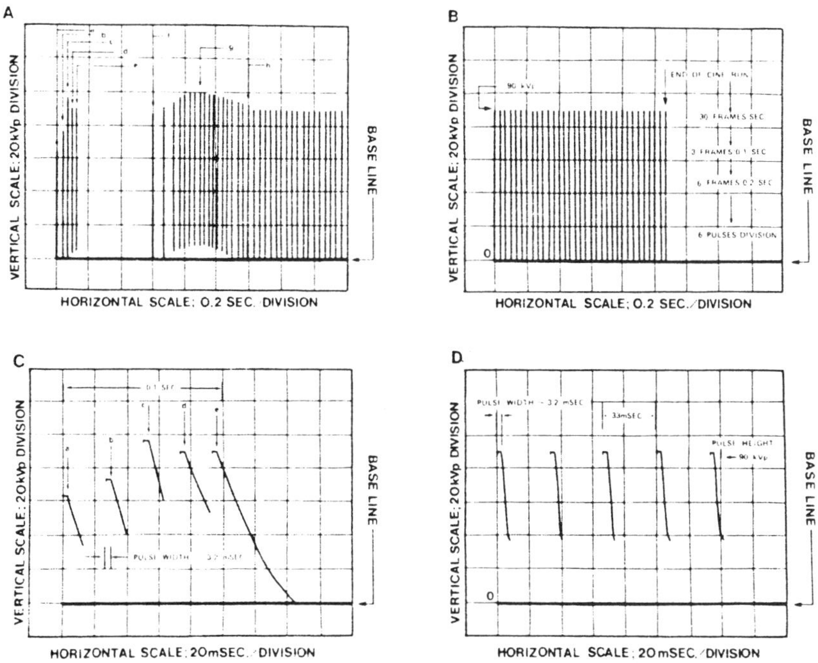

Figure 3. The X-ray Tube Potential Wave Forms for Cine Camera System Operation/Automatic Optimum Tube Potential Search Logic.

Figure 3 confirms that the filming rate and the pulse width are AS PRESCRIBED. Evidently, the same testing must be performed for all available modes of cine operation to verify the preprogrammed technique factors.

The x-ray tube current wave forms should also be verified for the five measuring pulses; in other words the tube current should be at a value one half the maximum permissible tube current at 125 kVp as designed and to be programmed into the automatic circuits for the five measuring pulses.

The Iso-watt Loading Logic

In order to maintain the constant light output with the lowest possible x-ray potential (the maximum tube current), the x-ray tube is energized at a constant power input. This is accomplished by the iso-watt loading logic. It should be noted that the x-ray tube potential is varied to obtain a constant light output but is riding on the iso-watt loading logic at the same time.

By placing various thicknesses of copper phantom in the x-ray beam, and monitor the x-ray tube potential and the x-ray tube current during cine exposures for each phantom thickness, a set of data is obtained as shown in Table II.

TABLE II DATA FOR ISO-WATT LOADING VERIFICATION

Copper Phantom Thickness (mm)	Tube Potential (kVp)	Tube Current (mA)	Power (kW)
1	59	150	8.9
2	66	355	23.4
3	75	310	23.3
4	84	270	22.7
5	93	245	22.8
6	102	225	23.0
7	111	205	22.8
8	120	195	23.4

Figure 4 is obtained by plotting the data in Table II. The cine power rating of 26 kW is that determined from the installation manuals of the generator for this particular x-ray tube. The actual power loading is approximately 23.5 kW, shown as the A curve in Figure 4. Curves B and C are for other cine programs where their calculated cine power ratings are 16 kW and 19 kW, respectively. The actual power loading for these two curves B and C are 14 kW and 17 kW, respectively.

The calculated cine power rating and the actual power loading should be in agreement within $\pm 15\%$ of the calculated value. Overloading of the x-ray tube

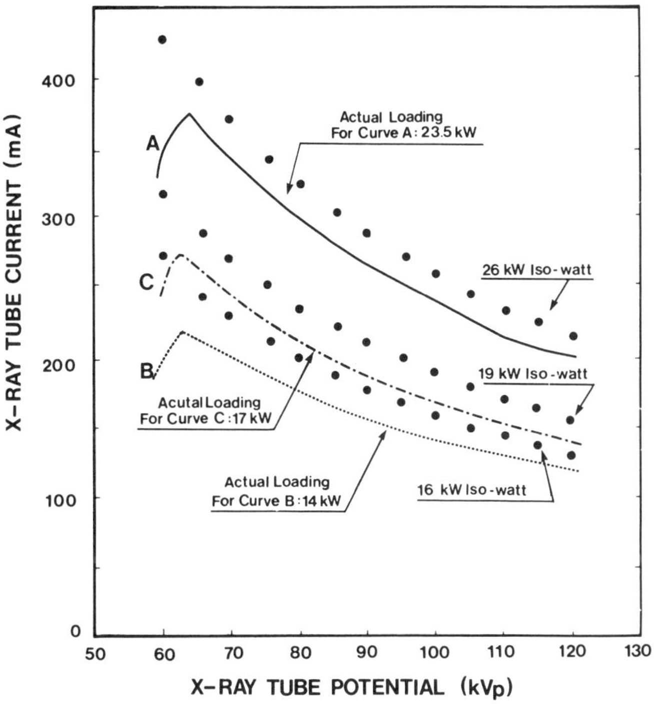

Figure 4. The Iso-watt Loading of X-ray Tube.

may result in serious damages to the x-ray tube. Fifteen percent overloading is permissible depending on the tolerance given to the original calculation chart supplied by the manufacturer. Significant underloading results in a higher x-ray tube potential for a given phantom thickness, therefore, a loss in cine film contrast occures when compared with the optimum (lower) tube potential that could have been employed. (See refernce 1.)

The Cine Film Sensitivity

In the calibration of the cine imaging system one of the most important factors is the radiation exposure required for cine filming. The cine filming sensitivity can be measured with an ionization chamber attached to the input face of the image intensifier. The grid is removed for this measurement.

The exact measurement protocol has been discussed and is being formulated by the Cine Task Force of the Diagnostic Radiology Committee of AAPM under the Chairmanship of Perry Sprawls. The cine filming sensitivity or the cine input exposure rate may be measured under a clinical geometry simulation as depicted in Figure 2. A set of data can be obtained as shown in Table III and **Figure 5**.

TABLE III CINE INPUT EXPOSURE RATE DATA (6" MODE)

Copper Phantom Thickness (mm)	Tube Potential (kVp)	Tube Current (mA)	Input Exposure Rate (μR/frame)
0.8	55	360	21.2
1.6	64	310	18.2
2.4	77	250	17.8
3.2	84	230	20.0
4.0	96	200	23.3
4.8	102	194	25.5

Since the input exposure rate is not constant over the range of the phantom thickness or the x-ray tube potential, the acceptance testing is conducted to verify the initial calibration set-up of the installation as prescribed by the manufacturer. The initial calibration is typically obtained at 80 kVp tube potential and set at the cine AEC sensitivity of 10 μR/frame for 9" mode and 20 μR/frame for 6" mode operation of the image intensifier. From Figure 5 (see next page) it can be said that this experimental data confirms the initial calibration of cine input exposure rate of 20 μR/frame.

Summary

The physical parameters involved in the cine camera system and the automatic operations of the cine AEC system have been described to assist in the formulation of the acceptance testing procedure. Some examples are shown for the automatic features provided to the generators specially designed for cardiac catheterization procedures and/or angiographic examinations.

There are many other automatic features that must be evaluated and verified for proper functions and operations. The emphasis of the acceptance testing for special procedures equipment is placed on the operational logics of automatic and/or preprogrammed features. Therefore, the acceptance testing must be carefully taylored to the specific generator under testing. To logically

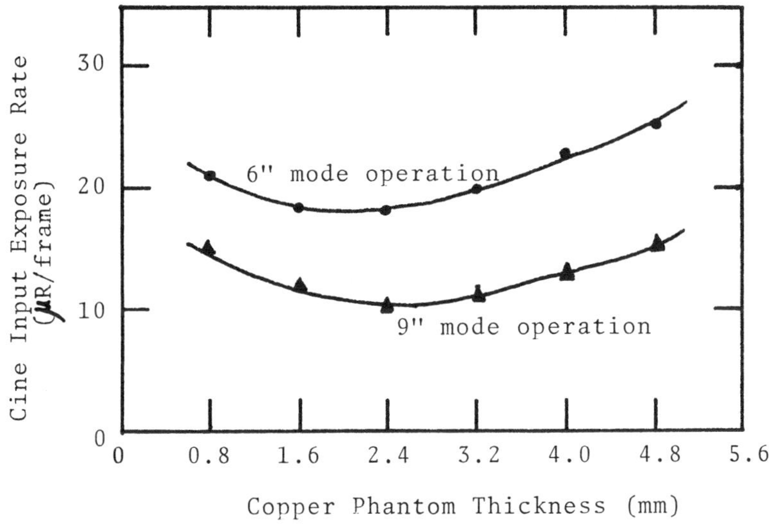

Figure 5. The Cine Input Exposure Rate Curve

evaluate the correct functioning of these automatic features, understanding of the equipment logics and design principles are essential and necessary.

Of ten special procedures suites (five cardiac catheterization and five neuro/general angiographic suites) that the author had involved in the acceptance testing, various miscalibrations and/or misprogramming have been found in addition to numerous malfunctionings. This is due to the complexity of the equipment and the high level of expertise required on the part of the installation engineers to accomplish the sophisticated tasks.

The acceptance testing should certainly be conducted for these specialized high-technology equipment to ensure proper equipment operation so that the full capacity and capability of the equipment can be utilized at its fullest extent.

References

1. Optimus M200 Generator: Book 1 and 2. Philips Medical Systems, Inc. Shelton, Connecticut.

2. P.P. Lin: "Selection of Phantoms and the Measurement Protocol for Cine Photofluorographic Imaging Chain Sensitivity", presented at the 22nd Annual Meeting of AAPM, July, 1980.

3. P.P. Lin: "Equipment Testing Procedure: Cardiac Catheterization Laboratory", presented at the Spring Meeting of the Midwest Chapter of AAPM, February, 1981.

4. P.P. Lin and S.N. Meyers: "Performance Evaluation and Acceptance Testing of Automatic Exposure Control Functions for Optimus M200 Generator Cine Photofluorographic Operation", Medicamundi (IN PRESS).

Acceptance Testing of Image Intensifier Imaging Chain

Pei-Jan Paul Lin
Department of Radiology, Northwestern University Medical School
Chicago, Illinois 60611

Abstract

The imaging characteristics of the image intensifier and its associated imaging chain/devices are measured and evaluated as part of the acceptance testing for radiological imaging equipment utilizing the image intensifier. A description and measurement procedures for such an acceptance testing of the image intensifier are discussed.

I. Introduction

Image intensifiers are employed for fluoroscopic and photofluorographic imaging operations in various radiological studies. The imaging properties of an image intensifier can be characterized by a series of physical parameters/quantities such as the conversion efficiency, flux gain,veiling glare, in addition to the resolution characteristics.

The acceptance testing which is primarily concerned with the performance evaluation of the equipment/devices in the clinical environment (or the field). It cannot encompass such elaborate measurements as the conversion efficiency, the flux gain,etc.without sophisticated instrumentation similar to a laboratory arrangement, nor is it practical to conduct field measurements of the same. Hence, the acceptance testing of an image intensifier in the field is somewhat limited but should include the physical factors that affect the clinical application of the complete imaging system.

The following factors are the principal parameters that should be included in the acceptance testing assuming that the image intensifier is equipped with a 100/105 photospot camera, and a television camera/monitor chain:

1. Input exposure rate for fluoroscopic operation.
2. Input exposure/frame for photospot camera operation.
3. The resolution in the center as well as periphery of the image intensifier coupled with the photospot camera and the television chain.
4. The distortion.
5. The low contrast visibility.
6. Function of automatic brightness control (ABC).
7. Rapid Sequencial Exposure Operation of Photospot Camera.
8. Function of automatic exposure control (AEC) for photospot camera.

Items 6 and 8 have been discussed and can be referred to a publication by the author in AAPM's Monograph No. 4. Item 7 is similar in nature to that of the acceptance testing for a cine camera system and discussed elsewhere in this proceeding. Hence, items 6 through 8 will not be covered in this paper.

II. Acceptance Testing Procedure

The acceptance testing procedure will be described in the order of the list presented in the previous section. Appropriate instrumentation and test devices will be introduced as required in the process of describing the test procedures.

II-1 Input Exposure Rate

The input exposure rate at the input phosphor of the image intensifier is important because the image intensifier can be looked upon as a converter. The image intensifier simply converts the x-ray photons to light output through its electron-optical system. The initial calibration of the image intensifier input exposure rate thus determines the amount of radiation required to properly represent the image at the output phosphor for viewing by its subsequent imaging chain. The initial calibration of the input exposure rate for a typical dual mode image intensifier is summarized in Table I. The calibration procedure differs from one manufacturer to another, but the x-ray tube potential may be set to 80 kVp with an appropriate phantom such as 1.6 mm of copper.

TABLE I THE INPUT EXPOSURE RATE CALIBRATION

	Fluoroscopic Operation (μR/sec)		Photospot Camera Operation (μR/frame)
	Low Level	High Level	
6"	80	160	160 (80)*
9"	25-50	50-100	100 (50)*

* For 70 mm Photospot Film

In the measurement of the input exposure rate, especially for ABC regulated fluoroscopy, the amount of copper phantom may have to be varied so that an interpolation of the input exposure rate can be determined at 80 kVp tube potential if the system is "kVp-variable" ABC. The acceptance testing of the input exposure rate may be conducted under an arrangement shown in Figure 1.

The grid installed at the image intensifier input window and/or in the

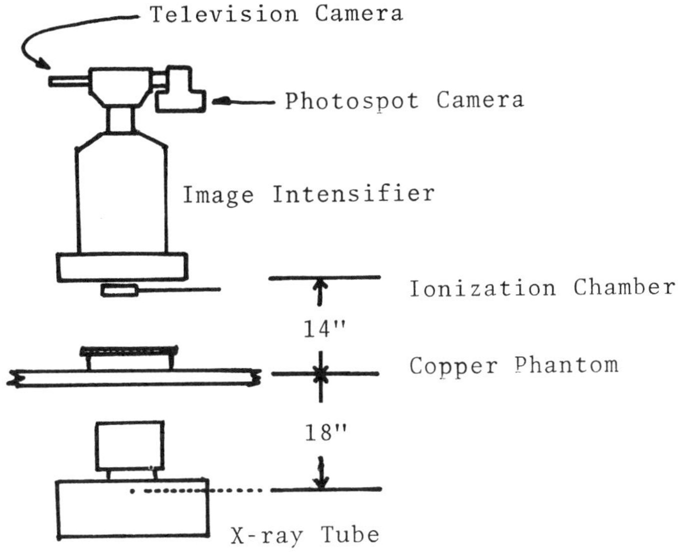

Figure 1. Schematic Illustration of Input Exposure Rate Measurements.

full size cassette spot filming device must be removed. The ionization chamber should be positioned as close as possible to the image intensifier input phosphor plane. Depending on the particular equipment some difficulties in achieving the ideal geometrical condition must be resolved. Alternatively, the values of the input exposure rate should be reassigned, taking into account the geometry and other modifications.

II-1-A Fluoroscopic Input Exposure Rate (μR/sec)

Figure 2 (see next page) depicts a typical input exposure rate curve plotted for a 9" mode operation of an image intensifier. The fluoroscopic input exposure rate varies with the phantom thickness (or the x-ray tube potential) due to the x-ray energy dependence of the image intensifier input phosphor material. For manual fluoroscopic operation and manual-kVp ABC operation, the tube potential can be set to 80 kVp - adjust the x-ray tube current such that the input exposure rate is approximately 50 μR/sec to 100 μR/sec.

II-1-B Photospot Camera Input Exposure Rate (μR/frame)

In the case of photospot camera input exposure rate measurements,the geometry remains the same as in Figure 1. The photospot exposure technique is set at an x-ray tube potential of 80 kVp and the tube current stations to 200 mA - 400 mA (fixed current system) or the falling load operation. For the phantom thickness in the range of 1.6 mm to 3.2 mm of copper the input exposure rate should be 100 μR/frame (±15%) for a 9" image intensifier operation with a

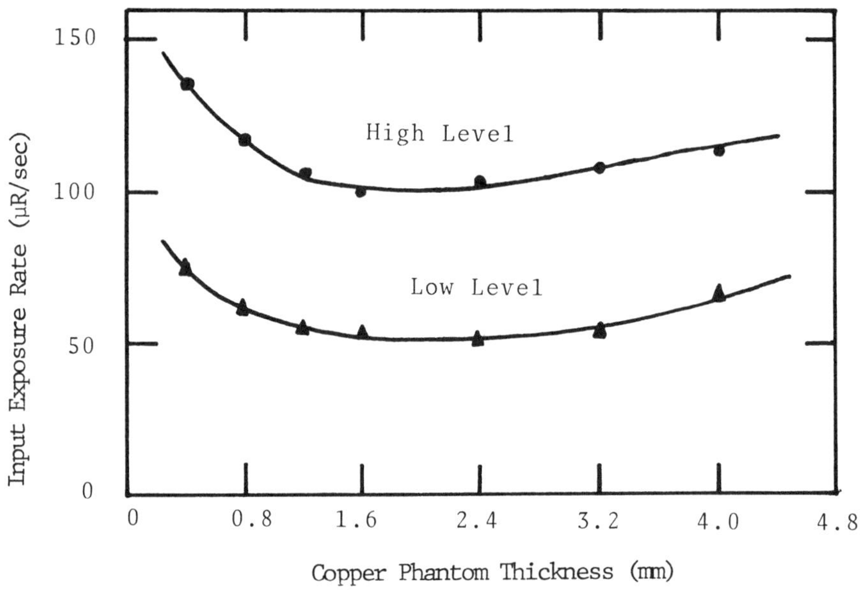

With a 1.6 mm copper phantom, the x-ray tube potential should be in the range of 75 - 80 kVp. When 3.2 mm copper pahntom is placed in the x-ray beam, the tube potential should be in the range of 90 - 95 kVp.

Figure 2. Fluoroscopic Input Exposure Rate vs. Copper Phantom Thickness. (The ABC is of kVp-varial type).

100/105 mm photospot camera. Thereafter, the photospot camera film density is adjusted to an optical density (O.D.) of 1.3 $\pm$ 0.10 utilizing the iris in front of the photospot camera lens.

II-2 Resolution Measurements

The resolution as measured by the line-pair test target is best accepted by the manufacturer, although the mesh plate has become widely accepted in the past five years. Since the line-pair test target is a high contrast object, the resolution measurements indicate the high contrast response performance of the image intensifier imaging chain.

II-2-A Fluoroscopic Resolution

The geometry for resolution measurements is similar to that of Figure 1. The grid is again removed. The line-pair test target should be positioned such that the line pairs appear on the television monitor at a angle of 45° to the television scanning lines. The x-ray tube potential should be set to the lowest possible value such as 50 kVp and adjust the x-ray tube current accordingly so that the image intensifier is protected from excessive radiation bombardment. The line-pair test target should be positioned at the input phosphor plane or as close as possible. The distance between the x-ray source and the image intensifier should be at its maximum to further reduce the influence of the x-ray focal spot size.

II-2-B Photospot Camera Resolution

Under the same condition as the fluoroscopic resolution measurement, place a phantom of appropriate thickness and material (such as 0.8 mm copper or 3/4" aluminum plate) so that the photospot camera exposure can be properly obtained.

The results of the resolution measurements should agree with the original specifications published by the manufacturer. A typical acceptance criteria or the specification for a dual mode image intensifier is listed in Table II.

TABLE II IMAGE INTENSIFIER RESOLUTION SPECIFICATION

Image Display	Image Intensifier Operational Mode*			
	6" Mode		9" Mode	
	Center	Periphery	Center	Periphery
Television	1.8	1.5	1.5	1.2
100/105 mm Photospot	3.5-4.0	3.2	2.5-3.4	2.2
35 mm† Cine Camera	2.2-2.5	1.8	1.5-2.0	1.2

* For a high quality CsI 6"/9" dual mode image intensifier.

* The unit is line pairs/mm.

† The resolution performance for a 35 mm Cine Camera is supplied for reference.

It should be noted that the best resolution of an image intensifier imaging chain can only be obtained when all imaging components are correctly focused. The focusing of the television chain including components such as the television monitor focusing and the photospot camera lens must also be correctly adjusted to yield the optimum resolution of the entire imaging system.

II-3 Distortion

The pincushion distortion can be measured with a rectangular grid and/or a polar mesh. The retangular grid may be designed with a piece of plastic onto which a grid pattern can be inscribed and filled with lead wire. The polar mesh is similarily fabricated but the pattern on the plastic is a number of concentric circles with four straight lines passing through the center.

Figure 3 depicts the rectangular grid and the polar mesh plates and the corresponding images that might be seen on a television monitor or a photospot camera film. The pincushion distortion as measured by the rectangular grid plate will be described in this article. A detailed discussion on this subject has been presented by L.W. Casperson, et. al (see reference 3).

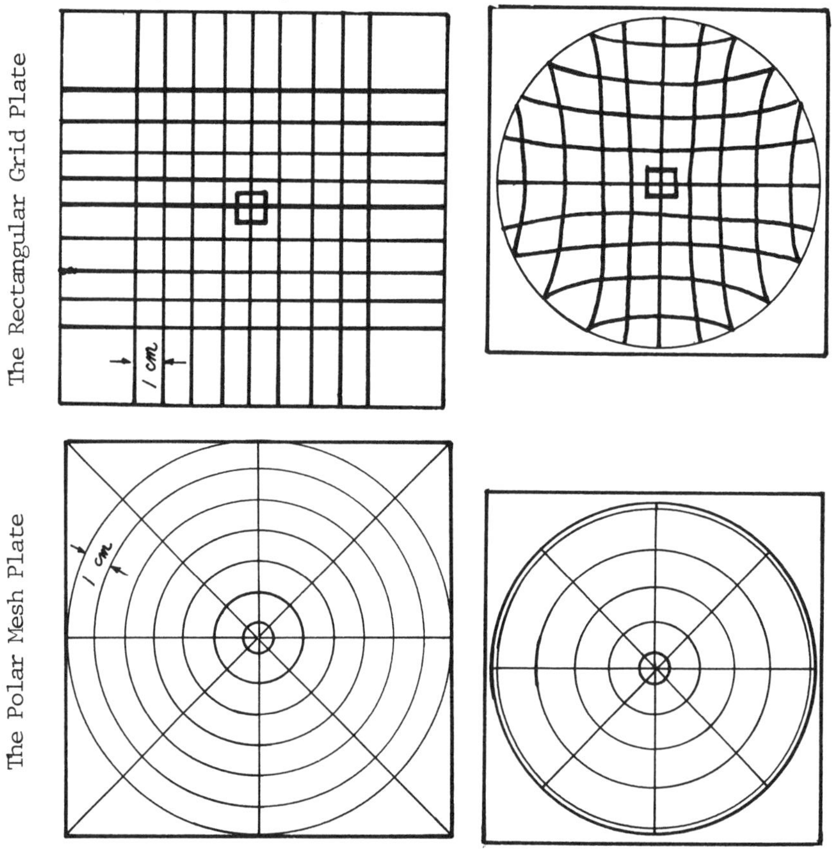

Figure 3. The rectangular grid plate and the polar mesh plate (left side) and their corresponding images (right side).

The percent distortion (P.D.) is expressed as: $P.D. = 100\left(\frac{L'}{L} - 1\right)$

Where L' is the actual length of the grid image measured from the photospot film or the television monitor at the point of interest, and L is the actual length of the grid inside the square shown in Figure 3. Therefore, the distortion at the center of the image intensifier is assumed to be zero; the distortion is measured as "normalized" to the center distortion. It should also be noted that the rectangular grid plate is attached as close as possible to the image intensifier input phosphor plane for P.D. measurement. The P.D. is normally specified as 10% or 15% without identifying the location the P.D. is measured. However, it is understood that the P.D. is measured at the points on a circle with a radius 80 - 90% of the image intensifier active input phosphor size. In other words, for the 9" mode operation the P.D. is obtained at points approximately 3.6" - 4" from the center.

The P.D. measured is the total percent distortion including, not only the image intensifier distortion, but also the distortion due to the television camera lenses and the television CRT tube distortion or the photospot camera lenses. The measured P.D. therefore is normally somewhat worse than the specified values.

II-4 Low Contrast Visibility

The low contrast visibility for the image intensifier can be determined by using the penetrameter method such as that described in AAPM report No. 4. The penetrameter consists of two aluminum plates (3/4" thick) and one aluminum sheet (1/32" thick). The 1/32" thick aluminum sheet has a series of holes drilled ranging from 1/16" to 1/4" in diameter.

A new fluoroscopic installation operated at 80-100 kVp with the entire penetrameter placed on the examination table should be capable of displaying the 1/8" hole. The 6" CsI image intensifiers should be capable of displaying the 1/16" hole on the television monitor. The focusing of the image intensifier and the television chain is assumed to be correctly adjusted prior to this test.

III. Discussion

The evaluation methods described in this article may not encompass all desired tests to verify the image intensifier performance as published by the equipment manufacturer. Measurements of more sophisticated nature such as the conversion efficiency or quantum detection efficiency, the flux gain, and the modulation transfer function (MTF) are not quite feasible in the field, although not impossible. The MTF measurements of the image intensifiers have been conducted by Lin, et. al. (reference 5) but cannot be routinely performed.

The functions of the ABC circuitry, the photospot camera, and the AEC circuitry for the photospot camera must be included in the acceptance testing. However, these items overlap with the acceptance testing of the generator, the ABC, and the AEC devices and are being discussed elsewhere in this proceeding.

References

1. P.P. Lin, "Acceptance Testing of Automatic Exposure Control Systems and

Automatic Brightness Stabilized Fluoroscopic Equipment", pages 10-27. Quality Assurance in Diagnostic Radiology, AAPM Monograph No. 4, American Institute of Physics 335 East 45 Street, New York, N.Y. 10017.

2. P.P. Lin and R.J. Kriz, "Variation of X-ray Tube Potential as a Function of Attenuator Thickness in kVp-Variable Automatic Brightness Stabilization Fluoroscopy", Application of Optical Instrumentation in Medicine VI, Proceedings Society of Photo-Optical Instrumentation Engineers, 127: 152-157, 1977.

3. L.W. Casperson, P. Spiegler, and J.H. Grollman, Jr., "Characterization of Aberrations in Image-Intensifier Fluoroscopy", Medical Physics, 3, No. 2, 103-106 March/April, 1976.

4. Basic Quality Control in Diagnostic Radiology, AAPM Report No. 4., Pages 28 and 29 under section O, automatic brightness control.

5. P.P. Lin, H.L. Neiman, R.A. Mintzer, and S.N. Meyers, "Performance Evaluation of Image Intensifiers Coupled with Photofluorographic Camera and Their Clinical Application". Proc. of SPIE and Optical Engineering. IN PRESS.

SHAKE-DOWN OF RADIOLOGICAL EXAMINATION ROOMS: ACCEPTANCE TESTING BY RADIOLOGIC TECHNOLOGISTS

Stephen Doerner, B.S.R.T., M.B.A., James Bova, B.S.R.T., and Pei-Jan Paul Lin, Ph.D.
Department of Diagnostic Radiology, Northwestern Memorial Hospital
and
Department of Radiology, Northwestern University Medical School
Chicago, Illinois 60611

Introduction

The purpose of acceptance testing for radiological installations is to verify the equipment performance, calibration and safety as defined in the purchase specifications. Although there is a variety of testing equipment and procedures which can be incorporated in the evaluation of equipment performance, two distinct stages exist in the acceptance testing process. The first and most intrinsic is the invasive testing performed by the radiologic physicist and/or in-house service engineer. The second stage in the acceptance testing process involves the evaluation of equipment performance through non-invasive techniques. Included in this stage is a process called "de-bugging" or "shake-down". This is defined as performance evaluation of radiological installations based upon simulated clinical applications.

The unique feature of the acceptance testing process developed at Northwestern Memorial Hospital (NMH) was that radiologic technologists assumed an active and vital role in the performance evaluation of new radiological imaging equipment, thus providing a complete viewpoint for practicality of clinical use. The objective of this paper is to discuss the role of the radiologic technologist in the acceptance testing process as performed at NMH. The step-by-step procedure employed to "de-bug" a typical radiographic/fluoroscopic room, tomographic room, and special application room such as automated chest unit will be discussed.

De-bugging and Shakedown

The objective of the "de-bugging" process is twofold. First, to optimize equipment performance prior to turnover to department personnel. Secondly, to establish equipment performance data which serves as a standard for future equipment performance. Successfully achieving these objectives results in the additional benefits of (1) greater overall department efficiency due to less down time, (2) higher staff morale from confidence in system equipment operations, and (3) reduction in patient radiation due to less films being repeated. At NMH the technologists involved in the acceptance testing were part of the quality assurance office which is under the auspices of the diagnostic radiological physicist. The diagnostic radiological physicist at NMH is soley dedicated to the diagnostic radiology department, and provides technical direction relative to the test procedures. The technologists are, in turn, responsible for performing the tests applicable to each specific room. However, most hospitals are not able to realize this approach. Acceptance testing is often performed by radiation therapy physicists, or nuclear medicine physicists, (if performed at all) who are not always as familiar with the practical or clinical operations of the equipment in diagnostic radiology. Therefore, the incorporation of radiologic technologists in the acceptance testing process to perform "de-bugging" and "shakedown" will be highly beneficial for such hospitals where the benefit of having their own dedicated departmental physicists cannot be realized due to the small size of the department.

Radiographic/Fluoroscopic Room

The most typical room in a diagnostic radiology department is the radiographic/ fluoroscopic unit which consists of a x-ray generator, a ceiling mounted tube, and a table with fluoroscopic carriage.

Generator

The "de-bugging" process begins with the x-ray generator. The technologists are responsible to verify the performance of each generator based upon the manufacturers specifications as viewed from clinical applications. In addition, each indicator light in the generator panel will be tested for proper functioning. The automatic exposure control (AEC) selection switches must be tested to verify that when, for example, the wall bucky is selected, the wall bucky AEC system is indeed activated and all other systems are not active. Furthermore, individual chamber selection switches must be tested that when, for example, the left outer cell is selected other AEC cells of the same AEC subsystem remain inactive, and only the left outer cell is functioning. Proper functioning of the back-up timer and tube limit circuits must be verified. These indicators should be checked against the tube-rating chart and the 600 mAs exposure limit as established by BRH (1). The anode speed rotation control must be tested to insure that when in the automatic mode, the anode speed of rotation will properly adjust itself to the radiographic technique selected. The mA and/or mAs meters must be checked for proper calibration. The anode heat calculator must be checked to verify that when the overhead, lateral, or fluoroscopic tube is selected, the calculator will function properly and retain the heat units of the other tubes in memory so that when the other tube is selected the display should show the pertinent heat units in percentage accordingly. The fluoroscopic timer must be tested for an audible alarm or buzz after five minutes of fluoroscopic radiation has elapsed. Verification of these functions and any additional functions of the generator should be based upon information supplied in the manufacturers' specifications.

Ceiling Suspended Tube

The most frequent test performed is the sizing of the positive beam limitation (PBL) device. A variety of testing procedures and devices exist which all have the objective of insuring that the size of the radiation field does not exceed the size of the image receptor. While the tolerance established by BRH is 2% of the source-to-image receptor distance, most manufacturers are able to fine tune the PBL system to within 1% of the SID. Included in this compliance testing for PBL is the verification of the override systems in the manual and semi-automatic modes. Another pertinent test performed is light radiation congruency. The objective of this test is to assure that the light-radiation is within 1% of the source-to-image receptor distance. This must be checked at 6"x6" and 17"x17" to maintain congruency at all sizes. The eight-penny test or a commercially available device may be used to perform and verify congruency. At NMH, a PBL testing device has been developed and will be discussed elsewhere in this proceedings (2). The half-value layer of the radiation beam is determined by using a digital dosimeter. The exact procedure for performing this test is outlined in the AAPM report - No. 4 under section "Half-value Layer" (3). The filtration must equal or exceed the minimum values required by BRH for use in the diagnostic range of 50 to 150 killivolts.

Focal spot size measurements are determined by using star test pattern. A 1°pattern is used on focal spots with nominal size smaller than .6 mm, a 2° pattern is used on focal spots sizes from .6 mm to 1.5 mm. Non-screen film is used to record the star test pattern image under geometry which provides a magnification factor of approximately 2.0X to 2.5X. All focal spot size measurements should be performed at 75 kVp tube potential and one half the maximum rated tube current (mA) of each focal spot (4).

A light meter must be used to inspect the collimator field light intensity to insure compliance with BRH regulations. The minimum acceptable allowance is 15 foot candles at a 100 cm source-to-image receptor distance. In addition, all patient identifications such as I.D. cameras for film cassettes and automatic I.D. flashers including 100 mm photospot cameras should be adjusted to produce an optical density of approximately 1.2 on the properly developed radiographs or photographs.

All locks provided with the x-ray tube housing mounting and ceiling suspension, i.e. longitudinal, vertical, horizontal, and angulation must be tested to insure proper operations and release under all clinical conditions. The cable drape should be set-up so that it does not interfere with any examinations performed. The tube support should be tested for smoothness of motion from minimum to maximum distances in height and the in-room movements. In multi-tube rooms, the ceiling rails of each tube must not interfere or inhibit the mobility and operation of the other tube and suspension column.

Examination Table

The initial test performed on the table was to check the alignment of the x-ray tube and the center of the bucky and grid. This was accomplished by placing a 10" by 12" cassette in the bucky tray. A 1·1/2"X 1/8"X1/8" lead bar was attached to the center line of the grid. The radiation beam is collimated to 6" by 6". A single exposure is taken using low tube current such as 25 mA or lower and a long exposure time.

Figure 1 demonstrates correct alignment and proper grid reciprocation. If A is not equal to A', the tube is not centered correctly to the bucky tray in the transverse direction. If B is not equal to B', the grid is not reciprocating about the central axis. If C is not equal to C', the tube is not centered correctly to the bucky tray in the longitudinal direction. During this test, the grid is verified for proper ratio and focusing distance. If a radiograph of the grid is not provided by the manufacturer, one should be taken to evaluate grid conditions. A visual inspection is made to verify simultaneous operations of the table tilt and tabletop drive. Special attention is directed to all locks and potential binding problems as the tabletop is extended over its full range of travel. Notice that the tabletop must be loaded with the appropriate weight to account for the patient weight. As described previously, the AEC device in the table is tested to insure proper balance and exposure termination when selecting any combination of chambers. Proper balance of the AEC devices is accomplished by using a 1 mm copper filter positioned on the tube collimator. Radiographs were taken at 80 kVp of each cell and then all combination of cells. All radiographs should exhibit optical densities equal to individual cells as well as any combination of cells within ± 0.1 O.D. Adjustments to produce proper diagnostic film densities for the various film-screen combinations is performed as well as testing of the major and minor density selections to verify percentage of change according to manufacturers' specifications. The center line of the tabletop must be aligned with the center line of the wall bucky system when the x-ray tube support is integrated with the table.

The evaluation of the fluoroscopic imaging system including such tests as the automatic brightness stabilization, image intensifier performance, photospot camera focusing and alignment, and radiation field size are performed. Testing procedures and expected results are outlined in references (2) and (3). Therefore, discussions with respect to this subject will not be included here.

The multi-format film divider is checked for proper operation as well as compliance with PBL requirements by performing collimation tests using all available framing modes such as; 2 on 1 and 4 on 1 modes (refer to reference 3 for detail). All movements and interlocks of the overhead carriage tested as well as all accessory equipment including patient handles, footboards, shoulder braces and compression bands for installation, movement, and overall safety.

Tomographic Room

All of the previously mentioned procedures performed on the generator and tube collimation and support systems are performed for the tomographic examination rooms. One of the most complete material available for the conventional testing of tomographic equipment for linear tomographic equipment can be found in HEW publication (FDA) number 80-8096 entitled "Quality Assurance for Conventional Tomographic X-ray Units" (5).

The shakedown procedure for tomographic operations begins by setting-up the unit in the tomographic mode. A visual check is performed for smoothness of motion, of tube column and film carriage, and stability of pivot points i.e., at the x-ray tube support, the cut-level selector and bucky connection. The cut-height indicator and cut-thickness are evaluated using a commercially available tomographic test phantom. Tests are performed using all available arcs, and several cut heights applicable to diagnostic procedures (3).

Automatic Chest Unit

All tests previously mentioned relative to the generator and the tube support collimation are performed on the automatic chest unit. In addition to these, complete automatic exposure control device testing is performed, as well as verification and inspection of the grid condition; e.g. grid ratio, grid focal length, etc.

If the unit is equipped with an automatic drive, the film changer should be moved over the complete range of travel paying specific attention to binding problems that may exist at the maximum and minimum heights. The servo-drive mechanism which aligns the tube to the film changer must be checked at all usable heights. The film transport system must be tested for smoothness of operation through the entire film con eyer mechanism. There shall be no light leakage in the entire system. A variety of automatic processors are available and the optimization of processing conditions must be established in order to maintain the optimum quality of radiographs. Specifics of these parameters should include optimizing the developer temperature and replenishment rates to the film and chemistry used. Continuation of this process should be implemented in daily quality assurance activities.

One of the most important tests to be performed on an automatic chest unit is film-screen contact. This is done utilizing a brass wire mesh phantom which is placed in intimate contact with the front panel of the automatic film changer. An exposure is made to produce a background density of approximately 1.0. The film is best viewed at about a 6ft distance in order to evaluate the uniformity in sharpness of the wire images. Provision should be made to clean and inspect the intensifying screens at regular intervals of 3 to 4 months.

Ancillary Equipment

Upright Wall Bucky: the vertical wall bucky is visually inspected for its general appearance and installation as it applies to its clinical function. All vertical buckys should be checked for proper counterbalancing and operation of its vertical lock. Additional features such as rotation, horizontal or pivoting movements are tested along with their respective locks. The interlock system which prohibits exposures should both the wall and table bucky trays be engaged must be checked. The grid is also inspected to insure that it is of the proper ratio and focusing distance. The alignment of the grid and the bucky tray to the tube and collimator are tested as well as the proper grid reciprocation. Specifics to these procedures can be found under the table section.

Radiographic Illuminators: radiographic illuminator brightness is tested to insure uniformity and consistency throughout the entire department. The procedure involves measuring the brightness in the center and each of the four quadrants of the illuminators in foot candles using an electronic flux meter (6). All lightbulbs should be checked to see that they are of the same brand and type. An additional feature of some viewboxes is an activator switch to illuminate the viewer when a film is placed onto it. Verification of the performance of this feature in all viewboxes should also be conducted.

Conclusion

The experiences discussed in this paper have proven that radiologic technologists in quality assurance can provide significant contributions to the entire realm of acceptance testing of radiological imaging equipment. Their involvement and input constitute a dimension of practicality sometimes overlooked by physicists and engineers. However, here at Northwestern Memorial Hospital, the support and direction demonstrated by the Department's Chairman, Administrative Director, and Diagnostic Radiological Physicist allowed this dimension in acceptance testing to become a successful reality.

References

(1) Regulations for the Administration and Enforcement of the Radiation Control for Health and Safety Act of 1968. HEW Publication (FDA) 76-8035. January, 1976.

(2) PP Lin, RJ Kriz, and R Stortzum, "An Alighnment Testing Device For Compliance Testing Of PBL Automatic Collimators", In this proceeding.

(3) AAPM Report No. 4, Basic Quality Control In Diagnostic Radiology. Published by the American Institute of Physics for the American Association of Physicists in Medicine. New York, New York, 1978.

(4) NEMA Standards, Test Method for X-ray Equipment XR5-1974. National Electrical Manufacturing Association, Washington, D.C.

(5) Wm. Hendee, and RP Rossi, "Quality Assurance For Conventional Tomographic X-ray Unit," HEW Publication (FDA) 80-8096, 1979.

(6) Panlux Electronic Luxmeter, Manufactured by Gossen GMBH, Postfach 1780, D-8520 Erlangen, West Germany.

Figure 1.

Grid Reciprocation And X-ray Tube/Bucky Alignment Test.

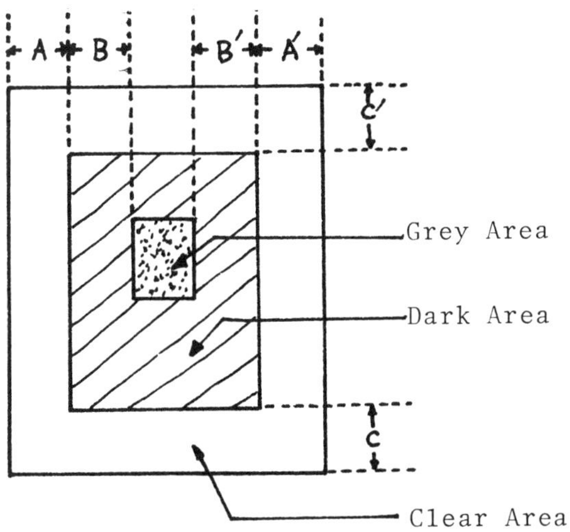

X-RAY ROOM BARRIER ACCEPTANCE TESTING

Keith J. Strauss, M. S.
Department of Medical Phyics
Michael Reese Hospital And Medical Center
Chicago, Illinois 60616

Introduction

This paper describes a practical method which allows detection (qualitative checks) of radiation field locations outside room barriers followed by quantitative measurements of the radiation field exposure levels. The quantitative measurements allow one to estimate the expected radiation dose to individuals in the vicinity of the room. The protocol for completing the barrier analysis includes the following sections: investigation, visual inspection, qualitative measurements, quantitative measurements, data interpretation, record keeping, and recommendations. Appropriate choices of qualitative and quantitative survey meters are discussed. Mathematical corrections for errors introduced by the quantitative survey meter are listed. Since the barrier analysis protocol uses the radioisotope Tc-99m as the radiation source for qualitative measurements, a radiation protection program designed to minimize radiation hazards to personnel in the vicinity of the radioisotope is discussed. A typical example of barrier analysis data is presented in the concluding section.

Unintentional radiation exposure of the patient, the general public, and radiology personnel due to x-ray fields inside and outside a diagnostic x-ray room must be controlled. National advisory groups, such as the National Council on Radiation Protection and Measurements (NCRP), have published extensive recommendations designed to protect individuals from x-ray exposure (1,2,3,4,5). State or local governments determine and enforce regulations based on the advisory group's recommendations. If state jurisdiction does not exist, the Joint Commission on Accreditation of Hospital's (JCAH) or Occupational Safety & Health Administrations's (OSHA) regulations control the use of x-rays with respect to radiation hazards (6).

The State of Illinois' regulations concerning the control of x-ray radiation hazards address the following (7):

1. X-ray field limitation and alignment
2. Beam filtration
3. Timer operation
4. Exposure switch operation and location
5. Minimum source skin distance (SSD)
6. Clinical exam entrance exposure levels
7. Use of personnel protective devices (aprons, etc.)
8. Use of patient protective devices (gonad shields, etc.)

9. Interlocks on primary barriers incorporated into equipment
10. Tube housing shielding
11. Personnel monitoring and recording
12. Room barriers

All of the above protection parameters except room barriers must be surveyed periodically during the lifetime of the equipment; usually an annual frequency is stipulated. While all 12 parameters affect patient or personnel exposure, the scope of this paper is limited to the analysis of room barriers.

Complete analysis of an x-ray barrier's effectiveness requires two types of measurements. A qualitative analysis of all sections of shielding, joints, and voids is completed to detect the shape and location of radiation fields outside the room barriers. Measurement of the exposure levels of the detected radiation fields follows. Knowledge of all the x-ray field locations from the qualitative analysis allows the surveyor to measure the MAXIMUM exposure levels outside the room while minimizing the number of quantitative measurements and subsequent heat load on the x-ray tube.

Description

The State of Illinois' Regulations (7) state that "Each installation shall be provided with such primary barriers and/or secondary barriers as are necessary to assure compliance" with the personnel dose levels in Table I:

Table I

Allowed Radiation Dose Levels

	rems/Calendar Quarter		
	Restricted Area	Restricted Area 16-18 yr old	Unrestricted Area
Whole Body	1.25	0.125	0.125
Hand, Forearms	18.75	1.875	1.875
Feet, Ankles	18.75	1.875	1.875

A qualitative analysis of barriers should be initiated prior to the installation of shielding; a quantitative analysis is required after the installation is complete. The following protocol describes the complete survey and indicates the time when each step is most easily completed. It is adapted from recommendations of the NCRP (5). The discussion is limited to installations using lead, but could easily be adapted to concrete installations. Two different radiation sources and survey meters are needed, one of each for qualitative and quantitative measurements.

Barrier Survey Protocol

Investigation

Prior to the installation of specified shielding, basic information should be collected concerning the facility. This information allows one to plan survey measurements and to evaluate collected data. Architectural drawings, the installation's shielding design, and personnel in the Radiology Department should be consulted to learn how the x-ray machine will be utilized. Specifically, one should determine:

1. Number of x-ray tubes
2. kW and maximum kVp rating of each tube
3. Orientations of each tube when in use
4. Maximum radiation field size for each tube
5. Projected workload of each tube
6. Occupancy factors outside barriers
7. Surrounding area classification (ie. restricted or nonrestricted)
8. Other planned operational controls
9. Planned physical safeguards such as barriers, interlocks, or warning devices.

The above information should be used to verify that the specified thicknesses and types of barrier materials (4,9-16) in the installation's shielding design are appropriate with respect to the planned use of the room.

Visual Inspection

During the installation of the lead shielding, progress should be visually inspected frequently to insure that the physical placement of the shielding follows the installation's shielding design specifications. Normally, the lead is bonded to the dry wall (Figure 1). This drywall layer is installed with the lead adjacent to the studs. Strips of lead are glued over the drywall joints. Lead tabs are glued to the drywall surface to cover screw or nail piercings. A "finishing" layer of drywall is glued over the first layer and secured with a minimal number of screws. This "finishing" layer of drywall allows normal taping and surface finishing procedures.

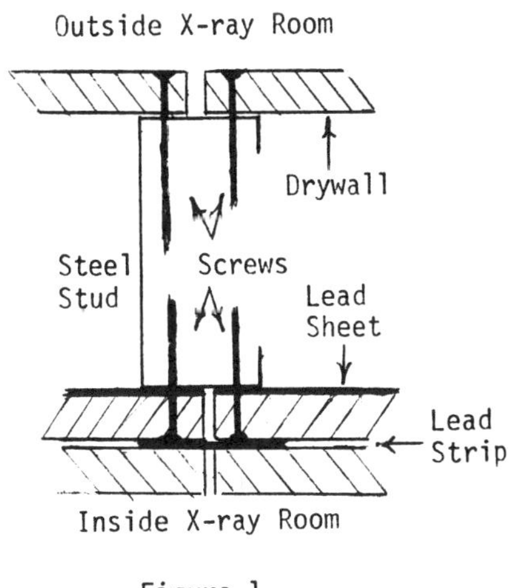

Figure 1

The visual inspection of lead construction must occur BEFORE the placement of the finishing layer of drywall and BEFORE the placement of drywall on the opposite side of the studs. This must be coordinated with the contractor. Specifically, the following must be checked with respect to the installation's shielding design specifications:

1. Thickness of lead bonded to drywall
2. Thickness of plate or lead glass in view window
3. Thickness and location of lead in door jams
4. Thickness of lead in doors
5. Thickness of lead in film pass boxes
6. Thickness of lead strips used for joints and shielding of electrical boxes, etc.
7. Thickness of lead used in baffles
8. Location and installation of interlocks.

The physical placement of required lead strips, tabs, and wrappings must be checked. The strips must adequately cover the joints in the dry wall. Since the strips are separated from the lead sheets by the 5/8" drywall (Figure 1), the strips should be at least 2 inches wide and centered on the joint. Lead tabs should cover the heads of all screws and nails which pierce the lead barrier. The lead wrapping of electrical boxes and other voids must minimize cracks and gaps. Since the lead strips, tabs, and wrappings are frequently glued in place, secure bonding must be checked.

At this stage of construction, lighting in the facility is usually dim. A bright light can be located inside the room at the location of the patient table. This light illuminates cracks or gaps in the lead placement when viewed from outside the room. These illuminated cracks or gaps must be eliminated by repositioning the lead strips or wrappings.

After the visual inspection, documentation of the survey should begin. A simple sketch of each wall is drawn which illustrates the locations of joints and other voids in the shielding. As shown in Figure A of the Appendix, this sketch is a view of the barrier as it appears to an observer outside the room.

Qualitative Measurements

Qualitative measurements follow the visual inspection. The qualitative measurements detect the location and GEOMETRIC size of radiation fields which exist outside the room barriers. These radiation fields require quantitative analysis at a later date. Detected radiation fields with large cross sectional areas and elevated readings suggest inferior materials or faulty installation. Replacement or repair of problems is simple at this time since the lead barriers are still uncovered.

The qualitative measurements are made by placing a radioactive source, Technecium-99m (Tc-99m), inside the room at the future patient location. The radiation emitted by the radioisotope models the direction and origin of primary radiation (emitted from the x-ray tube) and scattered radiation (emitted from the patient) which impinge on the barriers during clinical exams.

Since the radiation from the radioisotope is emitted continuously, one utilizes a qualitative rate survey meter (ie. Geiger counter) with an audible meter to detect radiation fields outside the room. Each section of shielding is scanned in areas away from joints and voids; this radiation level is noted as the section's base level. This is followed by scanning all joints and other voids wrapped by lead. Since the lead shielding is uncovered, the joints and voids are easily located. A Geiger tube of small cross sectional area is used to scan and define the geometrical dimensions of the detected radiation fields resulting from barrier voids and joints. The location, geometry, and relative qualitative level of the detected radiation fields are documented on the sketch shown in Figure A of the Appendix. The relative qualitative level is defined as the ratio of the Geiger counter reading at a gap or void with respect to its section's base reading.

The use of a radioisotope provides the opportunity to evaluate the lead installation with radiation when problems can easily be remedied. A portable x-ray machine should not be used at this stage of construction because dust free conditions do not exist. While the radioisotope facilitates the barrier evaluation, STRICT RADIATION PROTECTION GUIDELINES MUST BE FOLLOWED WHEN HANDLING A RADIOISOTOPE to insure that personnel radiation exposure is minimized. The choice of qualitative radiation sources, radiation protection concerns, and the choice of qualitative survey instruments are discussed in the sections: "Qualitative Radiation Source", "Qualitative Radiation Source; Radiation Protection Program", and "Survey Equipment; Qualitative Instrument" respectively.

Quantitative Measurements

The quantitative barrier evaluation should be completed after the x-ray equipment is calibrated and the room construction is completed. The quantitative measurements are required to determine the maximum radiation exposure levels outside the barriers as a result of leaks and the barrier thickness.

Before quantitative measurements are made, a final visual inspection is required. Any electrical service boxes or other voids created since the qualitative survey must be noted on the form illustrated in Figure A of the Appendix. The operation of protective devices such as x-ray "on" signs, door interlocks, and tube orientation limits should be checked. The lead thickness of doors hung after the qualitative checks must be inspected.

Clinical x-ray beam orientations and x-ray field sizes which result in maximum exposure levels outside the room barriers (4) are used for quantitative checks. X-ray transmission through primary barriers is measured with the largest x-ray field area selected and directed at the barrier without a phantom in the beam. Secondary barrier transmission is measured with a suitable phantom in the patient position, with the largest x-ray field area selected and with the beam directed in the most common clinical direction. A 5 gallon processor chemical container containing water makes a suitable phantom.

Measurements at four to eight locations usually are sufficient. These measurements must include one measurement for each wall at the location where the largest radiation field was detected during the qualitative checks. Additional quantitative measurements may be required if:

1. Regions of shielding exist with no designed margin for error (ie. plate glass view windows or areas utilizing only conventional building materials) or
2. Electrical service boxes or other voids were added after qualitative measurements.

The results of all measurements are recorded on the data form in Figure B of the Appendix.

The quantitative survey meter must be chosen carefully. A large volume ion chamber is required to provide adequate sensitivity to low level radiation fields. Since the radiation field may have a small cross sectional area, the entire cross section of a large volume chamber may not be uniformly covered. Thus, the meter reading may be low. The required correction for this and other errors introduced by the meter are discussed in the section: "Survey Equipment; Quantitative Instrument".

X-ray tube exposure techniques must be chosen carefully to avoid tube damage while providing suitable x-ray fields for quantitative measurements. While the barriers should be evaluated using the maximum energy x-ray beam from the x-ray tube (4), the selected kVp should be at least 10 kVp less than maximum tube rating to allow for possible kVp calibration error. The selected mAs must produce a large enough radiation field outside the barrier to allow an integrated meter reading within acceptable uncertainty limits. Feldman (17) has developed a procedure for calculating the required mAs for meters with analogue displays. A sample reading is usually sufficient for meters with digital displays. Since the required mAs and kVp are large, the tube rating chart must be consulted; low mA values must be used to avoid tube damage. Using a quantitative survey meter in the "integrate mode" avoids "long" exposure times required by long response times of rate survey meters. Since each exposure contributes large heat loads to the anode, the anode and tube housing must be allowed to cool between exposures. Equation (1) or (2) can be used to calculate the total heat units (HU) produced by each exposure. The additional factor of 4000 HU accounts for heat generated by the rotor. From the total heat unit calculation, one can calculate a conservatively long cooling interval for most tubes by assuming a cooling rate of 250 HU/sec.

$$HU = (kVp \times mA \times time) + 4000 \quad \text{(single phase)} \quad ---(1)$$

$$HU = (kVp \times mA \times time \times 1.35) + 4000 \quad \text{(three phase)} \quad ---(2)$$

Interpretation of Data

The radiation dose levels listed in Table I are the maximum radiation doses individuals may receive per calendar quarter. The Maximum Expected Dose per quarter (MED) (4) is estimated from the measured dose/mAs by using Equation (3).

$$\frac{\text{Dose/mAs} \times WQ \times U \times T}{1000} = \text{MED/QTR} = \frac{\text{rem}}{\text{qtr}} \quad ---(3)$$

where Dose/mAs = measured dose/mAs outside barrier (mrem/mAs)
WQ = machine workload (mAs/qtr)
U = Use Factor
T = Occupancy Factor

The dose/mAs is calculated from the survey meter readings collected during the quantitative analysis by applying the correction factors discussed in the section, "Survey Equipment; Quantitative Instrument". WQ is the total projected mAs of all exposures during a calendar quarter of clinical use of the x-ray machine. Both U and T have values between 0 and 1. T is the fraction of time during an eight hour shift one individual is assumed to occupy the location in question. U is the fraction of the total workload that the beam is directed toward a primary barrier; U is equal to 1 when a secondary barrier is evaluated. The appropriate values for U, WQ, and T are determined during the investigation step of the barrier analysis.

Acceptable MEDs for barriers have been a subject of much debate. The MED values allowed by State of Illinois' law are listed in Table I, but these values may be excessive with respect to the NCRP's ALARA concept of maintaining radiation exposures to individuals as low as reasonably achievable. Braestrup and Wyckoff (8) argue that if reduced radiation levels of 0.125 rem/qtr are reasonable for members of the general public in unrestricted areas, the same level should be reasonable for radiation workers in restricted areas. These authors suggest that using 0.125 instead of 1.25 rem/qtr as the MED for occupational workers in restricted areas increases the material costs of shielding for an average size room by only 25%.

Until this debate is settled, acceptable MED levels must be those assumed in the installation's shielding design provided they do not exceed the values in Table I. The designer of the shielding must discuss with the owner the costs and benefits of designing barriers to result in MED levels below maximum lawful levels. The owner must be informed that due to the conservative nature of NCRP recommendations, a shielding design utilizing NCRP #49 recommendations (4) normally results in MED levels significantly below maximum legal limits.

Quantitative Record Keeping & Recommendations

The form in Figure B of the Appendix is used for recording the quantitative barrier analysis data. The identification and function of the room is listed at the top of the page. The date of the survey, phantom utilized for secondary barrier measurements, environmental conditions, and identification of quantitative survey meter are required. Basic data and exposure techniques are listed for each measurement: Location, Radiation Type, Tube #, Beam Orientation, Phantom, kVp, mA, Time, and Survey Meter Reading. The meter reading is corrected to dose/mAs outside the barriers by utilizing Equation (4) found in the section: "Survey Equipment: Quantitative Instrument". Correction factors used in Equation (4) are listed at the top section of the form in Figure B: Calibration Factor, Energy Response, Directional Response, Intensity Response, and mrem/mR Factor. The required assumptions used to estimate the MED in mrem/qtr, WQ, U, and T are listed. This data allows one to estimate MED from dose/mAs values at a later date if operational conditions change.

Under the section of "Conclusions and/or Recommendations" the adequacy of the barriers is stated. If the estimated MEDs exceed acceptable values, corrective measures must be taken. While these corrections could include (4):

1. Occupancy restriction
2. Reduced machine workload
3. Installation of beam orientation restraints
4. Increased barrier thickness,

increased barrier thickness is usually the only practical choice. Occupancy restriction is difficult to enforce. Restricted workloads or beam orientations limit the clinical usefulness of the x-ray machine. The barriers must be resurveyed if corrections 3 or 4 are completed.

Qualitative Radiation Source

Technecium-99m (Tc-99m) is the radioisotope of choice for use as a qualitative radiation source for barrier analysis. Figure 2 (18) illustrates the decay scheme of Molybdenum-99 to the metastable state of Tc-99m. Tc-99m decays to Tc-99 with a half life of 6 hours by emitting a 140 kev photon. The 140 kev photon emitted by Tc-99m models the penetrating ability of the x-rays in a 140 kVp x-ray spectrum despite the spectrum's approximately 60 kev average energy. This is true because all the low energy x-rays in a 140 kVp beam are attenuated by a barrier designed to stop the majority of the beam's high energy x-rays.

Tc-99m is a good radiation source for qualitative barrier analysis for practical reasons. Since it is extensively used for Nuclear Medicine studies, it is readily available and inexpensive. A 200 mCi generator can be purchased for approximately $120 and can be used for one week. The International Commision on Radiological Protection (ICRP) (19) classifies Tc-99m in the group of lowest radiotoxicity due to its short half life and minimal emission of particles. Therefore, the radiation protection procedures required to minimize radiation hazards are manageable.

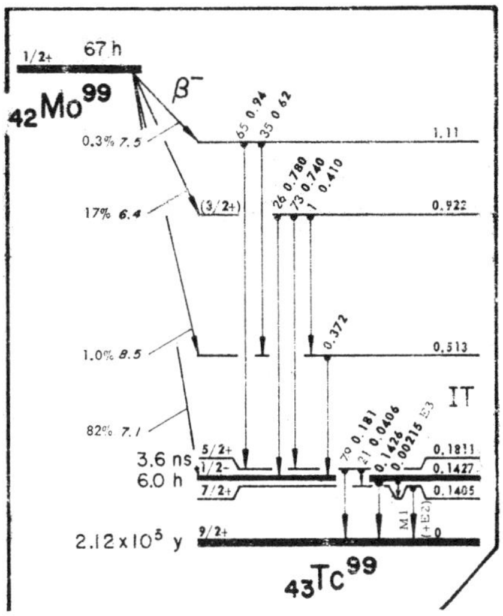

Figure 2 (18): Tc-99m Decay Scheme (Courtesy of C.M. Lederer et.al, Wiley, 1967)

A large activity of Tc-99m is required to qualitatively analyze most diagnostic x-ray room barriers. Approximately 200 mCi of Tc-99m is required to produce a radiation field of 0.05 mR/hr above background radiation levels outside a 1/16" lead barrier of a normal size x-ray room. Thus, a comprehensive radiation protection protocol must be followed to minimize radiation exposure and avoid radiation contamination to the handler and other individuals in the vicinity of the Tc-99m.

Radiation Protection Program

Radiation exposure from the photons emitted by Tc-99m is minimized by minimizing the time of exposure, maximizing the distance between the source and oneself, and using shielding. One must use the room barriers as shielding by remaining outside the room except when handling the radiation source. When the source is not needed, it must be placed in a lead pig with walls at least 3/16" thick. This lead pig also allows safe transport of the material. During testing when the Tc-99m is removed from the lead pig, the pig must be located so that it does not block the radiation directed toward the barrier to be tested.

Distance is used to minimize exposures by remaining outside the room whenever possible. Using tongs to handle the source increases the distance between the source and handler. Since the exposure rate from a source is proportional to 1/(distance squared), using 10" long tongs as opposed to direct handling of the unshielded source reduces the exposure to the fingers by a factor of 500.

Personal exposure to radiation is directly proportional to

the time spent in the vicinity of the source. Handling time can be minimized by practicing procedures with the tongs and a dummy source. Also, careful planning prior to entering the room can minimize exposure time while near the unshielded source.

Figure 3

Care is required to avoid contamination of objects in the room or of personnel. Contamination is defined as direct contact with the radioactive chemical. The vial containing the Tc-99m must be placed in a nonbreakable, thin walled vessel with closed top. Absorbent material should be placed under the vessel to catch spills. A buttoned lab coat and disposable gloves should be worn when handling the vessel. Any cuts or other skin breaks must be bandaged before putting on other protective layers. After handling the source, hands must be scrubbed with soap followed by a scan with a survey meter to check for contamination. Finally, smoking, eating, or drinking is prohibited during periods when the source vessel is handled.

Other individuals which may enter the vicinity of the radiation source must be alerted of possible radiation hazards. Signs similar to the one in Figure 3 must be posted on all doors of a room while the radioactive source is unshielded within. Ideally, the doors should also be locked to prevent entry. When stored, the Tc-99m should be placed in its lead pig with a warning label shown in Figure 4. The pig and source should be stored in a locked cabinet. Since Tc-99m has a short half life, decay reduces 200 mCi after a week to levels which can safely be placed in the regular garbage. If the Tc-99m is transported on the highway, it must be placed in a lead pig, wrapped in absorbent material, and labeled according to the regulations of the U.S. Department of Transportation (25). Personnel monitoring badges must be worn at all times by individuals involved in the qualitative scans to verify that handling procedures have minimized personnel exposure.

Figure 4

The above discussion covers routine radiation protection procedures. One must also be prepared to handle nonroutine problems such as a spill. If personnel are contaminated, protective clothing must be removed and the contaminated skin thoroughly scrubbed. Care must be used to insure the skin is not damaged or broken. Contaminated clothing must be secured to allow decay before further handling. Decontamination of the floor or other surfaces inside the room is not possible if the surface is porous (ie. concrete). In this case, the room must be placarded with the warning in Figure 3 and locked. Entry must be prohibited until the spilled radioisotope decays, eliminating hazardous radiation exposure levels. Three days allows sufficient decay time for a 200 mCi spill within the room.

Licensure of Radioative Material

Since Tc-99m is Byproduct Radioactive Material, the use of activities greater than 0.1 mCi (20) is controlled by the rules and regulations of the U.S. Nuclear Regulatory Commission (NRC). These rules(21,22) require that a radiation protection program be established by the licensee and approved by the NRC. Institutions with NRC licenses allowing Nuclear Medicine studies usually are licensed to possess larger activites of Tc-99m than their clinical needs require. Therefore, if the institution authorizes the surveyor to handle Tc-99m, it can supply the Tc-99m. The surveyor is then bound by the radiation protection procedures of the institution and by the procedures spelled out on his application submitted to the institution.

If the institution can not be utilized as the supplier of radioative material, the room surveyor must be licensed directly by the NRC before the Tc-99m generator can be purchased from a radiopharmaceutical supplier. The appropriate NRC license is an Industrial Use License of Small Quantities of Byproduct Material (23). A guide (24) is published by the NRC which describes the information which must be submitted in the license application. This information includes the type and activity of the Byproduct Material, its specific use, a detailed description of the radiation protection program, and the resume of the prospetive licensee. The resume must indicate sufficient experience to insure that the radioactive material will be handled properly. The current fee (26) for this type of license which requires renewal every 5 years is $110. Since the Tc-99m will be transported, the U.S. Department of Transportation guidelines must be consulted. These regulations (25) specify appropriate packaging, labeling, placarding of vehicle, and shipping papers.

Survey Equipment

Quantitative Equipment

An ionization chamber with a sensitive volume of a liter coupled to an electrometer with integrating capabilities is a suitable instrument for measuring weak radiation fields of short duration. The integrating mode eliminates the problems of measuring radiation fields which exist for shorter time intervals than the response time of the meter in the rate mode. The following factors must be considered when quantitative measurements are completed: energy response, directional response, intensity response, chamber cross sectional area corrections, and meter calibration.

The response of most quantitative survey meters to the spectrum of x-ray energies found in a 140 kVp x-ray beam is not constant. Figure 5 (27) illustrates that the energy response of the survey meter used for this study introduces an error requiring correction at most x-ray energies. To estimate this correction factor, one must assign an effective energy to the range of energies present in the measured x-ray fields. Since the majority of the low energy x-rays are filtered out of the beam during transmission through the barrier, one assumes an effective energy of 100 kev x-rays at the chamber for an x-ray field generated at 140 kVp. This suggests a correction factor of ind/act = 1.08 for radiation incident on the chamber's front side. For the chosen meter, uncertainty in the 1.08 correction factor is minimal since the meter's response to 100-140 keV x-rays is essentially constant. However, since the curve in Figure 5 is not a result of data collected using the actual chamber, a +/- 10% error is assumed.

Figure 5 (27): Energy Response of Survey Meter

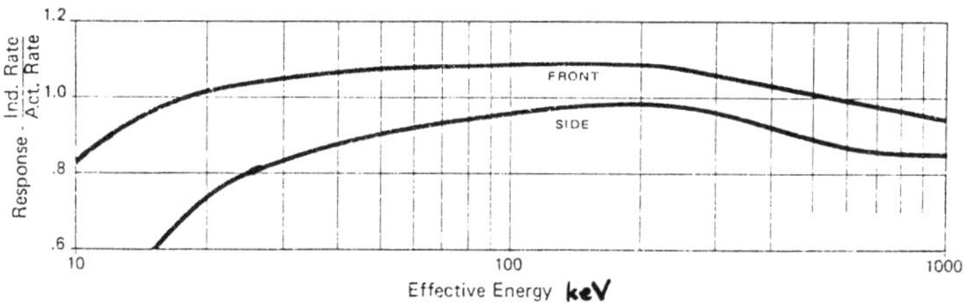

The directional response of the survey meter's ionization chamber must be considered. Some instruments have ion chambers mounted in a case with the electrometer. Since these chambers are not isolated from attenuating materials, these instruments exhibit directional responses similar to the instrument in Figure 6 (5) which requires correction. These corrections can be simplified, however, if the ion chamber is external to the electrometer. The instrument in Figure 5 is such a meter; Figure 5 illustrates energy response curves for radiation incident on the front or side of the ionization chamber which incorporates directional response. Therefore this meter eliminates the need for directional response corrections in addition to energy response corrections if the chamber is carefully oriented and radiation fields from multiple widely spaced sources are not simultaneously measured.

Quantitative survey meters accurately measure exposure levels over a limited range of x-ray intensities. At high x-ray intensities, ions recombine before collection; a low reading results. The x-ray intensity at which ion recombination corrections become necessary depends on the applied voltage and the geometry of the collection plates within the chamber (28). The chamber of Figure 5 collects 95% or more of the produced ions for x-ray intensities below 11 R/hr (27).

Large volume ion chambers (about 1 liter) are required to allow sensitivity to x-ray fields of low intensity. These chambers have cross sectional areas of 50-100 cm squared. Incomplete exposure of the chamber's cross sectional area results in a low reading due to averaging across the chamber's area. The ratio of the chamber's cross sectional area to the x-ray field's cross sectional area provides a first approximation correction factor. Qualitative measurements taken during the surveys may be used to estimate the x-ray field cross sectional area.

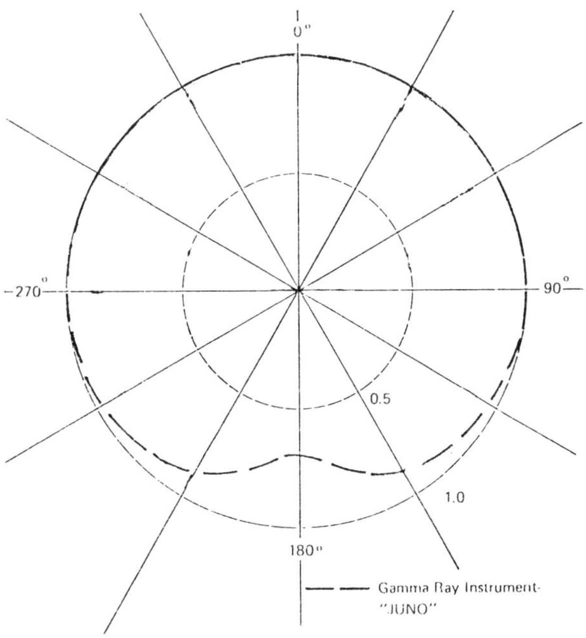

Figure 6 (5): Survey Meter Directional Response (Courtesy of NCRP, Report #57.)

The quantitative survey meter must be calibrated. This calibration should be traceable to the National Bureau of Standards through regional calibration labs. At the time of calibration, the user should establish a constant reproducible radiation field from a radioisotope with a long half life such as Cesium-137 or Radium-226. This field should be measured periodically with the quantitative survey meter to confirm that the calibration factor has not changed.

By including the correction factors which are required to correct for the errors introduced by the unique design of the quantitative survey meter in question, one may calculate the dose/mAs from the meter reading by utilizing Equation (4):

$$\text{Dose/mAs (mrem/mAs)} = R \times C \times ER \times DR \times IR \times AC \times (1/WE) \times [(T+273)/295] \times (760/P) \times ED \quad \text{---(4)}$$

where R = reading from meter (mR)
C = calibration factor of meter = 1 +/- 5%
ER = energy response correction = 0.93 +/- 10%
DR = directional response correction = 1
IR = intensity correction = 1 +/- 5%
AC = area correction = 1 except for x-ray fields of small cross sectional area
WE = mAs utilized to obtain meter reading
T = room temperature (Centigrade)
P = barometric pressure (mm Hg)
ED = exposure dose correction = 0.95mrem/mR +/- 5% (28)

The values for each term shown above assume an effective energy x-ray spectrum of 100 kev. The second through sixth terms correct for the characteristics of the chosen survey meter. The term ED converts the exposure in mR to an equivalent dose in mrem. The eighth and ninth terms convert to standard temperature and pressure. The corrected survey meter readings (dose/mAs) are assumed correct to +/- 15%.

Qualitative Instrument

A Geiger-Muller (GM) tube connected to a rate meter with audible indicator is a suitable qualitative survey meter for detecting the presence of radiation fields. The GM tube should have a small cross sectional area to allow determination of the cross sectional area of the radiation fields outside barriers. The meter's response time of a couple of seconds is not a problem since the radiation field from the radioisotope is continuous. The GM tube's sensitivity to low level radation fields is ideal for detecting x-ray fields which are one or two multiples of background radiation levels.

A GM tube and rate meter cannot be used to quantitatively measure x-ray exposure. X-ray exposure is dependent on both the number and energy of the x-rays in the field. A GM tube can only detect the number of ionizing events. It cannot differentiate between the various energies of x-rays in the x-ray spectrum which caused the ionizations.

X-ray fields of high intensity cannot be detected with a GM tube and rate meter. The GM tube has at least a 200 microsecond resolving time which is defined as "the minimum time that must elapse after the detection of an ionizing event before a second event can be detected." (29) This resolving time causes a GM tube to "jam" and register a count rate of zero when placed in a high exposure x-ray field. Thus, GM tube survey meters are capable of DETECTING x-ray fields with intensities less than 50 mR/hr (29).

Conclusions

This paper recommends a protocol suitable for acceptance testing of x-ray room barriers. Radiation from a radioisotope is used to produce radiation fields outside the room barriers. Qualitative measurements of these radiation fields are done while the lead installation is in progress which allows easy correction of inadequate construction. When the construction of the room and installation of equipment is complete, quantitative measurements are done. These consist of measuring the exposure levels of only the maximum radiation fields identified by the qualitative measurements. Since only the quantitative measurements use the x-ray tube as a radiation source, this measurement protocol allows analysis of all barrier joints and voids with minimal workload and wear on the x-ray tube.

Figure B in the Appendix is the quantitative data from a typical radiographic fluoroscopic room survey. One quantitative measurement is made adjacent to each wall where the maximum radiation field existed during qualitative checks. Additional measurements are completed inside an adjoining toilet which contains no lead shielding. Since personnel do not occupy areas above and below the room, measurements of x-ray transmission through the floor and ceiling are unnecessary. The calculated Maximum Expected Dose (MED) values are all significantly below the lawful limits in Table I. Thus, the barriers are more than adequate for the current planned workload. The inclusion on the data form of all assumptions, correction factors, and measurement techniques used to calculate the MED values makes further quantitative measurements unnecessary unless the barriers are physically changed.

Using Tc-99m as a qualitative radiation source requires licensure by the NRC and a comprehensive radiation protection program. Approximately 2 hours/room are required to complete the qualitative inspections, measurements, and documentation. Approximately one half hour of additional time per room is required to prepare the Tc-99m source and calibrate the quantitative surevy meter provided 2 or more rooms are to be surveyed. However, these extra steps are required to insure that the exposure levels of the MAXIMUM radiation fields outside room barriers are measured.

References

1. Medical X-ray and Gamma-Ray Protection for Energies Up to 10 Mev - Equipment Design and Use. NCRP Report #33, 1968.

2. Dental X-ray Protection. NCRP Report #35, 1970.

3. Basic Radiation Protection Criteria. NCRP Report #39, 1971.

4. Structural Shielding Design and Evaluation for Medical Use of X-rays and Gamma Rays of Energies up to 10MeV. NCRP Report #49, 1976.

5. Instrumentation and Monitoring Methods for Radiation Protection. NCRP Report #57, 1978.

6. Fields T: Radiation Control Surveys. AAPM 1975 Summer School: The Expanding Role of the Diagnostic Radiological Physicist, 200-204, 1975.

7. Rules and Regulations for Protection Against Radiation. Illinios Department of Nuclear Safety, Springfield.

8. Braestrup C.B., Wyckoff H.O: Shielding Design Levels for Radiology Departments. Radiology 107:445-447, May 1973.

9. Trout E.D., Kelley J.P., Lucas A.C: Broad Beam Attenuation in Concrete for 50 to 300kVp X-rays and in Lead for 300kVp X-rays. Radiology 72: 62-67, 1959.

10. Trout E.D., Kelley J.P., Lucas A.C: Conventional Building Materials as Protective Radiation Barriers. Radiology 76:237- 244, 1961.

11. Kelley J.P., Trout E.D: Broad Beam Attenuation in Lead for X-rays from 50 to 300kVp. Radiology 104:171-175, 1972.

12. Trout E.D., Kelley J.P., and Larson V.L: The Use of Plate Glass as Shielding Material in Diagnostic Radiologic Installations. Journal of Canadian Association of Radiology 25: 173-176, 1974.

13. Trout E.D., Kelley J.P., Herbert G.L: X-ray Attenuation in Steel 50-300kVp. Health Physics 29:163-169, 1975.

14. Glaze S.A., Schneiders N.J., Bushong S.C: Use of Gypsum Wallboard for Diagnostic X-ray Protective Barries. Health Physics 36: 587-593, 1979.

15. Christensen R.C., Sayeg J.A: Attenuation Characteristics of Gypsum Wallboard. Health Physics 36: 595-600, 1979.

16. Wohni T: Broad Beam Attenuation in LECA for 50-140 kVp X-rays. Health Physics 40: 205-209, 1981.

17. Feldman A: Factors and Strategies in Selections of Instruments for Radiation Surveys Around Medical Installations. Health Physics 38: 679-681, 1980.

18. Lederer C.M., Hollander J.M., Perlman I: TABLE OF ISOTOPES. New York: John Wiley and Sons, Inc., 1967.

19. Handling and Disposal of Radioactive Materials in Hospitals and Medical Research Establishments. ICRP Committee V Report, 1964.

20. U.S. Nuclear Regulatory Commission: Title 10, Chapter 1, Code of Federal Regulations, Part 30.

21. U.S. Nuclear Regulatory Commission: Title 10, Chapter 1, Code of Federal Regualtions, Part 20.

22. U.S. Nuclear Regulatory Commission: Title 10, Chapter 1, Code of Federal Regulations, Part 34.

23. Cooper J.R: Private Communication.

24. U.S. Nuclear Regulatory Commission: Guide for the Preparation of Applications for Licenses for Laboratory and Industrial Use of Small Quantities of Byproduct Material. Regulatory Guide 10.7. Washington, 1979.

25. U.S. Department of Transportation: Title 49, Chapter 1, Code of Federal Regulations, 1980.

26. U.S. Nuclear Regulatory Commission: Title 10, Chapter 1, Code of Federal Regulations, Part 170.

27. Victoreen Product Data on Model 660 Survey Meter.

28. Johns H.E., Cunningham J.R: THE PHYSICS OF RADIOLOGY. Springfield: Charles C. Thomas, 1969.

29. Cember H: INTRODUCTION TO HEALTH PHYSICS. New York: Pergamon Press, 1969.

QUALITATIVE BARRIER SURVEY

PRELIMINARY ACCEPTANCE

Institution: Michael Reese Building: Blum Room #: 205 Date: 2 - 20 - 80

Activity of Tc-99m: 137 mCi Survey Meter: Wm. B. Johnson GM Tube

Legend: Scale: 1/4" = 1'

Electrical Boxes: ☒ View Window: VW

Plumbing Voids: PL Door: D

Heating & Air Conditioning Duct Voids: ALL WALLS SKETCHED FROM OUTSIDE OF ROOM.

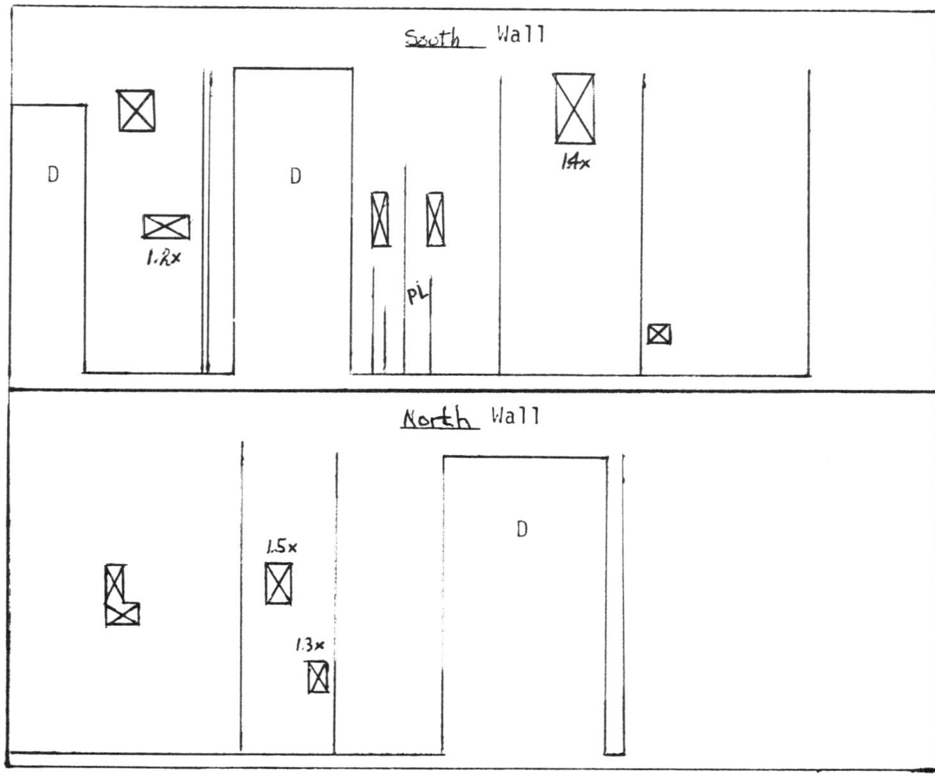

The Preliminary Acceptance of barriers consists of a visual inspection and qualitative radiation tests of the barrier thickness and quality of construction of:

1. Sheet Lead
2. Door Jams
3. Doors
4. View Windows
5. Electrical and Plumbing Voids
6. Film Passboxes
7. Baffles

Final Acceptance is based on measurements of the radiation field exposure levels after the installation of equipment and the completion of the facility.

PRELIMINARY ACCEPTANCE:

Signaure: Keith J. Strauss

Date: 4-23-80

Figure A

QUALITATIVE BARRIER SURVEY

PRELIMINARY ACCEPTANCE

Institution: Michael Reese Building: Blum Room #: 205 Date: 2 - 20 - 80

Activity of Tc-99m: 187 nCi Survey Meter: Wm. B. Johnson GM Tube

Legend: Scale: 1/4" = 1'

Electrical Boxes: ☒ View Window: VW

Plumbing Voids: PL Door: D

Heating & Air Conditioning Duct Voids: ALL WALLS SKETCHED FROM OUTSIDE OF ROOM.

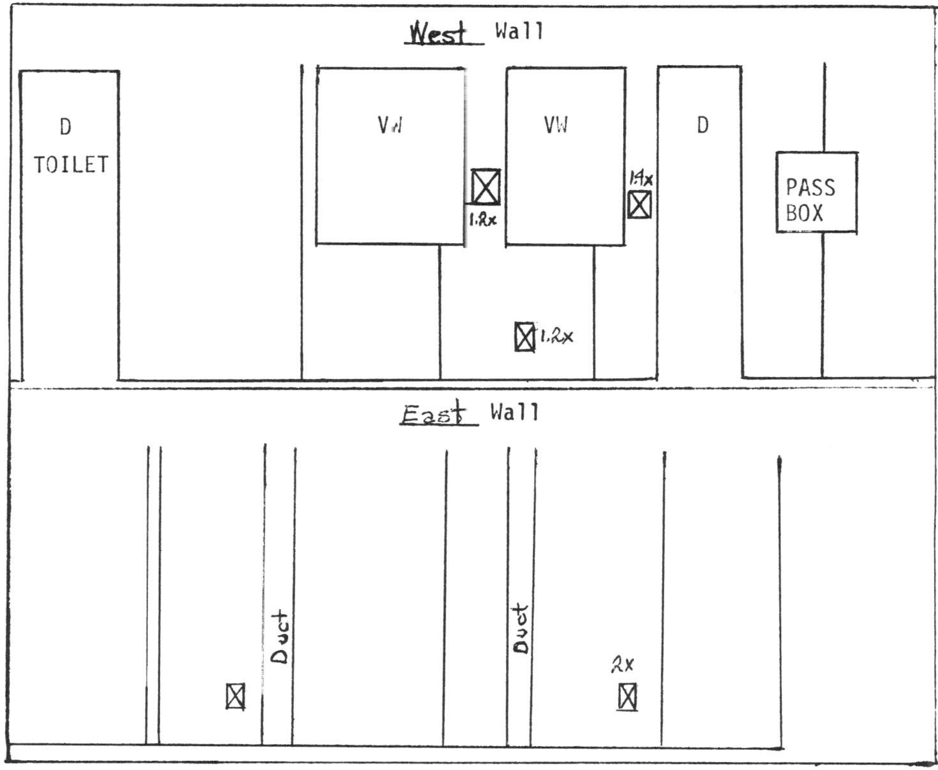

The Preliminary Acceptance of barriers consists of a visual inspection and qualitative radiation tests of the barrier thickness and quality of construction of:

1. Sheet Lead
2. Door Jams
3. Doors
4. View Windows
5. Electrical and Plumbing Voids
6. Film Passboxes
7. Baffles

Final Acceptance is based on measurements of the radiation field exposure levels after the installation of equipment and the completion of the facility.

PRELIMINARY ACCEPTANCE:

Signaure: Keith J. Strauss

Date: 4-23-80

Figure A

QUANTITATIVE BARRIER SURVEY

Code: -208 / BM / R Institution: Michael Reese Hospital Building: Blum Room #: 208

Equipment Identification: GE R & F Room Date: 5-30-81

GENERATOR: Manufacturer: GE Model #: MSI 850

Phantom: Water Temperature: $24^{o}C$ Pressure: 750.5 mm Hg

Dosimeter: Keithley 660 Directional Response: 1

Calibration Factor: 1 +/- 5% Energy Response: 0.93 +/- 10% Intensity Response: 1 +/- 5% mrem/mR Factor: 0.95 +/- 5%

LOCATION	RADIATION TYPE	TUBE #	BEAM ORIENTATION	PHANTOM THICKNESS	TECHNIQUE KVP	TECHNIQUE MA	TECHNIQUE TIME (sec)	SURVEY METER READING ('mR')	VOLUME CORRECTION	CORRECTED READING (mrem/mAs)	QUARTERLY WORKLOAD MAS/QTR	T	U	AREA TYPE	MED/QTR
Toilet	Primary	1	North	0"	140	300	0.35	0.203	1	1.7E-3	2.5E 5	1/16	1/4	Unrestricted	6.6E-3
Toilet	Secondary	2	Down	7"	140	300	0.5	0.003	1	1.8E-5	2.5E 5	1/16	1	Unrestricted	2.8E-4
North Wall	Secondary	2	Down	7"	140	300	0.35	0.001	1	8.6E-6	2.5E 5	1/4	1	Unrestricted	5.4E-4
South Wall	Secondary	2	Down	7"	140	300	0.35	0.007	1	6.0E-5	2.5E 5	1/4	1	Unrestricted	3.8E-3
Remote Control West Wall	Secondary	2	Down	7"	140	300	0.25	0.052	1	6.3E-4	2.5E 5	1	1	Restricted	1.6E-1
Control Room West Wall	Secondary	2	Down	7"	140	300	0.25	0.060	1	7.2E-4	2.5E 5	1	1	Restricted	1.8E-1
East Wall	Secondary	2	Down	7"	140	300	0.7	0.006	1	2.6E-5	2.5E 5	1/4	1	Unrestricted	1.6E-3

CONCLUSIONS &/or RECOMMENDATIONS:

Barriers are acceptable. Doses beyond North, South, and East Walls are at least a magnitude less than Unrestricted lawful limits in Table I. Doses beyond View Windows (West Wall) are 7 times less than Restricted lawful limits in Table I.

Signature: Keith J Strauss

 Figure B

ACCEPTANCE TESTING OF GEOMETRICAL TOMOGRAPHIC EQUIPMENT

Raymond P. Rossi, M.S.
Department of Radiology
University of Colorado Health Sciences Center
Denver, Colorado 80262

Abstract

Procedures and instrumentation requirements for the evaluation of the performance of geometrical tomographic equipment during initial acceptance testing are discussed. Consideration is given only to those aspects of equipment performance which relate to the tomographic mode of operation. Evaluation of stability of motion, exposure uniformity, location of the objective plane, flatness of the objective plane, objective plane thickness, exposure angle and resolution are considered. Major emphasis is placed on unidirectional systems, however discussion of pluridirectional systems is also presented. A brief discussion of criteria for acceptable performance is also given.

Introduction

Tomography is the now accepted general term used to specify a class of radiologic procedures which employ body section radiography. While tomograhic equipment ranges from simple to complex, all units share the basic principle of motion between the x-ray source and image receptor or patient during exposure. The result of this motion is that structures within a given plane in the patient remain in reasonably sharp "focus" while under and overlying structures are "blurred" out allowing improved visualization of structures not well seen during conventional radiography. (2,3)

The fundamental principle of tomography may be understood by reference to Figure 1 in which a simple linear tomographic system is illustrated. Differential motion blurring is achieved when the x-ray tube and the image receptor move in opposite parallel directions. A mechanical lever pivots about an axis B (fulcrum) in the plane containing the desired anatomical informational (objective plane). The image of point B on the axis is imaged to the same point in the image receptor plane during the entire exposure. All other points in the plane of B parallel to the image receptor plane are similarly sharply imaged. However, images of points such as A and C in planes other than the objective plane are imaged over many points within the image receptor plane during the exposure and are thus blurred. The motion of points, A, B and C in the image receptor plane is given by:

$$M_A = \overline{A''B''} + \overline{B'B'} = d\ \{[\tan(\Theta/2)]D/(a-d)\}$$
$$M_B = 0$$
$$M_C = \overline{B''C''} + \overline{C'B'} = d\ \{[\tan(\Theta/2)]D/(a+d)\}$$

where:

D = the source-image receptor distance
Θ = the tomographic exposure angle

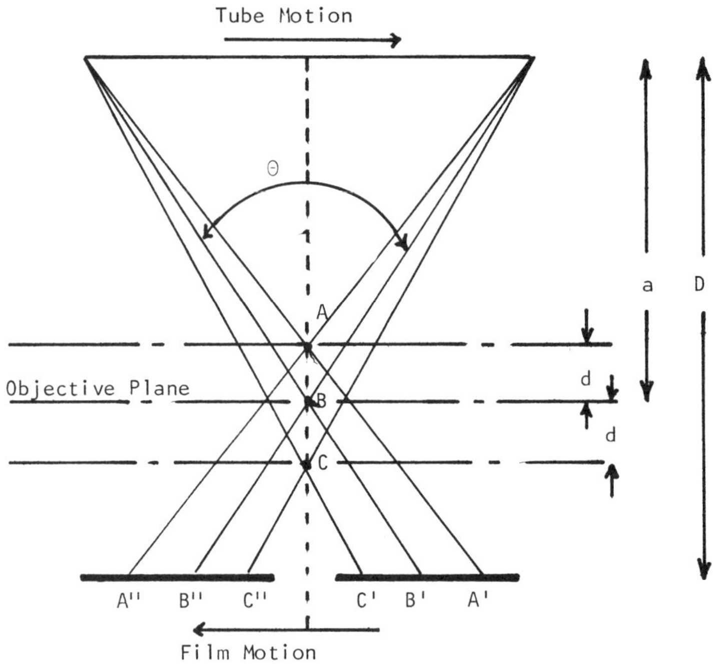

Figure 1: Principle of geometric tomography. Differential blurring occurs as the result of opposite parallel motions for x-ray source and image receptor about axis B. Images not only of B but also of all other points in its plane travel synchronously with the film. As there is no image-image receptor motion there is no blurring. Images of points outside of the plane of B move, respectively more or less than the image receptor and thus are blurred.

a = the objective plane-source distance
d = the distance of A or C from B.

Tomography can be carried out with a wide variety of geometrical motions including linear, circular, spiral, hypocycolidial and random with the choice of motion being dictated by the anatomical structure to be imaged. The magnitude of the exposure angle can be varied to vary the thickness of the objective plane with the thickness being inversely related to exposure angle.

The following nomenclature is generally used in discussing tomographic systems:

- Fulcrum: A fixed adjustable axis about which the mechanism which interconnects the x-ray source and image receptor rotate.

- Tomographic scale: A means for indicating the height of the fulcrum above a given reference plane (usually the plane of patient support). This scale indicates the location of the objective plane.

- Objective plane: The plane in which all points will be sharply imaged into the image receptor plane; the plane of maximum focus; the plane of the depth of cut.

- Source-image receptor distance: The distance between the x-ray source and image receptor.

- Source-plane distance: The distance between the x-ray source and objective plane.

- Plane-image receptor distance: The distance between the objective plane and image receptor.

- Section Thickness: The effective thickness of a tomographic section.

- Operative angle: The angle through which the tomographic apparatus moves.

- Exposure angle: The angle through which the central ray moves during exposure.

Tomography by its very nature results in an image which is degraded compared to conventional radiography. Thus it is necessary to verify that the mechanical aspects of the system are in proper working order to yield optimum results. In order to assess the performance of geometrical tomographic equipment it is necessary to conduct tests to determine the mechanical stability of the tomographic motion, the uniformity of exposure during the tomographic exposure, the accuracy and linearity of the tomographic scale with respect to the actual objective plane, the flatness of the objective plane, the thickness of the objective plane, the accuracy and symmetry of the exposure angle and the resolution in the objective plane. These tests are performed in addition to other system evaluation procedures carried out during initial acceptance tests.

Required Test Equipment

A variety of test objects (phantoms) are available for the assessment of tomographic equipment performance ranging from the very simple to extremely elegant. A representative list of the equipment required is given in Table I and should be sufficient for the evaluation of the performance of the majority of tomographic systems. The major requirement for tomographic test objects relates to the need for accurately calibrating scales whenever quantitative measurements are to be made from the images of these test objects.

Table I
Instrumentation for the Assessment
of the Performance of Geometric Tomographic Equipment

Test Object	Application
Aperature Plate	Exposure uniformity, motion stability
Inclined Scale	Objective plane location, flatness of objective plane, exposure angle, thickness of cut
Incline Mesh Resolution Pattern	System resolution

Evaluation Procedures

Procedures for the evaluation of many aspects of radiologic imaging equipment seem to be generally agreed upon; the same cannot be said for evaluation of tomographic procedures. While the procedures outlined below seem to be the most commonly employed, alternative methods of evaluation are also noted.

Mechanical Stability and Exposure Uniformity

The most common method to evaluate the mechanical stability and exposure uniformity of the tomographic unit is to employ an aperture plate of lead which contains a 3 mm beveled pin hole (1, 6, 7). With the x-ray tube perpendicular to the image receptor the aperture is located in the central ray and positioned approximately 12 cm above the set objective plane. This arrangement is illustrated in Figure 2. Tomographic exposures are then made for the various tomographic motions, exposure angles, etc. which are available utilizing radiographic techniques which will yield an image density of approximately 1.2. The recorded image will describe the geometric form of the obscuring movement as illustrated in Figure 3. In a properly functioning system the density of the image over the complete trajectory should be uniform and for pluridirection systems proper closure of the described geometric motion should be exhibited. Mechanical instabilities will be demonstrated as jerkiness in the recorded image and non-complete exposures will be demonstrated as non-closed trajectories (1). When evaluating tomographic equipment equipped with single phase generators and pulsed nature of the beam will be seen in the resultant image and is to be expected (7). One must also always verify that the time of exposure set on the generator is longer than that required for the tomographic motion. If a spot exposure is made with the x-ray source in its perpendicular position then the pin hole image will allow assessment of the symmetry of the exposure angles (1).

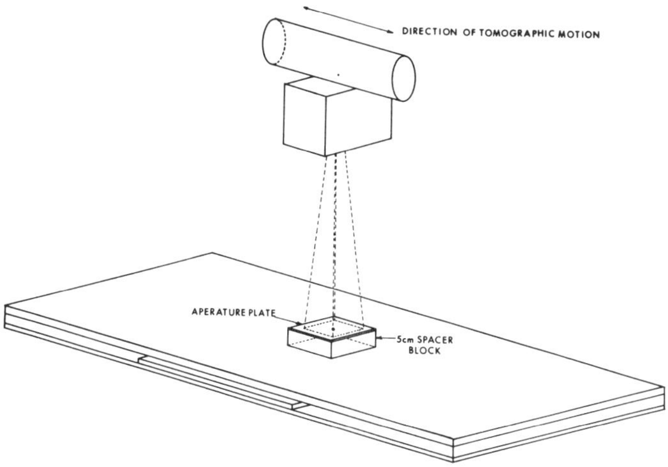

Figure 2: Arrangement for determining exposure uniformity and mechanical stability.

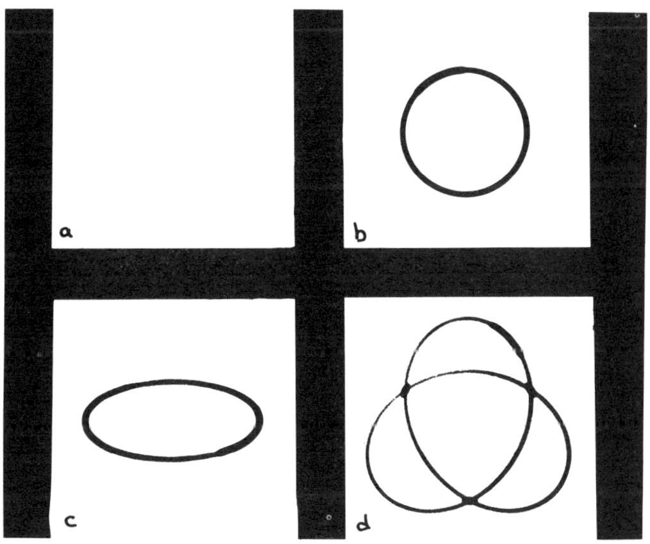

Figure 3: Typical test films for different tomographic motions from which uniformity of tomographic exposure and mechanical stability are determined: a) linear; b) circular; c) elliptical; and d) hypocycloidal.

Tomographic Scale Accuracy & Linearity

The accuracy with which the tomographic scale indicates the actual objective plane may be evaluated using a calibrated inclined wedge as illustrated in Figure 4 (6). For linear system the wedge should be oriented so that the scale is perpendicular to the direction of tomographic motion.

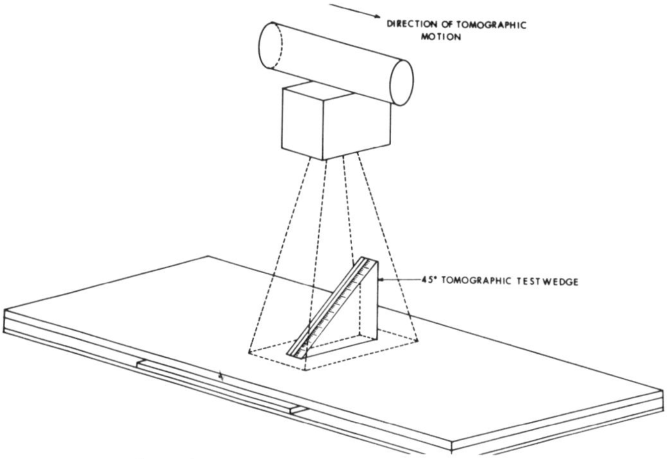

Figure 4: Arrangement for determining accuracy of tomographic scale with respect to objective plane.

Tomographic exposures are made for the various tomographic motions, speeds and at different settings of the tomographic scale, typically 5, 10, 15 and 20 cm. The resultant image will appear as indicated in Figure 5 from which the location of the objective plane is readily determined. As section thickness decreases with increasing exposure angle interpertation of these images is aided by using large exposure angles. Linearity of scale indication is assessed by noting the correspondence of measured objective plane location with scale indication at the various levels.

Flatness of Objective Plane

The flatness of the objective plane may be evaluated by using the same procedure as outlined for determining the accuracy of the tomographic scale with respect to the objective plane by making a series of tomographic exposures at each cut level with the inclined wedge being positioned in different quadrants of the field for each exposure (6). The resultant images should indicate the same level of objective plane regardless of location of the wedge. A large coarse wire mesh located in the objective plane may also be used for these measurements (5).

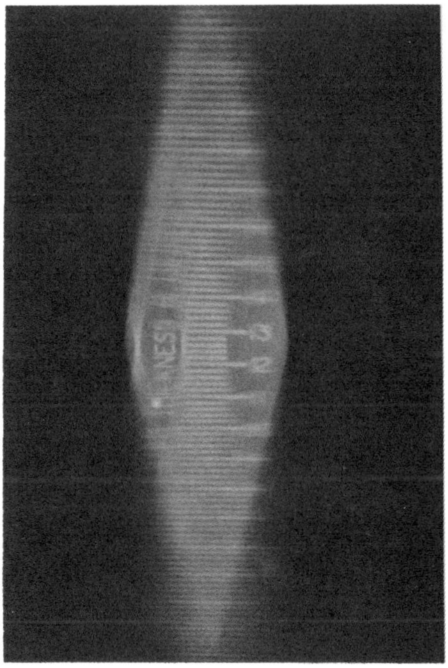

Figure 5: Typical tomographic image from which accuracy of tomographic scale is determined.

<u>Thickness of the Objective Plane</u>

The thickness of the objective plane is inversely related to the exposure angle. It may be assess using the inclined wedge as illustrated in Figure 6 (6). Resultant images will be similar to those shown in Figure 7 for various exposure angles. The section thickness is given as the region over which the image of the scale is in sharp focus. The measurement of this region is somewhat subjective but nontheless can provide useful information. For linear systems the inclined wedge should be orientated perpendicular to the direction of motion.

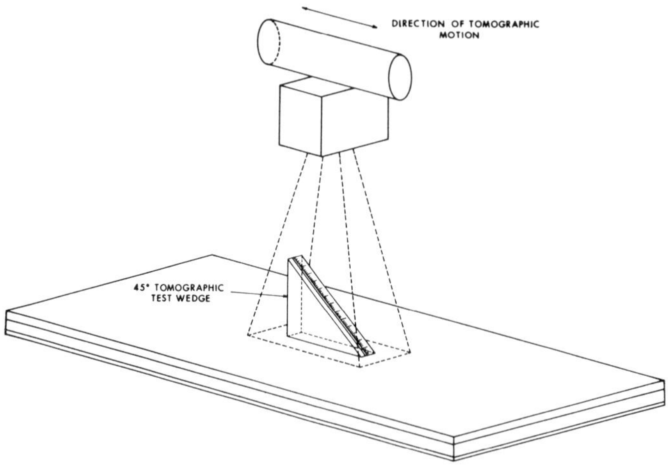

Figure 6: Arrangement for determining tomographic section thickness.

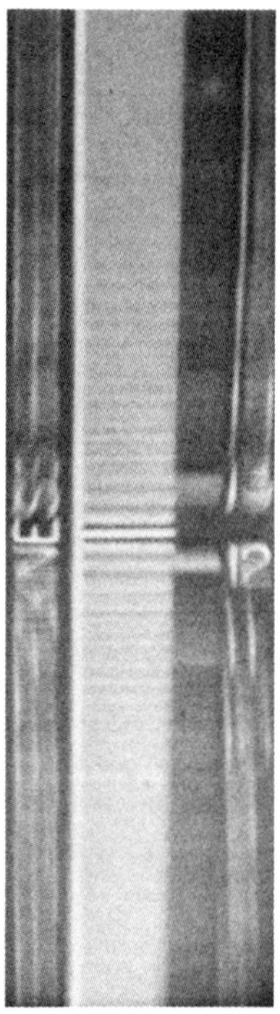

Figure 7: Typical tomographic image from which section thickness is determined.

Section thickness can also be determined using the test object shown in Figure 8 (4). This test object consists of a cylinder 7 cm in diameter and 12 cm long wound with a 1 mm diameter wire to a pitch of 1 cm. A wire runs coaxially to the cylinder. The cylinder is placed with its long axis normal to the objective plane with the objective plane intersecting the cylinder. A tomographic exposure is made and the resultant image will show a portion of one of the wire turns more distinctly, and the arc can be related to 360 degrees and the 1 centimeter pitch.

Section thickness = arc x pitch/360.

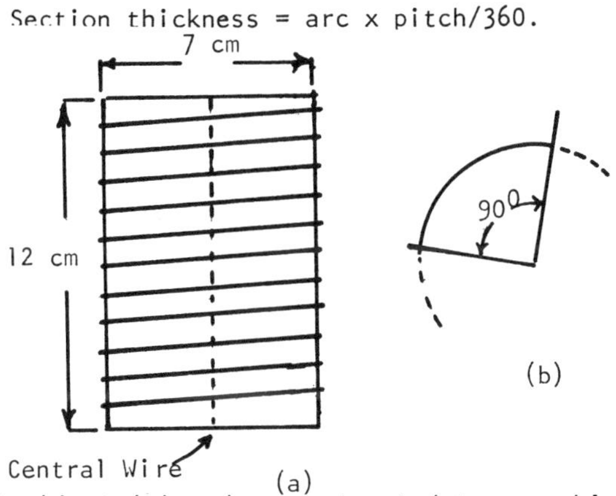

Figure 8: Test object (a) and reconstructed tomographic image (b) which may be used to determine section thickness.

Exposure Angle

The tomographic exposure angle can be assessed by use of the inclined wedge using the same arrangement shown in Figure 4 (6). For linear systems the wedge is orientated perpendicular to the direction of motion. The tomographic scale is set for an objective plane of approximately 12 cm and tomographic exposures made for the various motions and exposure angles. An example of the resultant tomographic image and necessary reconstructions to determine the exposure angle are shown in Figure 9. For a wedge angle inclination of 45°:

$$\tan \beta_1 = c_{11}/b_1$$

$$\tan \beta_2 = c_{21}/b_1$$

for points above the objective plane and

$$\tan \beta_1 = c_{12}/b_2$$

$$\tan \beta_2 = c_{22}/b_2$$

for points below the objective plane where c and b are measured in the tomographic image using the same scale.

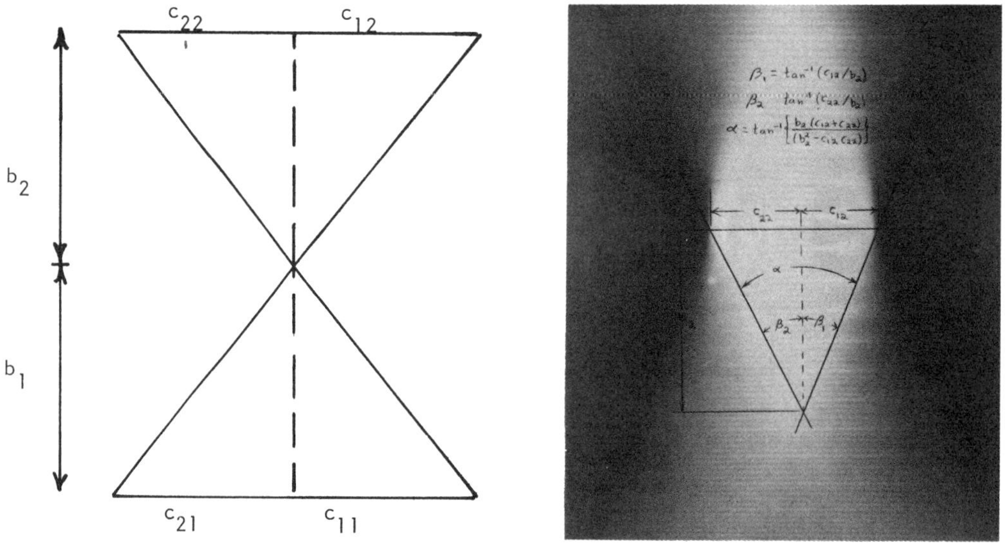

Figure 9: Reconstructed tomographic image from which exposure angle is determined.

For the case in which the exposure angle is symmetrical, the exposure angle is given by:

$$\alpha = 2 \tan^{-1} (c/2b)$$

In the general case in which the exposure angle is not symmetrical

$$\alpha = \beta_1 + \beta_2$$

or for points above the objective plane

$$\tan \alpha = \frac{b_1 (c_{11} + c_{21})}{(b_1^2 - c_{11} c_{21})}$$

and for points below the objective plane

$$\tan \alpha = \frac{b_2 (c_{12} + c_{22})}{(b_2^2 - c_{12} c_{22})}$$

The exposure angle may also be determined utilizing the test device illustrated in Figure 10 (4). The test object consists of a .5 mm thick lead foil at an angle of 45° which contains a narrow slit marked off with radio-opaque wires at 1 cm intervals and horizontal lead foil containing a single hold in line with the slit. The x-ray tube is positioned so that the central ray is perpendicular to the objective plane. The test object is placed on the table with the hole in the horizontal lead foil directly under the central ray and a non-tomographic exposure made to record an image of the hole and slit. The x-ray tube is then moved to any angle, the hole covered with lead and another non-tomographic exposure made.

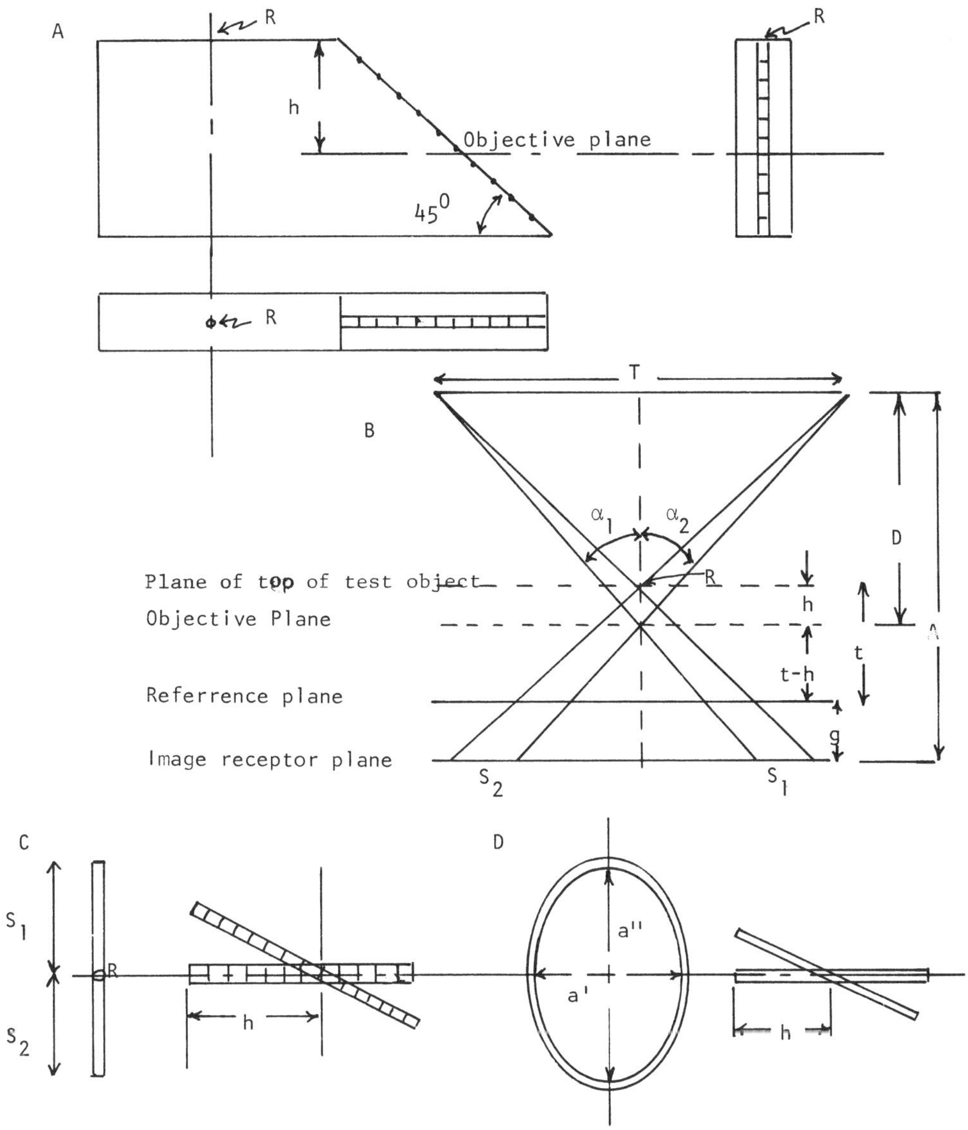

Figure 10: Alternate test object for determination of exposure angle (A). Geometry necessary to determine exposure angle (B) and reconstructed tomographic images for linear (C) and elliptical (D) motions. A is the source image receptor distance, D is the source objective plane distance, h is the distance to the objective plane from the top of the test object, t is the height of the test object, g is the image receptor-reference plane distance, T is the distance moved by the source and S_1, S_2 are the excursions of the central ray.

Finally, the slit is covered, the hole uncovered and a tomographic exposure made. The resultant tomographic image will appear similar to those shown in Figure 10. These images may be to determine the location of the objective plane and the exposure angle may be derived from the excursion of the image of the hole to each side of the central dot according to the relationships:

$$\tan \alpha_1 = (D - h)\ S_1/Ah$$

$$\tan \alpha_2 = (D - h)\ S_2/Ah$$

Another method of determining exposure angle utilizes the technique of recording the image of a narrow slight afixed to the collimator of the x-ray tube and a stationary film oriented at a slight angulation with respect to the central ray. The exposure angle may be directly measured from the recorded image. Details of this procedure may be found elsewhere (5).

Resolution

The ability of the tomographic system to demonstrate anatomical detail depends not only on the mechanics of the given tomographic system but also upon the type of image receptor employed, the size of the x-ray tube focal spot and the magnification employed to form the image. Tomographic resolution may be assessed using the arrangement shown in Figure 11 (6, 7).

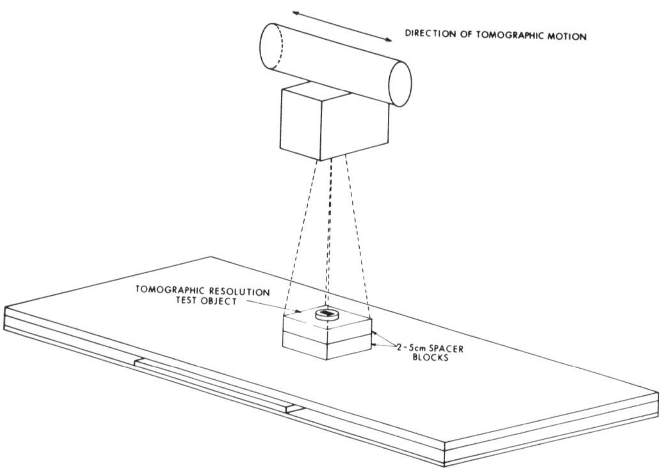

Figure 11: Arrangement for determining tomographic resolution.

The test object consists of a suitable resolution pattern which passes through the objective plane at a slight angle (e.g. 15°) and may be either a series of wire mesh patterns ranging from 10 to 50 mesh or a bar pattern of spatial frequencies from .5 to 4.8 lpm. When using a bar pattern for linear systems the bar groups should be orientated at right angles with respect to the tomographic motion. The resultant tomographic image will appear similar to that shown in Figure 12. The highest resolvable mesh or spatial frequency is noted and is the resolving capability of the system including the influence of the focal spot size, magnification and image receptor employed.

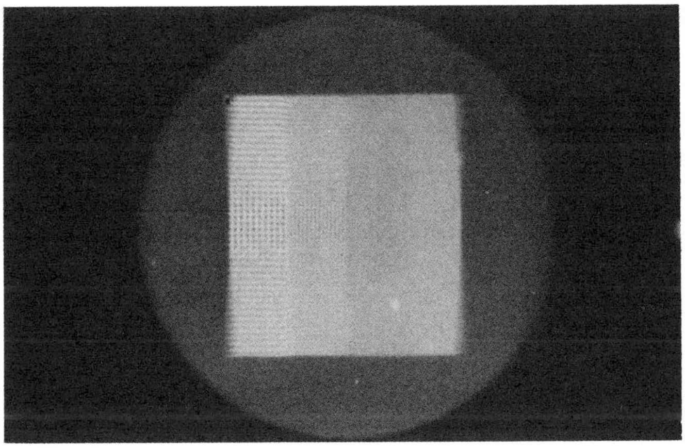

Figure 12: Tomographic image from which system resolution capability is determined.

Criteria for Acceptable Performance

In evaluating the performance of geometrical tomographic equipment for purposes of initial acceptance testing the criteria for acceptable performance should be that agreed upon in the equipment purchase contract. Guidance as to realistic levels of performance levels to be expected may be found in a number of publications. While no longer in effect, the standards of NEMA, summarized in Table II, may also be used as guides which would be applicable to any manufacturers equipment.

Table II
Performance Values for Geometrical Equipment
(NEMA Pub. No. XR3)

Parameter	Performance Level
Objective Plane	The measured objective plane shall coincide with the machine indicated objective plane within ±.25 centimeters.
Exposure Angle	The measured exposure angle shall coincide with the machine indicated exposure angle (if any) within ±3 degrees.
Flatness of Plane	The measured flatness of plane shall be in accordance with the manufacturer's specified tolerance for flatness of plane.
Tomographic Resolution	The measured tomographic resolution shall be in accordance with the manufacturer's specified tomographic resolution capabilities.
Section Thickness	The measured section thickness shall be within ±50 percent of the section thickness specified by the manufacturer.

Summary

Test equipment and procedures for the evaluation of geometrical tomographic equipment have been described. When carefully utilized these procedures will yield data which will serve for the evaluation of equipment for purposes of initial acceptance tests and for future life performance testing. Guidance as to acceptable levels of equipment performance is also given.

References

1. Litteton, J. T. , A Phantom Method to Evaluate the Clinical Effectiveness of A Tomographic Device, Amer. J. Roen., Vol. 168, No. 4, pp. 847-856, April 1970.

2. Litteton, J. T., Winter, F. S., Linear Laminagraphy, A Simple Geometric Interpretation of Its Clinical Limitations, Amer. J. of Roen. Rad. Ther. and Nuclear Med., Vol. 95, No. 4, pp. 981-991, Dec. 1965.

3. Kieffer, J., Analysis of Laminagraphic Motions and Their Values, Radiology, 33, pp. 560-585, Nov. 1939.

4. NEMA Standards Publication No. XR3-1970, Test Methods for X-Ray Equipment, National Electrical Manufacturers Association, New York, 1970.

5. ICRU Report 10f (NBS Handbook 89), Methods of Evaluating Radiological Equipment and Materials, International Commission on Radiological Units and Measurements, 1962.

6. Hendee, W.R., Rossi, R.P., Quality Assurance for Conventional Tomographic X-Ray Units, HEW Publication (FDA) 80-8096, USDHEW, Oct. 1979.

7. Basic Quality Control in Diagnostic Radiology, AAPM Report No. 4, 1977.

ACCEPTANCE TESTING OF COMPUTED TOMOGRAPHIC EQUIPMENT

Robert J. Kriz, M. Sc.
Department of Radiology
University of Illinois Medical Center
Chicago, Illinois 60612

INTRODUCTION

Invasive measurement does not give definitive information on image quality for Computed Tomography (CT) as it does for other radiologic imaging equipment. The reason is that the "image" exists only as a sequence of numbers stored in computer memory and more importantly, is generated from computer manipulation of collected data. To be sure, the use of invasive measurement can aid in determining the accuracy with which data is collected, but this does not insure image quality. One must then "fall back" to the use of first level testing procedures, in particular the use of phantoms, to determine image quality.

Since a large number of variables relate to image quality, a large number of specialized tests are needed. A series of these tests are described below, as well as their relationship to image quality. In addition, since image quality is related directly to dose, methods for defining and determining dose are described. Most manufacturers specify their best performance for a given parameter for scans which optimize that parameter. For example, high contrast resolution will be determined using the smallest reconstruction circle and an algorithm selected to optimize high contrast resolution. This may be very different from the algorithm and even the reconstruction diameter used for routine scans. It is therefore very important to determine the parameters used for a "routine" scan and measure all values for the image quality variables selected for that scan. Thus, image quality will be defined for the routine scan as well as for "optimum" conditions.

HIGH CONTRAST RESOLUTION

High contrast resolution can be measured by several different methods. The most common are resolution test patterns, edge response and line spread functions. All of these methods have both advantages and disadvantages. The line spread function and edge response function both give complete information on the high contrast resolution of a system, but computer analysis is needed to generate the Optical Transfer Function (OTF) or the Modulation Transfer Function (MTF). In addition, because of the discrete nature of the CT image due to finite pixel size, accurate line spread functions are difficult to obtain.

Holes drilled in plexiglass and filled with air or water have been suggested as a resolving-power test (1). Caution should be used to insure that the spacing of the holes is such that the center to center spacings are equal to twice the diameter. See Figure 1.

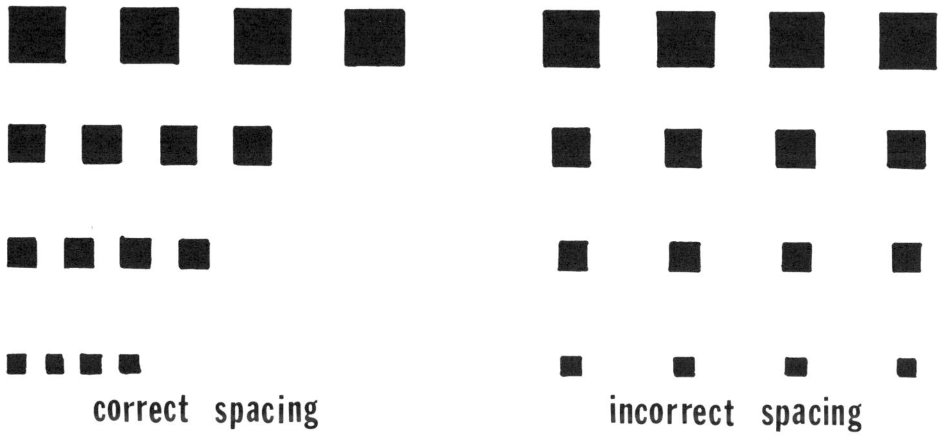

Figure 1: Correct Vs Incorrect spacing of resolution test objects. "Correct" is understood to mean agreement with Square Wave Response (SWR) or MTF cut off resolution.

This is very important because a single hole is easy to resolve in comparison to resolving each hole of a set. Square holes can be directly compared with resolution tests made from a star test pattern as they produce a square wave response. Also, for this reason, square holes have less loss of contrast than round holes of the same size (diameter compared to width). One should also understand that registration (or mis-registration) of square holes with pixels can affect resolution.

The disadvantage of a set of line pair resolution patterns is that actual cut off frequency can only be estimated. But as will be shown, because of the large amount of data to be taken, the simplicity of their use far outweighs this disadvantage.

It should also be realized that resolution is maximized when the pattern is at an angle of 45 degrees to the horizontal.

Resolution should be measured at the center and at least one other location usually at one-half the radius centered in one of the four quadrants. Measurements should be made for each scan diameter (head mode, body mode, etc.) or if this is continuously variable, for a large enough set of scan diameters to insure that the system operates as expected. Resolution should be checked for all reconstruction algorithms.

In addition, the combination of scan diameter and algorithm resulting in the highest resolution should be used to check resolution as a function of scan speed and number of views, if this is also variable.

LINEARITY

Linearity refers to the ability of the system to accurately assign CT numbers to a given material depending on the linear attenuation coefficient.

In principle this is relatively straight forward to check. One need only scan various materials of known linear attenuation and see if the CT numbers obtained are correct.

In actual practice the problem is greatly complicated by the fact that the linear attenuation coefficient is a function of the atomic number of the absorber and x-ray energy. The effective x-ray energy of a polychromatic x-ray beam at some depth in a phantom is difficult to determine. This problem can be overcome using a method suggested by Millner (2).

Linear attenuation coefficients for various materials that might be used in a linearity phantom are not available. In addition, the mass attenuation coefficients for many materials are not available and must be calculated from the mass attenuation data available for the elements according to the method published by Hubbell (8).

A further complication arises when one considers the energy ranger over which data is required and the increments which are possible. The effective energy of an x-ray beam used for CT ranges from 60 to 80 keV. It should also be understood that because of beam hardening effects the effective energy of the beam changes with depth in the phantom. All pins in a linearity phantom should be at the same depth (4). Values of mass attenuation coefficients published in the National Bureau of Standards (NBS) report number 29 are given for 50, 60, 80 and 100 keV (8). If other values are required a method of interpolation must be used. See Figure 2. Cacak (3) has suggested using $M = A \ln E + B$ where M is the mass attenuation coefficient. E is the energy and A and B are constants found by least squares fitting the data. When this is done the difference between the calculated values and the listed values runs from .25% to .8% for hydrogen and carbon to 1.4% to 2% for nitrogen and oxygen. Hubbell's data is only accurate to 1%, and therefore the curve fit is good.

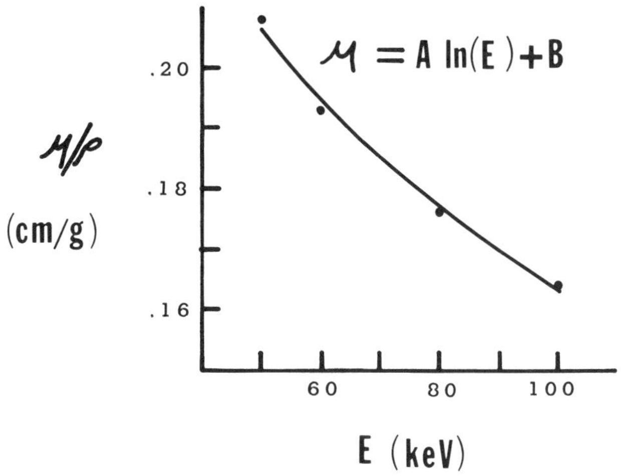

Figure 2: The mass attenuation coefficients of Polymethyl Methacrylate is plotted as a function of energy. Plotted points are data from NBS #29.

After the mass attenuation coefficients have been determined for a compound, the density of the compound must be determined. This can be done using a graduated cylinder and a good scale or balance. Note that a 1% error in the density will cause at least a 1% error in the linear attenuation coefficient since the linear attenuation coefficient is the product of the mass attenuation coefficient and the density. If one uses an average of approximately 1% for the error in the mass attenuation coefficient and assumes a 1% error in determining the density - the total error becomes 1.4%. This corresponds to a change of approximately 3 keV in the energy used to determine the linear attenuation coefficient of water. It is therefore reasonable to construct a set of tables for linear attenuation coefficients for various materials from 60 to 80 keV in 3 keV increments. This table can then be used to determine the linearity of the CT system. For a given energy, one uses a least square technique to fit the data (CT number Vs attenuation coefficients) and records the correlation coefficient. This is required for different energies. A plot of correlation coefficients vs energy is then made and the energy corresponding to the correlation coefficient closest to 1.0000 is selected as the effective energy of the system. All points on the CT number vs linear attenuation coefficient curve for this energy should lie within two standard deviations of the curve for acceptable linearity. In addition, the slope of the curve determines the contrast scale.

NOISE VS EXPOSURE

In CT scanning, noise is defined as the statistical fluctuation of CT numbers. It can be measured by calculating the standard deviation of CT numbers about their average. This implies that the region over which the standard deviation is measured is homogeneous. The standard deviation is determined by the equation given below.

$$\sigma^2 = \frac{\Sigma (x_i - \bar{x})^2}{n - 1}$$

Since the noise is largely due to counting statistics and the number of phontons is dependent on the attenuating medium, the mAs and kVp, it is recommended that noise be measured utilizing a water phantom having a minimum diameter of 8" for head modes and 12" for body modes. In addition, noise should be determined both at the center and at the periphery of the phantom. Measurements are made at all available current stations and two plots of σ vs mAs made for each usable tube potential, (one for each measurement location) on log paper.

Brooks and Dichiro have shown that noise is proportional to the inverse square root of the dose (5), therefore the slope of the curve should be -0.5. A lower value indicates additional noise is present. See Figure 3.

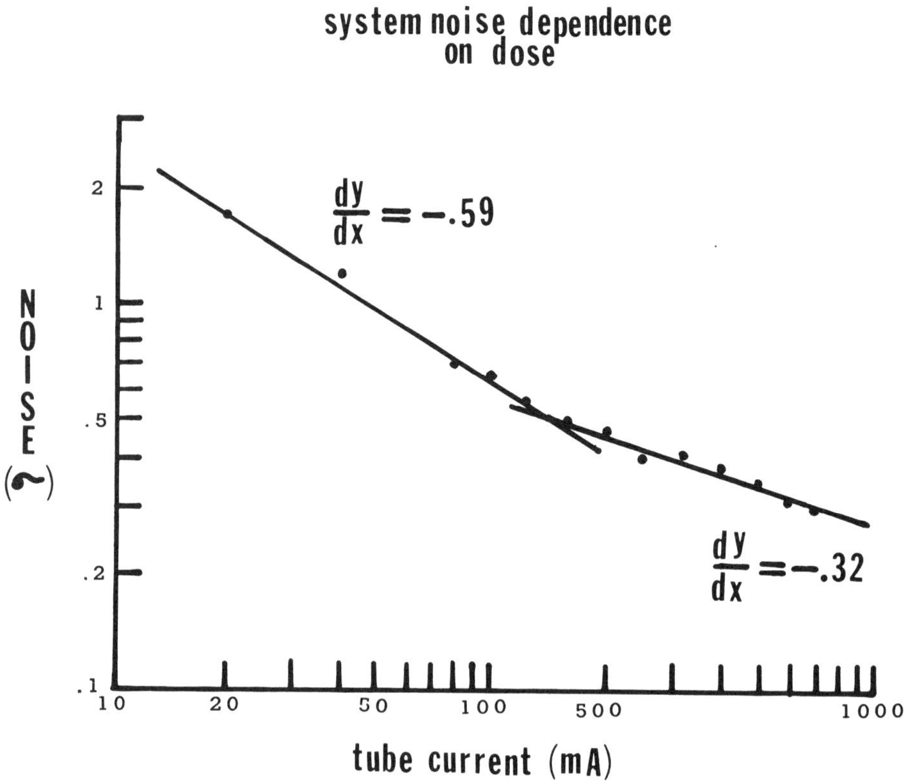

Figure 3: Noise is plotted as a function of dose. Noise is the standard deviation of CT numbers.

Noise limits should be as specified by the manufacturer. Also at this point it is important to recall the statement made at the end of the introduction; that one determines all values for selected variables of a "routine" scan. The manufacturer normally specifies noise under what he considers "best" conditions. Hence, it will probably be necessary to request specifications for noise for "routine" scans in writing before purchasing a system.

MACHINE INDEPENDENT NOISE

CT scanners use different scales for their CT numbers. Some systems use number ranging from $\pm$ 500 and others $\pm$ 1000 and still others use different scales. The contrast scale (CS) referred to earlier related the CT numbers to the linear attenuation coefficient. It can be determined by scanning a phantom of acrylic and water (1). Since the difference in attenuation coefficients between acrylic and water is constant one can calculate the CS from:

$$CS\ (cm^{-1}/CT\ \#) = 0.024/CT\ \#\ PLEX\text{-}CT\ \#\ H_2O)$$

or use the inverse slope of the linearity curve (a plot of CT numbers vs linear attenuation coefficient).

The contrast scale can then be used to determine the standard deviation (noise) with respect to the linear attenuation coefficient of water. This can be expressed as a percentage by using the following equation:

$$(PAC)_W = \sigma \times CS \times 100/\mu_W$$

where (PAC) is the percent attenuation coefficient of water and μ_W is the linear attenuation coefficient of water (1).

The method of dividing by the range of CT numbers is not as accurate as the method described above. But, it is much simpler to use and most manufacturers do use it. Therefore, much care should be used when comparing "noise" or "sensitivity" as stated unless the definition of the term or its measurement method is also stated.

SENSITIVITY PROFILE

The sensitivity profile is a description of the systems response in the axial direction through a slice. A loss of sensitivity is due to the scan geometry. The x-ray beam diverges and this results in volume elements (Voxels) that are short on the tube side of scan and tall on the detector side. Thus, the partial volume effects are different for the same voxel scanned from different directions. This effect is greatest near the edges of the scan circle and minimum at the center.

The sensitivity is not constant across a slice and the sensitivity profile is not identical in all locations in the plane of the slice. Therefore, it is useful to measure the sensitivity profile in more than one location. Usually it is adequate to make the measurement in the center and at one location off center e.g. at a distance of one-half the scan radius along the vertical axis. However, if the CT scans other than a 360 degree rotation, additional measurements may be required because the sensitive volume may not be circularly symmetric.

A method for determining the sensitivity profile was described by Brooks and Di Chiro (6). One can use the aluminum ramps in the AAPM phantom for this purpose (7). Some care must be taken to insure that the ramps are inclined at 45 degrees to the scan axis or error will result. Once scanned, the CT numbers can be printed out along the ramp and the sensitivity profile plotted. The Full Width at Half the Maximum (FWHM) is a measure of the slice thickness and should be compared with the nominal slice thickness. The method does not work well for slice thicknesses less than approximately 5 mm. Also, if the ramps do not appear to be perfectly vertical or horizontal the angle to the horizontal axis must be determined and corrections made. One advantage of the AAPM phantom is that it contains 3 ramps and thus the sensitivity profile can be determined at three locations in a single scan.

The sensitivity profile should be measured for each slice thickness available within the limitations of the measurement method.

If a printer is not available, one must use care in determining the FWHM. The apparent length of a ramp will change depending on window width and level. Some scanners have an ability to display a plot of CT numbers vs position and the true FWHM can be determined from this graph. If this function is not available, one can use the cursor and search for pixel locations having the appropriate value.

In addition to measuring the single slice profile, the correct relationship between the slice profile and table indexing should be checked. This can be done allowing the CT to automatically index the table by the nominal slice thickness and determine a second slice sensitivity profile. It is recommended that the half maximum point from the two profiles should be no more than 10% of the nominal slice thickness or 1.0 mm apart (which ever is greater). However, this should be specified in the bid prior to purchase and some manufacturers will use different values.

DOSE PROFILE

The dose profile is the dose distribution through the slice in the axial direction. Ideally, it should be a square wave, that is zero outside of the slice and uniform inside of the slice. In reality, the dose profile can be far from the square wave shape. This is because of the same geometric situation as discussed in the sensitivity profile section. In addition, scattered radiation and penumbra cause non-zero dose to be present outside of the slice.

The dose profile can be measured using Thermo-Luminescent Dosimetry (TLD). The standard TLD 100, 1 x 3 x 3 mm chips can be used. If the chips are stood on edge, dose determination can be made in 1 mm increments. If so, it is wise to include at least 20 chips for a 10 mm nominal slice thickness. This is required because of the possibility of a rather long tail on the dose profile curve. The dose profile should be measured for the most commonly used slice thickness and in several locations along the scan circle. An 8" water phantom should be used to simulate both attenuation and scatter, although other phantom sizes and materials are acceptable. However, one should use the same or similar phantom when comparing different systems. Since the dose at the center of a slice is different than the dose at the edge of the slice, it is useful to measure the dose profile in the same location as the sensitivity profiles. However, this will require the use of a specially designed phantom to allow proper positioning of the TLD's. If this is not possible, the TLD's can simply be placed on the surface of a water phantom and consideration given to the difference in location between the two measurements.

The dose profile should be plotted and the FWHM determined. It is recommended that this FWHM should agree within one millimeter of the FWHM of the sensitivity profile. See Figure 4. It should be understood that these tolerances depend on the individual manufacturers and should be determined in the bids.

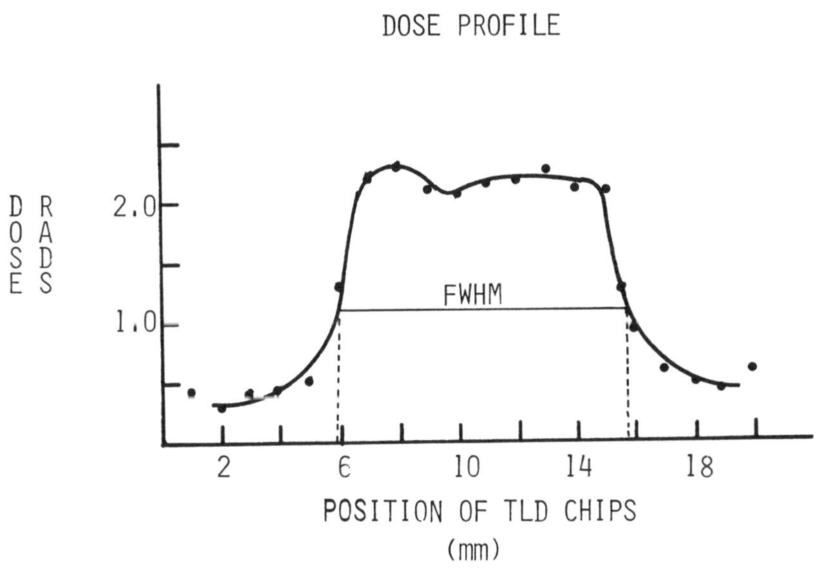

Figure 4: The dose in rads is plotted as a function of position. The direction is across the slice width.

It should be understood that if anything other than a 360 degree rotation is used, the dose profile will be different for different locations around the scan circle. In that case, the measurement should be made at a location where the highest dose is expected.

DOSE VS SENSITIVITY PROFILE

Since the sensitivity profile and dose profile have different shapes, some trade offs between dose and sensitivity are made. It is this relationship that is most important in determining the actual nominal slice thickness. If for example the nominal slice thickness is chosen such that the half maximum points of two adjacent sensitivity profiles are coincident it is possible and likely that the dose values will be much higher than 50%. Thus, dose far in excess of 100% will result at places where the nominal slices are adjacent. This is a major reason for very high multiple to single dose ratios. If on the other hand the nominal slice thickness is chosen such that the dose profile are brought together at the 50% point, lower multiple to single slice dose ratios result but at a cost of lost sensitivity between adjacent slices.

This result can cause a small, just barely detectable lesion to be lost by "falling between the cracks" in the scanned column. Perhaps what is worse, is that the appearance or detection now becomes dependent on the registration of the patient with the table. Thus, an image of a lesion may appear one day and disappear the next, leading to the false conclusion that the treatment has been successful.

Manufacturers do not currently specify the relationship between sensitivity and dose profiles. Thus, it is important to either specify tolerance when purchasing a system or to specifically request information to review during the bid evaluation. For existing scanners it is advisable to at least make the physicians in charge aware of any possible problems so that corrective action can be taken (e.g. overlapping slices when small lesions are expected).

As an example of ways one may specify tolerances, one can specify the separation between the two curves for various slice thicknesses at the 25%, 50% and/or 75% sensitivity profile locations. It is probably better however to simply ask the various manufacturers to specify what their system is capable of at these points and then judge the systems based on these results.

TABLE INDEXING

The ability of the table to accurately reposition itself should be verified. This is in addition to the test described in the sensitivity profile section. This can be done by fixing an indicator to the table (a pen or pencil works fine) and a stationary ruler mounted next to the table, a 200 or 250 lb weight should be placed on the table and the table moved a total distance of 100 cm with several changes in direction. The table should return to within 1 mm of its original position. This is especially important if radiographic like projections are used for selection of slice locations.

In addition, the accuracy of light localizer fields should be checked by taping paper pak film to a phantom. The location of the light field can be marked on the film using a pin or pocket knife. A scan is then made to expose the film to radiation. Congruence of the light and radiation fields should be better than one millimeter. The congruence should be checked at the top, left side and right side of the scan circle.

Another useful test is to expose the film to multiple scans with automatic table indexing. Agreement between the nominal slice thickness and table motion can thus be verified.

CT NUMBER DEPENDENCE ON POSITION WITHIN THE GANTRY

The dependence of CT numbers on position of a phantom within the gantry should be checked. This will not be specified by the manufacturer and is therefore not an acceptance test unless prior arrangement has been made. However, it is of some importance and should be measured.

The linearity sections of the AAPM phantom can be used. Use a large scan diameter (40 cm) and a small diameter (25 cm) phantom. The CT number of each material, as a function of position are plotted by scanning the phantom and then moving it in 2 cm increments to the right and left. The procedure should be repeated by changing the phantom position in the vertical direction. There should be no large change in CT numbers as long as the entire phantom remains within the reconstruction circle.

SPATIAL UNIFORMITY AND DISTORTION

Spatial uniformity is checked with a uniform phantom, usually a water bath. A cupping artifact (higher CT numbers at the edges of the phantom) should be at or below the level specified by the manufacturers.

In addition, a phantom constructed of two sets of equidistant holes or pins aligned in the vertical and horizontal directions can be used to check distortion. The distances between various pins can be measured with the system software, at the TV monitor and on hard copy images. Distances should be checked along 45 degree inclines. Cacak (4) has designed a phantom for this purpose.

LOW CONTRAST RESOLUTION

Low contrast resolution is a measure of the ability of a system to discriminate between two objects having nearly the same attenuation coefficients.

One method of describing low contrast resolution is by the use of C-D diagrams as has been suggested by Cohen and Di Bianca (9, 10). A C-D diagram is a plot of the contrast (defined as the difference in CT numbers) vs the minimum size hole that can be resolved. However, the usefulness of this technique has been contested (11). Since CT numbers are proportional to linear attenuation coefficients, they are dependent not only on the material used to construct a phantom but also on the kVp and filtration used by a particular CT system. Results are further complicated by beam hardening effects. Since perceptability is also influenced by the observing conditions, the author prefers not to utilize low contrast perceptability tests, but rather favors the use of noise measurement in conjunction with high contrast resolution.

REFERENCES

1. Edwin C, McCullough and J. Thomas Payne: Med. Phys. 4, 85 (1977).

2. Michael R. Millner et al: Med. Phys. 5, 543 (1978).

3. Robert Cacak: Private Communication.

4. Robert Cacak and William R. Hendee: Performance evaluation of a 4th generation CT scanner. Application of Optical Instrumentation in Medicine VII, p. 194-204, Proc. SPIE, V. 173.

5. Brooks and Di Chiro: Statistical limitations in x-ray reconstructive tomography. Med. Phys. 3:237-240, 1976.

6. Brooks and Di Chiro: Slice geometry in CAT. JCAT 1 (2) 1977.

7. AAPM Report #1. Phantoms for performance evaluation and quality accurance of CT scanners.

8. J. H. Hubbell: Photon cross sections, attenuation coefficients, and energy absorption coefficients from 10 keV to 100 geV. NBS #29.

9. Gerald Cohen and Frank A. Di Bianca: JCAT 3 (2), 189-195, April (1979).

10. Gerald Cohen: JCAT 3 (2), 197-203, April (1979).

11. B. H. Hasegawa , R. K. Cack, J. A. Muluany, W. R. Hendee: AJR, Vol. 138, p. 135-139, Jan. 1982.